Dearest Babe,

Kelly Farris Mazade
Kate Mazade

ISBN-13 979-8-9854753-0-2
eISBN-13 979-8-9854753-1-9

Published in 2022 by Kelly Farris Mazade and Kate Mazade through IngramSpark with Author's Note, Introduction, Chapter introductions and narratives, Afterword, Bibliography, and Indexes.

The text of this book was composed in Times New Roman
and Lucida Sans Typewriter typefaces.

Book Designer: Kate Mazade

www.dearestbabe.com

Printed in the United States of America

In loving memory of
Babe

table of contents

Author's Note

Because of my interest in genealogy, I have become the keeper of the "Family Archive." I have amassed hundreds of documents, photos, and stories from all branches of the family. This particular story came from a box of letters that were in my grandparent's house in Houston after they died.

I think it is important for us to know who our loved ones were. We have memories of them, but those memories are not always the whole story. When he was alive, I knew Joe as Happy, my grandfather who always wore old scrubs and his hat cocked over to the side. I knew him as a man who rescued bunnies for us and frequently ran out of gas on the side of the road. Writing this book changed the man I knew. The belief that people are more than what you remember of them brought me to this book, which is not really mine, but Joe and Babe's.

Babe always wanted Joe's story written and once told me about the letters that she had saved all those years. After she died, I read Joe's letters and realized that they represented the first five years of their life together. I spent the next two decades transcribing each letter and researching places and events. I even joined the Former Members of the 451st hoping to come across someone who knew Joe.

This book turned into a project that overwhelmed and frustrated me. It has made me laugh and it has, at times, broken my heart. There are still so many questions that have gone unanswered and many things I wish I had asked Babe. My hope is that you discover something new about the people that you loved—those who meant so much to you and helped shape who you are today. I hope I have made Babe proud.

Kelly

Introduction

This book started as three boxes, dusty old containers that changed my life. I pulled them out of my grandmother's closet a few months after her death in 2002.

The first box was a wide, flat file box that was filled with letters—almost five hundred soft, yellowing handwritten letters, folded neatly into their original envelopes. They were nearly in order, but some envelopes held two or three letters without dates. These were the letters that my grandfather wrote to my grandmother during the second World War. He wrote to her almost daily from June 23, 1943 to July 26, 1945.

The second box was my grandfather's old field medicine kit—a sturdy wooden case with a latched lid. Its scratched olive drab paint was stamped with the U.S. Army Medical Department's caduceus with first aid instructions and inventory still printed on the underside of the lid. Inside, the box held rolls of film negatives and a handful of fading photographs. There was also a small leather journal with vertical pages full of cramped writing.

The last box, a clear plastic storage container, was one in a set of 6 that held all of my grandparents' personal records and memorabilia—six decades of birthday cards and party invitations, medical journals and lecture notes, greetings and condolences, tax records and correspondence.

Among the pages was my grandfather's complete military personnel record—the only existing copy after a 1973 fire in the National Personnel Records Center destroyed the others, along with another 16-18 million official military personnel files. There were also two letters, from Robert Karstensen and Peter Massare, inviting my grandfather to join the Former Members of the 451st reunion group.

I took these boxes home, and over the next twenty years, started to go through their contents. I put them in order, slid each letter into a clear protective sleeve, and typed them up, so the other members of the family would have a chance to read our grandparents' story and learn about the history of our family.

* * *

My grandfather was Dr. Joe Wesley King. He was born in 1915, in Shannon, Mississippi but later moved to Memphis, Tennessee. He lived through the Great Depression, World War II, the Civil Rights Movement, and the Space Race. It was a generation of personal struggle and great opportunity. Joe saw many advancements in medicine during his time in the war and his career as an orthopedic surgeon at Houston Methodist Hospital. Like many other men of his generation, he believed that it was his duty to serve his country in whatever way he could.

When he was a boy, he injured his hand while roller skating. He was hanging on to the bumper of a milk truck and fell. When he went to the doctor, the physician gave him a generous discount on his care. Still, Joe had to pay the bill in installments, but the doctor's kindness left a lasting impression on him. From then on, Joe knew he wanted to be a doctor. At sixteen, Joe went to college at the University of Texas, hitchhiking to Austin each term because he didn't have a car. He finished both college and medical school in three years. He paid for his own education, working as a nanny during college and as the night shift medical personnel for a refinery plant during medical school. Joe received his medical degree from the University of Tennessee in 1938. He then completed his internship at John Gaston Hospital in Memphis, Tennessee.

My grandmother was Olive Madison Black, but everyone called her Babe. She was born and raised in Memphis, Tennessee. She went to Southwestern in Memphis (now called Rhodes College) and lived at home while she attended college. Babe was very involved in college activities, such as clubs, sororities, and campus events, where she was popular and well known. She graduated with a degree in education and taught 5th grade until she married Joe.

Joe and Babe had known each other for two years before they started dating, but their first date was an accident. Joe called Babe to arrange a double date. He was to go with another lady, while Babe was to accompany Joe's friend Dan Baldwin. When Joe and Dan went to pick up Babe, she thought Joe was her date. Joe stood up the other lady and took Babe out himself. That was the beginning. At the time of their courtship, Joe was practicing medicine in Helena, Arkansas and would travel back and forth to Memphis to see Babe. On May 31, 1940, after a short six-week courtship, Joe and Babe were married. In the first two years of their marriage, Joe built his own medical practice in Helena. Their first daughter, Kay, was born in January of 1942.

After the fall out of Pearl Harbor, Joe, along with tens of thousands of Americans, joined the military to serve their country. Because he was the only surgeon in west Helena, Arkansas, an under-served community, Joe was originally exempt from service. It took nearly nine months to have the exemption reversed and qualify for service, allowing him to enlist in the Army on September 15, 1942. He served as a flight surgeon in the Army Air Corps, which later became the U.S. Air Force. Joe was assigned to the 15th Air Force, where he served in Italy with the 451st Heavy Bombardment Unit. At twenty-seven, he was older than many of the enlisted men he cared for, who were in their late teens and early twenties.

Almost every day of his deployment, Joe wrote a letter to Babe. Some letters said almost nothing. They just told Babe that he was still alive and doing well. Some letters reflected the struggle and heartbreak he saw as the group doctor. He treated a variety of medical problems, some combat-related, some psychological, and some social. Years later, he told my father that the two biggest problems the men faced were anxiety about flying and VD.

* * *

When I first read Joe's letters, I didn't see the full story. They were just records of his daily life and glimpses of Babe's life at home with their two daughters. But as I continued to look into the history, to track down the people whom Joe mentioned, to research the unit's missions, the full picture started to fall into place. Now when I read Joe's words, I see the man standing next to him in the mud. I feel the vibration of planes rumbling overhead. I hear Kay chattering in the background as Babe opens letter after letter. And I know the stories that built our family and can share them with the people I love.

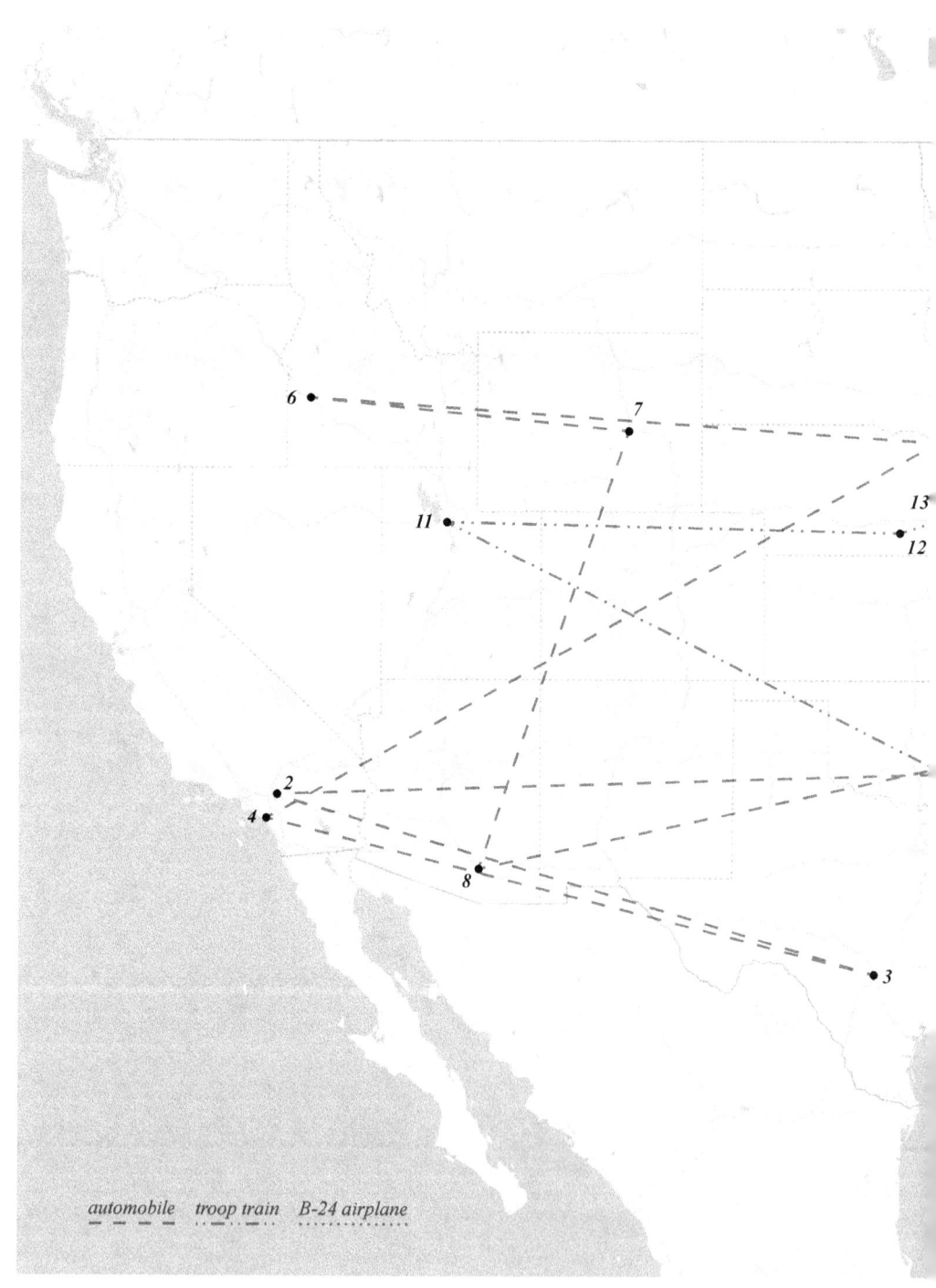

automobile   troop train   B-24 airplane

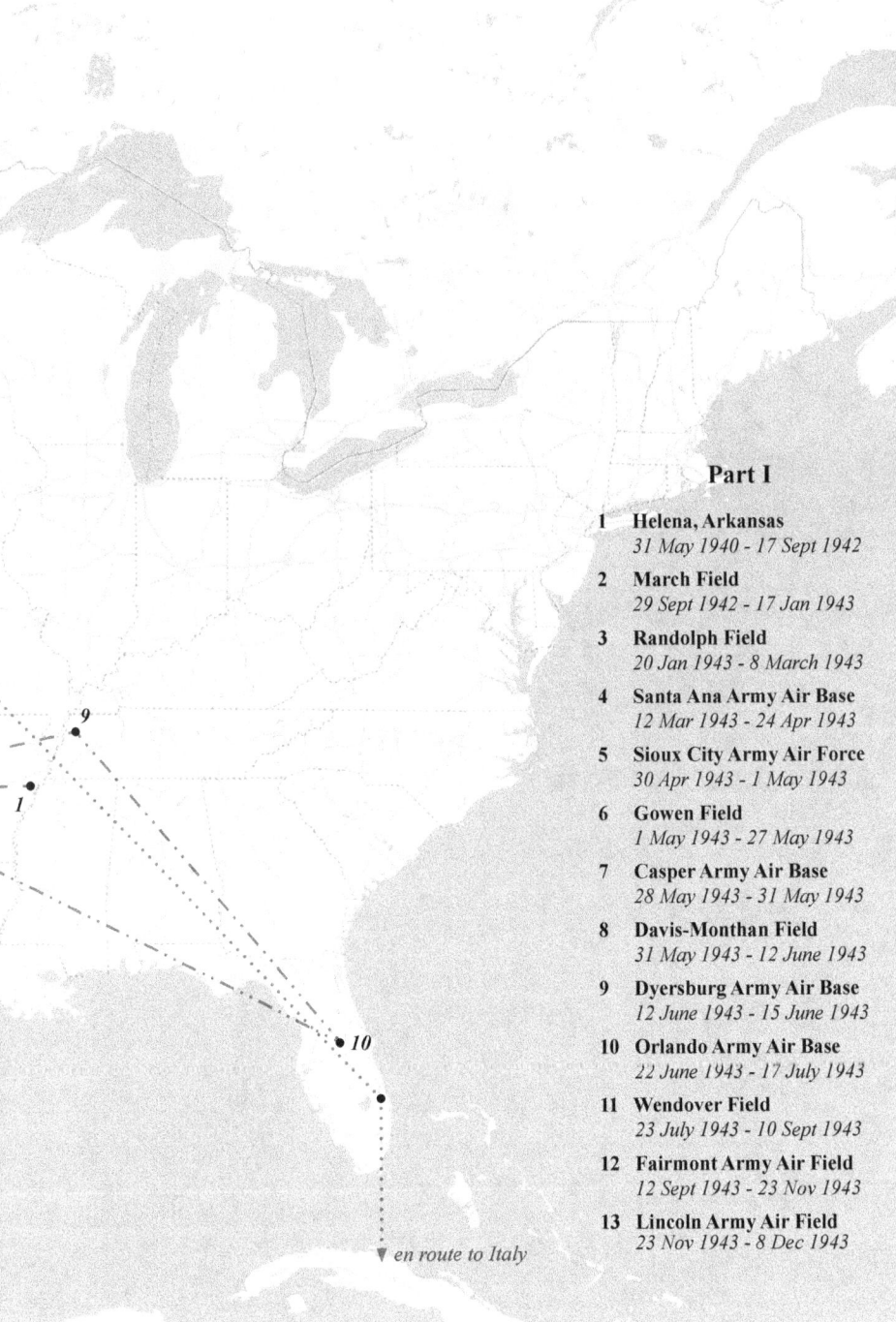

Part I

en route to Italy

Stateside Training

Joe enlisted in the Army on September 15, 1942 and had to report to March Field in Riverside, California by September 29 of that year to begin his training. Joe, Babe, and Kay packed their grey Chrysler convertible coupe and drove from Memphis to California. The three of them continued to travel from base to base during the first part of Joe's time in the Army.

When Joe's training in Riverside, California ended in January, he was sent to Randolph Field in San Antonio, Texas to go to the School of Aviation Medicine. Even though Joe was already a licensed and practicing doctor, he still had to complete the medical training he would need for the Army Air Corps. He was stationed at Randolph Field for six weeks before being sent to Santa Ana, California, where he completed his Aviation Medicine training. Joe received his Aviation Medical Examiner certification on April 22, 1943.

451st Heavy Bombardment Group Insignia

After short term assignments in Sioux City, Iowa; Boise, Idaho; and Casper, Wyoming, Joe was sent to Davis-Monthan Field in Tucson, Arizona to join the 451st Heavy Bombardment Group, which had been activated in April 1943 with only 11 men: ten officers and one enlisted man. Shortly after their activation, the unit moved to Dyersburg, Tennessee under the 346th Bombardment Group.

From Babe in Riverside, Calif. to her parents in Memphis

October 4, 1942

Dearest Mother and Daddy,

Here I think I'll be sure and write you so you can keep up with our trip, and I wrote you a card every single day until we got to Calif. Fort Worth, Carlsbad, Albuquerque, Grand Canyon, San Diego, Riverside. Hope by now you have gotten the rest of them.

Well, we have been very lucky. Friday, one of the real estate ladies called and said she had us a very nice little garage apartment. I got her to bring me out Friday afternoon, and we moved some of our things in Friday night and the rest yesterday. We have one large room with a studio couch, 2 big chairs, a big bed with a Beautyrest mattress and a utility table. The table has a 2 burner electric hot plate and the lady let me have a toaster and Silex coffee maker. They also furnished all dishes, pans, linens, etc and pay all the utilities — all for $30 a month. And in Riverside. That's something. Also have a large bath and dressing room combined, containing a chest of drawers and a large wardrobe. We keep Dad's ice box in here also — if it hadn't been for that we couldn't have considered the place. Very important too is that they are such nice people, named Ingham, and it's in a very nice residential section. Joe and I are both pleased to death to get it. We put up Kay's bed for her, and now she's very pleased too. Of course we haven't the vaguest idea of how long we'll be here, but it's a nice reasonable place to stay while we're here. There's a nice big grassy yard with a pomegranate tree, fig tree and lots of pretty shrubs. And Kay and I really enjoy getting out in the sun. I shall start fixing our meals here either tomorrow or next day and then it will really seem like home.

Joe had to be at March Field all day, but he gets Sundays off which is just wonderful. Now that he's gotten acquainted and used to things, he's just wild about it. They have classes all day, and he says it's just like being in medical school again and that he's really enjoying it. He's tired when he comes in and his feet hurt, but otherwise he's feeling fine.

This afternoon, Joe wrote his mother and dad, and I wrote Hugh a long letter since you sent me his address — it was only yesterday I got your letter saying where he was. Please write me Air Mail (6¢) and I'll get it within 2 or 3 days, otherwise it's from 4 to 6 days

Well, honey, will write more later just wanted y'all to know we'd found a place and we're very satisfied and getting along fine. Take care of yourselves and write again soon.

Love,
Babe

January 10, 1943

Dearest Mother and Daddy,

Got your sweet letter yesterday and was so glad to hear from you. We also enjoyed Kay's birthday letter. Miss Kay has gone to sleep in one of her new dresses and with her new shoes on — neither bothers her rest as she is nearly sawing logs. She's had her bath and shampoo, and I guess it is rather tiring to drink a half a tub of water and sling the other half all over the room.

Well we expect to leave here either a week from today, 17th or 18th, as Joe's class at Randolph begins January 25 instead of February 1. We shall be glad to hit the road again, and we've a larger place to live in San Antonio. Either a bedroom, dinette, kitchen and bath or if we're lucky a living room in addition to the above. The address is Yucca Cottages, Austin Highway, San Antonio, Texas. We'll probably get there about the 22 or 23rd. I'll try to write you again this week and will send postcards while we're on the road.

They have gotten in fifty new doctors here, so Joe has gotten to come home 4 nights this week and I'm expecting him again tonight. Friday night, we went out to dinner and to see Bette Davis in "Now Voyager" — it was grand. I wore a new outfit given to me by Mrs. Ingham. It was an old blue handknit suit of hers that had gotten faded. I took it to a good cleaners, had it dyed black, blocked to fit me — put new silver buttons on the jacket and bought me a beautiful white lace collar to wear with it — Joe said I looked "knockout" — however, he's sweet enough to say I look pretty when I get up in the mornings! I wore my new pearls with it and they made the outfit!

Mrs. Ingham has also made Kay a poke bonnet out of the blue corduroy to match the coat and lined the brim with pink taffeta — it really does look cute on Kay and as soon as we can get some more film we'll send you some more pictures. Mrs. Ingham says she just can't stand to think of Kay leaving — says she'll miss us too (that's for politeness sake).

Grandmother and Papa sent Kay $2 for her birthday, which we thought was awfully sweet but I'm so afraid they denied themselves to send it. I wouldn't send it back cause I was afraid it would hurt their feelings — but Mother couldn't I send it to you and you could send it to them. What do you think?

Surely do appreciate you offering to send me some canned goods — however we can still get most everything in that line, and the folks that have already gone to San Antonio say that they have just plenty of meat, butter, etc, down there. If I find we can't, I sure will write to you.

I have a bouquet of narcissus on my table that came out of the yard. They smell so sweet and remind me of the ones you used to bring me from Lucy, Tennessee. The sweetpeas too are in bloom and tiny little winter iris that look like orchids. Well honeys, remember I love you both so very much and think of you a hundred times a day.

Lovingly,
Babe

Stateside Training

From Babe in San Antonio, Texas to her parents in Memphis

January 27, 1943
Wednesday, 7 a.m.

Dearest Mother and Daddy,

If the early bird really catches the worm — we should have a garden full. Joe's school day begins at 6:50 a.m. so we have to get up at 5 or 4:45 a.m. in order for him to get there on time. Ain't that somethin'!

Joe loaded the trailer Saturday afternoon, and we left for San Diego about 3 p.m. — drove to Gila Bend the first night — stayed the second night in El Paso — we blew into Texas with a "Norther" (as they call a cold wave — the temperature at El Paso was 8 degrees. We nearly froze. From there we drove to Del Rio and then drove down into San Antonio on Wednesday. Joe, Kay and I huddled together in the front seat the last two days and wrapped up in all her blankets and the lap robe we had. It was some trip — it's still cold here about 24 degrees — but the sun is out today and everything looked brighter anyway.

We stayed at some other tourist court Wednesday through Sunday as our place here wasn't to be vacated until Monday. We moved in Monday afternoon, and I was so thankful that Joe didn't have to report at Randolph until Tuesday, so he could help with Kay and unload the trailer.

We have run into two couples that Joe and I knew when we were in Memphis — Lt. and Mrs. Yates (he interned at the Baptist and she was a technician there) and Lt. and Mrs. Tatum. Both were in some of Joe's classes when he had his teaching internship at John Gaston. Also Lt. Alfred Page who went to Southwestern when I did — was Joe's pupil at Fort Sam Houston here. We all got together at the Court Sunday evening and really had a good gossip session. It really was wonderful seeing someone from home.

Joe started classes yesterday. They attended lectures from 7 a.m. until 5:30 p.m. with an hour off for lunch. He comes home — we have dinner — he has to study about 2 hours, as they have exams every other day — I wash dishes and attempt to keep Kay quiet. By that time we're all ready for bed and how!

We have a nice little cabin — old and nothing fancy — but comfortable and clean — and it's so nice to have a little more room to spread out in. The cabins are individual white cottages with green trim, scattered around a nice grassy lot. We have a front porch, large bedroom, big bath, and kitchen. I cook on a 3 burner gas plate but we have a nice big electric refrigerator. We're about 4 miles from town on the Austin Highway towards Randolph Field which is about 9 miles further out. Joe and several other Lts. pool their cars — so I do have the car about four days a week — course due to the gas shortage I only use it most occasionally. There are about 7 or 8 cottages here — all occupied but Dr.'s and their wives or families. We've met two of the couples, and they are very nice. We got a chance to be with the Craycrofts' one evening before they left for Santa Ana, Calif. We went out for dinner in shifts — Joe and Burr, while Jean and I kept the babies — then Jean and I went out while Joe and Burr kept the babies. When we got back — Joe and Kay were asleep on one bed and Burr and Carolyn on the other — Jean and I certainly did tease them about how much they enjoyed one another's company! It's too bad we can't be in the same place at the same time — but it does make it nice having someone get you a place to live wherever you go.

By the way — as for the food situation, Texas doesn't know there's a war going on — I never saw so much meat in my life, and they have plenty of canned goods too. Right now I have black eyed peas cooking — for dinner I'm having peas, pork chops, slaw, and corn pone — Joe's favorite meal just about. People in California don't eat black eyed peas. They think they're only to feed livestock! Can you hear that!

I surely would do my best to go see Hugh in February if I were you — regardless of whether he comes home in March or not — he wouldn't ask you to come if he weren't terribly lonesome and homesick to see you cause he knows what an expense and what a hard trip it would be — so I sure would try to go. You just don't know what it means to see a familiar face — much less your mother and daddy!

Honeys, I must stop as I have dishes to do — Kay to bathe and dress and then we're going to drive down the highway and get some gas. Kay hasn't been out since we moved Monday and I think the sunshine would do her good.

All our love now,
Babe

March 1, 1943
Monday, 11:45 a.m.

Dearest Mother and Daddy,

Your sweet letters received and so appreciated — two from Mother last week and Daddy's this morning. You really enjoy and look forward to letters from home when you've been away so long and so far.

My how I wish you could see Miss "Precious" — last Tuesday I had to go get our Ration Books for canned goods, so I asked Pauline the little Mexican girl to stay with Kay while I was gone. I was gone about 30 or 40 minutes. When I got back, Pauline says "Kay can walk" — set her down on the floor and Kay walked clear across the room to me. You can imagine my amazement! Joe was so tickled. When he came in that night, I set Kay down and she walked to the door to meet him. She looks like a combination of a "small plane" taking off and a baby duckling waddling around — and does she love it. She sits down so hard, but she must be made out of rubber she bounces up so quickly. It really is funny to watch her "taking off."

I can hardly believe that this is our last week here. The time has gone by like lightning. Joe is having final exams today, Tuesday and Wednesday. Thursday they don't have much to do. Friday, Saturday and Sunday he has to go on a 3 day and night bivouac. They sleep in pup tents, march, get tested for gas attacks and get a taste of real army life. I think Joe is rather looking forward to it, as it means no more studying. He has worked hard and really done well. 100 out of the 300 have been warned that unless they do well on these final exams, they will "flunk" out having already flunked some of the quizzes. Joe studied Friday afternoon and nite — Saturday evening and nite until about ten — then got up at 5:30 yesterday morning and studied all day. At 5:30, we went into San Antonio to the Poseys for dinner. I hadn't eaten off of a tablecloth and out of good china and silver in so long I hardly knew how to act. We really did enjoy a good home cooked meal. Particularly baked potatoes and biscuits! Joe and I are about to turn to "fried food."

Next Monday, Joe "checks out" at Randolph and we leave here around 10 or 11 — as soon as we can anyway. Joe wired his folks to come and drive out to Calif. with us. We haven't heard from them yet, so we don't know whether they will or not. Joe feels it would give him more chance to actually be with them, and I didn't have the heart to veto the idea, although, it will make it crowded for such a long trip. However, I did want them to come see us while we'd have so much room in Calif. cause goodness knows when we'll even have another place large and convenient. After this week my address will be:

General Delivery
Balboa Island, Calif.

It will probably take us 3 or 3 ½ days to make the trip as Joe has to report at Santa Ana on the 12th. I'm looking forward to the trip — no cooking — no diaper washing — I just sit and ride. Joe never lets me drive. He says taking care of Kay is enough. He does all the driving, packing and unpacking — so you see I really get a change and when we stop he helps with Kay.

We've left our car at the garage for a week to get a new top (old one was leaking) having everything checked over and any necessary repairs made. It should look and drive like new. It really has been a good car, and we've really enjoyed putting some mileage on it.

Love to my two honeys,
Babe

From Babe in Casper, Wyoming to her parents in Memphis

May 31, 1943

Dearest Mother and Daddy,

Your 3 gypsies have arrived safely in Casper, Wyoming. Joe received orders Tuesday to come here. We left Boise Thursday and arrived here Friday evening. We had a very pleasant trip — the weather was grand, and we didn't have a hard time finding places to eat and sleep. This part of Wyoming is very pretty; although the climate is most uncertain. We were greeted upon arrival in town by a hail storm, but the rest of the day was beautiful.

We looked for a place to live, but fortunately, we hadn't been able to find one, as Joe received orders to go to Tucson, Ariz. where he and three other doctors will be assigned to the 451st Bomb Squadron or Group. Those are the big four motored bombers. I expect this is his permanent assignment for the duration and wherever this group goes, he will go too. I think they usually get a couple more months training in the U.S. before going overseas.

This is so the doctors can get personally acquainted with the men in his group and know better how to take care of them mentally as well as physically. I really think it is a relief to Joe to have some definite goal in mind, instead of this being pushed from pillow to post with nothing in particular to do.

Joe is out at the Post now, signing out and getting our gas coupons and as soon as he gets back we shall leave. His duties at Tucson begin June 7. We'll probably get there on the 4th and have that extra time to look for a place to stay. However, it probably won't be long before you and Daddy will have a couple of "hearty-eaters" (me and Kay) on your hands — for as soon as this Group starts training, no families are allowed to go along — so don't eat up all those good things from the farm before we get there!

It's a little better than 1200 miles to Tucson, so we're looking forward to another nice trip down through Colorado, New Mexico, Arizona. We'll get to see parts of the Rocky Mts. that we haven't seen yet — even Pike's Peak. We only have to retrace about 100 miles of road that we have been on before — so it will all be new and interesting. We'll wire you as soon as we arrive in Tucson.

Kay is now getting ready for her morning nap and I must start getting things collected so I'll be ready when Joe comes. Still think I've got the swellest husband in the world!

Our best to you both,
Babe

P.S. Due to traveling so much we have more coupons than we need, too, honey, but thanks just the same — I'm saving the sugar ones for you.

* * *

By June 12, Joe and Babe had moved from Arizona to Tennessee. They were still together during the first week of Joe's assignment at Dyersburg. To prepare for his deployment, Joe helped settle Babe and Kay near her parents and in-laws in Memphis. He purchased a fully-furnished house on Spring Street, where Babe moved, while expecting their second child.

For the next six months, the 451st continued to grow, adding to its personnel and becoming four squadrons: the 724th, 725th, 726th, and 727th. They were stationed in Dyersburg, Tennessee; Orlando, Florida; Wendover, Utah; and Fairmont, Nebraska before staging in Lincoln, Nebraska.

Dyersburg Air Base was built in 1942 as a combat training center for B-17 airplanes. After the 451st arrived at Dyersburg, the unit was divided into two groups: the ground echelon and the air echelon. The ground echelon stayed in Dyersburg to finish building the 451st unit with men from the 346th Bombardment Group, while the air echelon went to Florida for Army Air Force School of Applied Tactics (AAFSAT) to learn how to fly B-24 Liberator airplanes.

During AAFSAT, the pilots and crews trained for their upcoming overseas duties, bombing missions, and simulated combat conditions. Joe accompanied the group to Florida and spent 4 weeks there. While previously in Santa Ana, Joe had completed a physical exam and requested flying status. His request was approved, and he was able to participate in flight training with the rest of the unit in Florida. Joe continued to log flight hours throughout his time with the 451st. With his flying status, Joe qualified for increased pay; however, even though he learned the basics of piloting, as a flight surgeon, he never participated in combat missions.

AAFSAT involved 10 days of classroom training in Orlando and 20 days of flight combat training in Pinecastle. Training in Florida was less than ideal; the men lived in tents during a very rainy summer, the airplane mechanic crews were short handed and lacked necessary tools. Four B-24 airplanes were sent to Pinecastle for training. However, three of the planes were inadequate for the missions assigned and were returned, causing a short delay in the schedule. In spite of the challenging conditions, the 451st excelled in its AAFSAT training and missions. After the completion of AAFSAT, the air and ground echelon were reunited in Wendover, Utah.

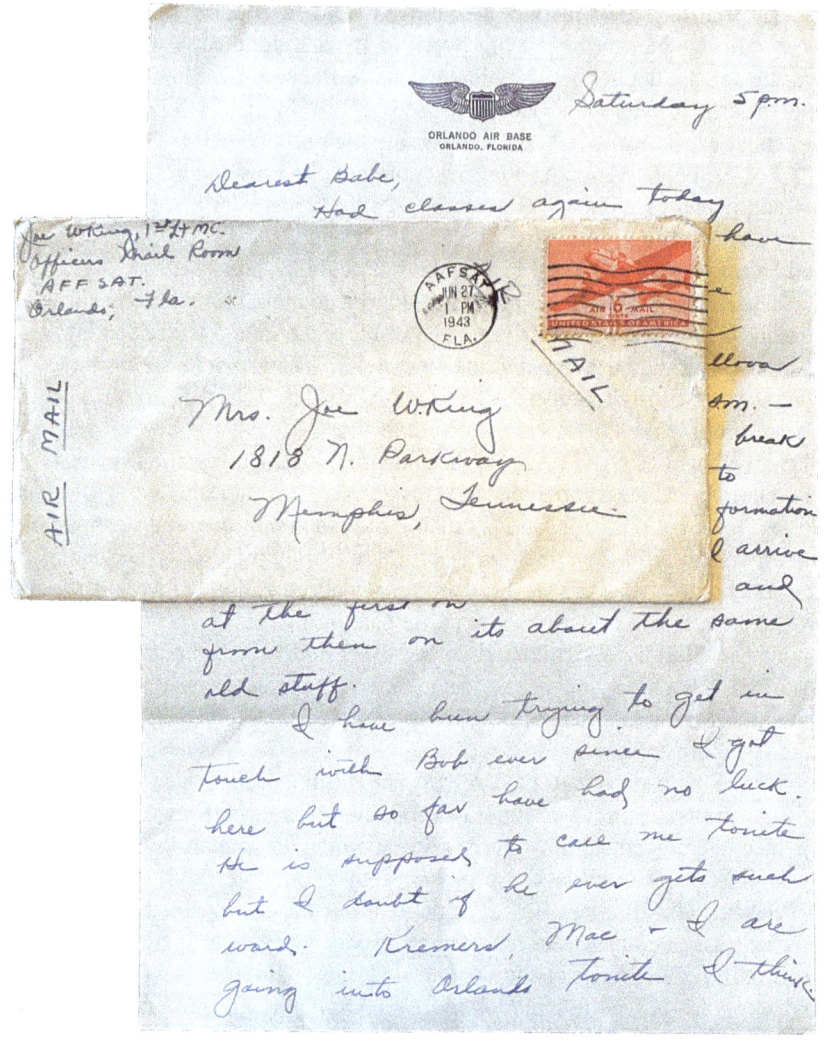

From Joe in Orlando to Babe in Memphis

June 23, 1943
Wednesday, 9:30 p.m.

Dearest Babe,

Dropped you a postcard this morning in Jacksonville. We pulled out there at about 10 a.m. Finally arrived in camp at 5 p.m. The whole trip was quite a mess. We started out with three tourist cars at Memphis, and when we got to Atlanta the following afternoon they took one of those from us. So we made it the rest of the way with only two.

Yesterday when we arrived in Atlanta, four of us went to one of the hotels and got a room just so we could take a bath. It was one of the nicest hotels in town, and they only charged us $2.00 for it — that is for the four of us. That was the dirtiest trip I have ever taken, even in my bumming days. All of us were so terribly filthy when we got here. When we got off the train in Orlando, it was raining and it continued to do so while we unloaded the cars and waited for transportation out to the post. Two of my cleaned shirts in the black suitcase got pretty wet, but I will be able to wear them anyhow. The rest of the stuff came through ship-shape.

By the way, we are seven miles from Orlando. We have to be on the base at 11 p.m. every night except Saturday, which is 2 a.m. We sleep in barracks — 30 men to a barrack. Make your own bed, etc. We have more outdoor latrines — 1 latrine to 4 barracks. Even more rugged than Dyersburg. No Officer's Club. I have been trying to either find a coca-cola or a cold drink of water ever since we hit here, but so far no such luck. I have now given that up until tomorrow.

Sorry, I couldn't be with you for a longer time at the station the other night, but I couldn't. They wanted to load us up and get us accounted for on the train before it pulled out. Too, it was a good thing in a way for I can't take it.

There is no use telling you how much I already miss you — much to your surprise. But telling you of such and wishing to be with you through Uncle Sam's mail just makes it worse for both of us, so I will try to refrain from mentioning it too much.

Now don't worry about me for I'll be O.K. We are starting to get in condition right away — we will have calisthenics every morning at 6 a.m., so you know I will be in good shape. Give my love and regards to everybody. Remember, honey, when you are trying to get everything straightened out up there and in your further trials and tribulations that I wish I could be helping you to get through it all. Too, remember how much I love you. Don't worry, everything's going to be O.K. in the end, and we will make up for being apart later.

Tell Kay hello for me and take care of both of you.

> I love you,
> Joe.

> June 25, 1943
> Friday afternoon, 4:45 p.m.

Dearest Babe,

Well things have been happening fast since we have been here, and today I started going to school again — wonder if I will ever get old enough to quit going to school? Yesterday was an off day. They let us rest up. Yesterday afternoon Kremers, Wagner (our group surgeon) and myself went into town, had lunch then went to a "double-feature" picture show (as usual, I have forgotten the names), then had dinner and returned to the barracks and to bed. None too eventful — with the exception that we saw a man and his wife on the main street immediately after he slashed her throat in an effort to decapitate her. Anyhow someone stuck their finger on the bleeding jugular and saved her life.

This morning, we had a talk by General Peabody, then attended lectures. This afternoon we had a demonstration of fire power. A piece of shrapnel from a 300 lb. bomb that was detonated 1000 yards from us caught one fellow in the thigh, and he was carried to the hospital — no other casualties. The above is all the blood shed that I have seen since I have been here. They should break us in gradually and realize that we medics haven't seen blood since private practice.

I received the Trust Deed for the house and the notes this morning and signed them and will send them back to George Abernathy when I send you this letter. I will not send them Air Mail, however, so they will take a couple of more days to get there. I hope that you continue to be as enthusiastic about the house and don't find some glaring fault with it when you move in that both of us should have noted right off. I also am interested to know what Mr. Black thinks about it when he gets a chance to see it.

Another thing — for goodness sake, don't you go over there and start moving things around, scrubbing, washing woodwork, etc. Not only must you realize that you are in no shape to do such, but you must also realize that you gotta stay in fair shape to look after Kay. There's just you and Kay for a while you know. It may not be a bad idea to be looking around for someone to help you out all the time.

Honey, Kremers and Wagner have just come by for me to go to mess with them. Since I don't like to eat by myself, I think I'll close and go eat. Kiss Kay for me and take care of yourself. Remember that I love you,

Joe.

June 26, 1943
Saturday, 5 p.m.

Dearest Babe,

Had classes again today but finished at 1:30 p.m., and have been laying around ever since. I suppose from now on however, there won't be a hellova lot of loafing. Get up at 6 a.m. — stand formation at 6:20 a.m. — go to mess — finish mess and fall in formation at 7:30 — march to classes and arrive at 7:50 a.m., and from then on it's about the same old stuff.

I have been trying to get in touch with Bob ever since I got here, but so far have had no luck. He is supposed to call me tonight, but I doubt he'll get the message. Kremers, Mac and I are going into Orlando tonight, I think to have dinner and go to a show. This is the dullest post ever without an officer's club.

By the way, the deputy commander of this outfit told me today that we wouldn't leave the U.S. until December or January, so perhaps we will get to live together some more before we pull out. Of course, it's best to take such stuff with a grain of salt because if they need us badly at any time, I expect they will pull us out after we have had some training. At least, I will be here when the next "little one" comes along if all these rumors are true. That will help some.

As yet, I haven't received any money from Dyersburg. As soon as it comes in, I will send most of it to you. It should be a considerable amount for all of that travel. I won't get last month's pay, however, until I get paid for this month. I will send you most of both when I get them, and you can either send them to Helena National Bank or use it as you see fit. By the way, I don't get paid for any travel from Dyersburg to here because it is considered a troop movement.

I just answered the telephone and am supposed to meet Bob at 6:45 downtown. I hope it won't be raining, but it will. It has rained at least twice every day since we have been here. Not like when you and I were down here — (It's raining already).

I almost envy you in fixing up the house. I wish I were there to help you. We should be able to make it a nice, livable, comfortable little home. Tell Kay that I love her — in fact, next to you, I love her most of all. Take care of yourselves.

All my love,
Joe.

June 27, 1943
Sunday, 7:30 p.m

Babe dearest,

I saw Bob last night — he looks about the same, only a little more sun tanned. He's griping about this place, and I expect he's got a gripe coming. I suppose we can stand all this pretty easily — think how much tougher combat is going to be. Anyhow, from all Bob Lewis says, we won't learn anything here.

We all went to town last night, had dinner and then went to the Officer's Club uptown (which isn't an officers club at all but just a dump) and had a couple of drinks and then came home. You have to go uptown to eat as often as possible for mess out here is the worst I have ever seen by far.

This afternoon all the M.D.'s in our outfit (all 5 of us) went to the picture show — Cary Grant in Mr. Lucky — which was fair. We then came by the Officer's Club at the Orlando Air Base, which is an entirely different outfit from this and is 3 or 4 miles from here. They were having a Bachelor's Officers party of some sort. We went in and had a couple of glasses of punch apiece, then went to dinner (which was excellent), and then came on back to AAFSAT. Seeing all the boys and girls at the party made me homesick. So I'm just sitting here writing and thinking of you and Kay. I have a new Reader's Digest so in a minute I will start on it and to bed later.

Next morning, 7 a.m.

Somehow or another we all got into a bull session last night, and neither did I get to finish this letter nor did I get to read my Digest.

Dearest Babe,

I'm enclosing $20.00 more in this letter. Maybe you won't run out. Maybe my travel checks will come in from Dyersburg pretty soon, and I will send them either to you or Helena National Bank. Perhaps with them, the $150 sent to the bank on the 1st and my salary, we can keep from going into debt. From time to time, let me know how the business comes out and how you stand with the world — I can't keep from thinking about it some.

Well honey, it's about time for me to start classes. Be a good girl and take care of yourself for me. Hope Kay isn't too cross while she's cutting these teeth. Tell her I miss her but most of all I miss you.

I love you,
Joe.

June 28, 1943
Monday, 5 p.m.

Dearest Babe,

Received a letter from you this morning. I sent you $20.00 more yesterday — let me know if you get it. I haven't written to the folks as yet — you can tell them I'm getting along O.K.

It's a good feeling to know that 1 week from today you will be living in your new home. I certainly hope that it proves as satisfactory as my anticipation. I can hardly see how that can be possible though. We should have bought a place long ago — I've had a lot of fun thinking about it even though I may not ever live in it.

Got a letter from the Office of Internal Revenue today acknowledging my Income Tax Return (believe it or not) saying that they had received it. So now I really don't know what to do. However, even so, as soon as I can, I am going to fill out an estimated return and send it in. I don't believe I will pay anything on it, however — just request exemption until after the war. There is no doubt that they will have to make some kind of provision for us guys after the war or they will have to send damn near everyone to jail. Anyhow I'll attend to that when I can get around to it.

The other boys have gotten their travel pay that was due them at Dyersburg, but so far mine hasn't come through — I wish to the devil it would, so I could get it on the road to you. This morning we went to some very dull lectures for four hours, and this afternoon we went on a field trip to inspect "RADAR" equipment that they have strung along the coast. It really is remarkable what that stuff will do. In fact, it will damn near tell you what color the plane is.

Well, I have just gotten back from seeing Clark Gable and Lana Turner in "Somewhere I'll Find You" again. Remember it? We go over to the show here on the post for a turkish bath. It's the hottest place I have ever been. We don't dare go over there unless you have intended to change clothes the next day.

This bunch we're in is really a bunch of screwballs. They seem to be pretty nice boys but are as screwy as they can be. We got one guy in here that was assistant professor of genetics and anatomy at Yale for 11 years — he wants to tell me about his past illnesses all the time. Another was a reporter on the N.Y. Telegram — he borrowed $10 from me this morning. Another's Mother and Dad has never been married — apparently just "common law" proposition — he doesn't mind at all telling about it and talking about it and wanting a 30 day furlough for he says he has "Navigator's Fatigue." There are many, many characters of all different dispositions and previous varying walks of life, and somehow they get along without too many arguments and are at least happy on the surface. Of course this is all uninteresting, but somehow you just wonder how they all live together in one room.

Honey, I appreciate all the letters — they are a hellova help. I shall love you and Kay always.

Joe.

June 29, 1943
Tuesday night, 10:30 p.m.

Babe Dearest,

Well the travel check came in this afternoon's mail, and it should be enough to help you out some. I believe if I were you, I would send it right on down to Helena National Bank. This $250, plus the $150 that goes to the bank tomorrow, would just about make up what is, or will be, overdrawn. Then when I get paid for the past 2 months, I will also send it to you, and that will provide you with about a $250 capital to operate on. You need at least that much. That, however, is going to damn near leave me flat. I'm not going to write a check on Helena National if I can possibly help it, so I may write you for money if I really have to have it. I'll let you dole it out to me, so I can have the embarrassment of asking you for it. Good enough for me since I've been doing you that way for 3 years. I suggest for a couple of months, that is until we can build up a little reserve for you to operate on, I would not spend anything that wasn't absolutely necessary for you need a little extra in the bank. I realize you are no child, honey, and you must forgive me for writing as if you were one. I don't mean any harm, and I know you will take care of it as good or better than I — I'm like an old woman and just wanta get in my two cents worth.

Tonight, the five of us Medical Officers went uptown to dinner again. We had a couple of rounds of scotch first and then felt pretty good. There was no meat on the menu but fish. Believe it or not, I ate fried scallops. They, to me, tasted like all other kinds of fish. I can't say that I enjoyed them too much. After that we came home. I then entered a bull session on airplanes. On talking to the boys, I found out that Tink gave me a pretty good course in Aerodynamics. Of course, I act dumber than I am about such.

I'm getting pretty tired of being without my wife. Living by myself is certainly everything but fun to me. Lord only knows what I will be like in 3 or 4 months. Save your money for I may be sending for you for a week or so even though I couldn't be with you very much.

It's after time to go to bed — Kiss Kay for me (if you can hold her long enough). I love you more than it seems possible.

Joe.

June 30, 1943
Wednesday, 10 p.m.

Dearest Babe,

Today is the first time that I haven't gotten a letter. I realize that Monday you were probably terribly busy putting across your deal. Hope you didn't have too much trouble and wish that I could have been there to help you.

I can tell right now that this letter is going to be very disconnected for there is the biggest argument going on around here about which is the better plane, a B-24 or B-17.

The following day — July 1

As you can tell it is impossible apparently for me to stay out of arguments. I got in the one above and didn't get to finish my letter. For that I humbly beg your pardon, and I will try to stay out of discussions when I should be writing to you, but for some reason or another they conglomerate around my bed for the bull sessions. I suppose that means I'm more full of bull than the rest of the guys. Anyway, I will try to get this letter in this afternoon's mail.

I received two letters from you today. One of them contained the abstract of your first business deal. I'm glad you didn't have to pay but $2100 more. I still believe it was a good deal — the more I think about it the more satisfied I am.

I went up for a spin in one of these huge C-47's yesterday. They can carry a couple of jeeps on the inside or be used as an ambulance plane and carry 18 or more beds. They are pretty big. In fact, our whole class went up in the thing at once. This is restricted stuff and shouldn't be told I guess.

I cut a class while ago and went and got a haircut — perhaps I got scalped instead. I don't have any sideburns for the first time since I was 8 or 9 years old and look rather peculiar. I just sat and let him cut away, and sure enough when he got through I didn't have but damn little hair left. It doesn't make any difference — it'll be grown back out on the sides in three or four days.

At this moment, honey, I'm having a hellova conflict. I'm just about to walk over to the telephone, call you and tell you to come down here. I don't know how much time we would get to be together — probably not much, but I certainly would like to see you.

Honey, I've sent you a check for $259 and two $10 bills lately. Let me know when you receive them. I gotta hurry so this will get in the afternoon mail. Love Kay for me. Tell the folks I'm O.K.

All my love,
Joe.

* * *

Joe and Babe had only been separated for a short time. However, Joe sent Babe a telegram on July 1, asking her to join him before the unit was transferred to its next assignment in Utah. She was in Florida from the beginning of July until Joe left around July 20.

From Joe in St. Louis to Babe in Memphis

July 20, 1943

Dearest,

Arrived here at 6:30 a.m. All is well. Leave here 8:30 for Kansas City. I hear that they are going to hook us behind a troop train, so I know what kind of a trip we are going to have — another cattle train engineer, I'll bet.

Only laid over about 1 hour in Evansville, Ind. Sorry town. Slept pretty good last night — don't know whether it was cooler, or I was just exhausted. Regardless, the results were the same.

I'm wondering how you got home. Bet 2 to 1 you had to ride a coach into Memphis and that you are just about getting in now. Hope you were able to get a Pullman, but I almost know you weren't. I want you to know how swell it was of you to come down and stay with me.

Don't work too hard with the new house until you have had sufficient time to rest up. Even then, don't work too hard. Take care of yourself and Kay. I miss you already.

All my love,
Joe.

* * *

Joe moved to Wendover Field in Utah, approximately 100 miles southwest of Salt Lake City. Wendover Field was another air base used to train heavy bombers: B-17, B-24 and B-29. However, Wendover presented the 451st with a new set of problems. The air echelon arrived first to find that the base was overcrowded, causing a severe shortage of living and office space. The 451st quickly put up tents to accommodate their men. To make matters worse, the group was split into 3 sections scattered across the base. The men also had to deal with high winds, dust, and haze. The winds blew down tents, causing the men to continually rebuild their living quarters and chase papers and supplies scattered across the salt flats. In the midst of all this, training began and new crews arrived.

Prior to World War II, upper level medical commanders wanted to create an area of medicine specifically for airmen because of the unique challenges associated with high altitude combat. In addition to the physicals, examinations, immunizations, and pre-flight checks that Joe mentions, his job was to monitor the physical and mental health of the airmen. After Joe completed the School of Aviation Medicine at Randolph Field, his title changed to Flight Surgeon, although surgery wasn't in his job description. His training covered medical problems specific to aircrews, including anoxia (lack of oxygen), aero-otitis media (ear problems), frostbite and extreme cold conditions, and psychological impacts of flying.

Starting in Wendover, Joe and the other medical officers — Wagner, McFarland, Quinn, and Kremers — worked out of a dispensary, a small medical tent with supplies, records, and exam space. Joe mentions performing 64 exams, which were a comprehensive evaluation for every crew member that included a physical and psychological exam and were required before the crews left the United States and again prior to their return.

From Joe in Wendover, Utah to Babe in Memphis

July 23, 1943
9:30 p.m.

Dearest Babe,

It's just going to be a note tonight, Honey, for I'm so damned tired I can hardly sit up. We finally arrived at this God-forsaken place (and I mean exactly that) this afternoon just after lunch. We are staying in barracks just like the ones at AAFSAT, i.e. 30 to a room only these are tar-paper. Wendover consists of one building, which contains the railroad station, cafe, and hotel. Nothing luxurious, I can assure you. It is on the salt flats. Looks just like snow and you have to wear dark glasses in the daytime or have snow blindness.

We laid over in Salt Lake City last night. Our troop train didn't get here until 8 this morning — even though it left 14 hours before us — a distance of 125 miles. That shows the swell service given to the troops. In Salt Lake last night, we all went to a night club and danced and drank. I missed you there too.

Upon arrival here we found out that we may, and probably will be here until October 1. By then we will all be "stir-crazy." I wouldn't be surprised if that wasn't changed however for there just isn't room here for us. There are three outfits here, and the place is built to accommodate about half that. We are going to have to put up a 16x16 tent for our medical dispensary. I know that is just going to be swell for 5 doctors and 40 enlisted men to work in — don't you think.

You know this army isn't so bad when you are along, but it ain't worth a damn when you aren't. Too, send me another picture of Kay — you know I haven't seen her in about 6 weeks. Pretty soon it will be hard for me to realize what she looks like at the present time — that's awful, ain't it? I love you two with everything I've got. Remember that.

Your husband and daddy,
Joe.

P.S. Send my foot locker please. Thanks.

July 24, 1943
Saturday night, 10:10 p.m.

Dearest,

Received a letter from you today and was awfully happy to hear from you. I had just about decided that I was the forgotten man when there wasn't a letter waiting for me.

I have worked today for the first time in I don't know how many months. Nasty work too — trying to beat all the supply sergeants out of supplies. I finally got 2 ambulances and a couple of tables, but still no chairs. Our dispensary for the whole group is 2 tents now instead of one. Getting started is going to be a job.

I wish you could see these damn salt flats. Just as far as you can see to the east of us is just as level as it can be. Not a blade of grass or a tree or a house or nothing — just salt and I mean plenty of it. There is not one weed around here. Not a bush or tree within sight. It's almost another wonder of the world — but it is not an ideal place to live. There is one thing that's good however, and that is the food is wonderful.

Sunday, 2 p.m.

Here it is Sunday afternoon, and we are doing just like we do every other day. You just don't keep up with the days. They had a dance out here last night. Imported girls from Salt Lake City for it. I stayed in my barracks and read. However reading to me is just about like going to a picture show lately — somehow I don't remember it.

By the way, I hear that we are going to get travel pay from Orlando. If and when we do, I will send you some money. I know you are going to be running short pretty soon.

I certainly am eager to find out how you are getting along living in the new house and how you are going to like living by yourself. Certainly do hope the dryness didn't ruin the grass or shrubs. I doubt that any damage could be done to those trees, as I remember they were pretty large and sturdy.

What are you going to name the new one when she or he gets here? We should have a name picked out and have some clothes for the baby. I believe if you look through what was left over from Kay, you won't have to buy much. Poor little thing — here I'm already suggesting that the baby take "left-overs."

A couple of us really got gypped last night. We walked up to the hotel to get a drink and found out that nothing but enlisted men could get a drink on Saturday night. So we walked another Utah mile down to a grocery store, so we could buy a bottle and take it to the Officer's Mess for a drink, and when we got there we found that we couldn't buy whiskey due to the fact that it was a holiday. Now what holiday July 24th is, I haven't found out. Anyhow, we came back to camp empty handed and when we went to chow, they had run out of steaks, which they had for supper, and we had to eat hamburgers. So we got gypped all around. You are getting gypped all the time here, anyhow, for whiskey is rationed, and you can only get one quart every two weeks. I suppose they have to do that to keep everyone living in these conditions from being sots.

Write me, honey, for I can look right into my empty mailbox from here, and I am eager to hear how you and Kay are faring. Tell Kay that I love her. You know I love you best.

All my love,
Joe.

P.S. Temperature at present is only 124. Yesterday it was 120 but you don't sweat. Orlando and Dyersburg really seemed worse. The boys send their regards.

July 27, 1943
Tuesday, 11 a.m.

Dearest Babe,

My squadron was granted a 48-hour pass today. Most of the boys have gone into Salt Lake City. There is no reason for me to take one for there is no place to go. No golf course and no place to fish. There are plenty of things I can be doing here, so it's just as well.

Well I'm through with sick call. I just got and read a letter from you — it sure is nice to hear from home. In your letter today, you mentioned how much you and Ernestine had been working. Again I want to insist that you take it somewhat easy. You have plenty of time to fix up your house.

You asked about bathing facilities. Yes, we can bathe daily if we want to now. However, if you ever get soap on you, it's hell to get off. Instead of having the usual kind of showers, they have showers that put out a fine spray. It takes 5 minutes to get wet under one. No doubt about it, they really conserve water. This place really isn't so bad though. Probably nothing like as bad as it sounds. I believe I like it better than Pinecastle. The food is better. It costs to live out here though. Food costs $1.50 per day. They charge to live in the barracks, cokes are 10¢ each. Every little thing you do costs something. They should be making a nice little profit at the Officers' mess. Everything is higher, I expect, because of the transportation problem. Supplies are hell to get. We can't even get one foot of lumber. I took my two ambulances and just drove up and down the streets of the post yesterday. I finally found a dump of lumber, so I stopped my ambulances and loaded them both up to the top. I gotta have some tables and benches in the dispensary, and as far as I know, that's the only way to get them. So now I have some tables and benches.

Sure was swell of Dad Black to build Kay a sand pile. I surely do appreciate it. I really believe the folks back home are going to look after my family for me and that helps a lot. How does she like it?

I'm glad to hear about Sis. I know she is happy about being pregnant again. Tell her I'm surprised that I hadn't heard about it before now for I'm sure she had already written and told everyone but me. I wrote her a note today, and I'm going to write Mother and Dad today or tomorrow.

Kiss Kay for me and try not to let her forget that she still has a daddy who would love to see what she looks like with a mouth full of teeth. Now, for the second time, take care of yourself and don't push yourself.

I love you and Kay and it (her),
Joe.

July 28, 1943
Wednesday, noon

Dearest,

I was very popular this morning. I received two of your letters. The travel pay hasn't come through yet, and I don't know now when it will. Write and tell me how the finances are doing. I'm worrying about that a little. If there is anything I can do, let me know.

You mentioned a new tire on the car. I suppose you are talking about the one the old man and I fixed in Roswell N.M. I surely hope something hasn't happened to where you had to get a new one. By the way, did you ever get those two slick ones retreaded? I don't imagine you have 'cause I know you haven't had any time.

Sure am glad to hear that Hugh is connected to a service center. That should be damn nice work in pleasant surroundings — i.e. as pleasant as is possible. I know Mr. and Mrs. Black feel a whole lot better about him. Really that's swell. I wonder if he would like to trade jobs.

I have been up to the hotel a couple of times, which is just outside the gates up on the highway. One night, they wouldn't let us in for it was enlisted men's night, and the other time I had the usual drink and came home. I have played a few games of pool at the Officer's Club and a few games of cards, but I'm not too crazy about this Officer's Club. It's like Dyersburg. There are too many officers' wives there. Here there is no other place for the officers' wives to go, and they have just about taken the club over. It's annoying — particularly when you are already lonesome — to walk in and see a bunch of guys and gals together. There have been no picture shows on the post that I have not already seen. At present, Abbott and Costello are in the picture that you and I saw in Orlando.

You are going to have plenty of time to fix up the house before I get to Memphis apparently, so don't hurry and wear yourself out. From all we can hear, we may be here quite a spell. September 1 is probably the earliest possible date that we can leave here but probably not until October 1. Then somewhere else probably in the U.S. — I certainly hope so. I, at least, want to be here when the little 'un gets here. I surely do love you and wish I could tell you that personally. Tell Kay hello.

All my love,
Joe.

July 29, 1943
Thursday, noon

Dearest Babe,

First, let me tell you about where to find the ticket about buying that new tire. It is in the compartment of the car — I'm not sure whether it is with the registration papers and such or not, but I'm quite sure that it is there. I have looked through my stuff to make sure, and, so far, I can't find it. Tell me whether you have been able to find it or not.

I received your letter and Jean's this morning. Will enclose the Craycrofts baby's picture in this letter. When I opened the letter and the picture of Carolyn fell out, I was amazed. At that particular time I was talking to a Major. He asked me if that was my baby, and I told him I guess it was but she certainly must have changed a hellova lot. It took me about another minute or so to realize it wasn't Kay.

I was going up on a flight to Boise, Idaho this morning, but when we got out to the hangar, the #3 motor wasn't acting right, so we didn't go. I have gotta get in some time before the end of the month, however, because one of these days I may get authorization for flight pay and sometimes it comes through retroactively, and I wouldn't like to miss that little extra cash.

From your letters, it sounds like the folks are dropping by quite often. I certainly hope they aren't annoying you. Probably as soon as the "newness" wears off, they will slack up a bit. I know that you like to have some time to yourself.

I have to go this afternoon and vaccinate and inoculate some of my men. In the morning at 5:30 I am going over and do a famous "short-arm" inspection of my men. Sounds like Boise, Idaho doesn't it.

By the way, Elliott Arnold, is going to get married in a few days up here to some girl. He has only known her for two weeks. Personally, I'd hate to have to start off a poor newly married girl in this hellhole, but you didn't get started off in a bed of roses, did you? Even so, this is worse than Helena. I haven't seen a civilian man in so long, I won't know what one looks like.

Went over and got a haircut yesterday and, boy, here they really cut it. When you sit in the chair, the barber doesn't ask you how you want it cut, he just starts cutting. Everyone around looks like a skinned rabbit. It doesn't matter because no one is going to see you anyhow.

Well, honey, I've got to shave before I go over to tend to the boys, and I haven't got too much time. I love you and Kay and the little one very very much. Please take it easy.

All my love,
Joe.

Dearest Babe,

My dearest,

As per usual, I received a letter from you this morning. Sure is swell to look over in that mailbox and see a letter from you there.

Some interesting things have happened since I wrote to you yesterday. Last night a couple of the boys and myself went to the picture show (it was lousy) and we returned to the Officer's Club, and they were paging all of the medical officers of our outfit. I answered and was told that there were two full colonels and a major in our dispensary tent looking for a medical officer. I immediately proceeded for that neck of the woods. When I got over there, it was Colonel Kinard and Colonel Smith, who are respectively the surgeons of the Second Bomber Command and the 15th Bomb Wing. It so happened that I had met both of them previously. Colonel Kinard is the doctor who was with the 19th Bomb Group in the Philippines when the Japs attacked. He talked to us at Randolph. I had met Col. Smith while at Boise. Anyhow, they were swell. Treated me like I was a doctor instead of a lieutenant. We discussed various problems confronting us and cussed AAFSAT, praised Randolph, and had a good talk in general. They promised to have us a building for a dispensary before Monday and told me if we didn't have one by then, to call them at Boise — so I imagine we will get it. After sitting over there for quite a spell, they took me to the Officer's Club and bought all of us 4 rounds of beer (can you imagine me drinking that stuff). Anyhow, it looks as if we are going to get something for a change — but that Colonel Kinard is one swell guy. He has the most charming personality that I have ever come in contact with and a ready smile. I am sure that you have heard me talk about him after he talked to us at Randolph and told us his experiences on Bataan.

Too, it seems that they believe that this outfit will ultimately be broken down into training units and that we may not go overseas until spring. It seems that there is not too much demand for O.T.U. outfits like ourselves at the present.

They didn't say all of this, of course, they just mildly intimated it — so, for gosh sakes, don't count on it. At least, all of it is interesting conversation.

I am going up for a 4-hour flight at 1:30. Don't know where to; probably just up to check instruments and such.

Enclosed I am sending you a little money. I will send you some more pretty soon. We should get our paychecks on the first. Hope you aren't too short. I know you have needed a lot with the moving. Sorry you didn't have it.

Seems as if my letters must take a pretty long time to get to you. Most of the time, it only takes 3 days for yours to come here. However mail leaves the post only once per day — at 2 p.m., and I know the train for Salt Lake doesn't leave until about 6 p.m. — guess it takes at least one day for it to get to Salt Lake.

You know, honey, I love you a lot. I'm so terribly glad you are satisfied with the house. Keep taking care of yourself and "ours" and one of these days we are all going to make up for lost time somehow.

I love you,
Joe.

Joe was required to fly every month in order to keep his flight pay. By the time Joe was discharged from the Army Air Corps, he had flown 178 hours.

July 31, 1943
Saturday, 1:45 p.m.

Dearest Babe,

I received the letter today with the newspaper clipping about the house. I was glad to get it. In fact, I had forgotten that there was so much shrubbery around the front of the house. The house looks pretty good in the picture, doesn't it? Too, the reporter gave you quite a write up. You must have made eyes at him to get all that space in the paper without paying for it. As it was I thought it was O.K.

Had a nice flight yesterday. We left at 1:30 — flew over the salt flats then over the Great Salt Lake and then to Salt Lake City. We turned north and went to Pocatello. I took the co-pilot's seat after we left Salt Lake and flew it (rather poorly). We flew up there with a 12 man crew. Really enjoyed it. As soon as the boys take a trip southward, I am going along and we are going to fly over the Grand Canyon. That should be sumpin' at 12,000 ft where you can see the whole thing at one time.

Last night we all went up to the State Line Hotel to give Arnold a stag party. We all just sat around, talked, and tried to drink up as much scotch as we could. Got back to the base at about midnight. That isn't enough sleep for me anymore. Today I'm dead tired.

You surely are getting smart putting up all the canned stuff — and real domestic too. The chili and tomatoes both sound good — wonder if they will last long enough.

We are working pretty hard now, sure enough. Getting everybody ready for overseas is a job. This week, we have been doing immunizations. Next week, we are going to start on overseas examinations. It's really going to take some

HOUSE, FURNITURE AND VACUUM CLEANER—Dr. and Mrs. J. W. King, formerly of Helena, Ark., bought from Mr. and Mrs. Oscar E. Wilkins everything in this home at 804 Spring but the Wilkins' clothes and dishes. There is even a Victory garden in the back yard. Stanton Abernathy of George M. Abernathy & Co. handled the sale.

* * *   * * *

Army Couple Buys Home And All the Furnishings

Dr. and Mrs. King and Baby Move In And Oscar Wilkinses Pack Only Clothes

Mrs. Joe W. King today was still poking about her new home at 804 Spring, finding new things she didn't know she owned.

Mrs. King's husband is an Army doctor. Mrs. King had been traveling about the country with him, until he finally was assigned to a place where Mrs. King could not go. So it was decided to settle Mrs. King and the Kings' 18-month-old daughter in Memphis.

"We saw an advertisement in The Press-Scimitar," said Mrs. King. "We drove right out to the house, looked at it for half an hour, then bought it."

Mr. and Mrs. Oscar E. Wilkins sold the house, stove and refrigerator, furniture, pictures on the wall and even the vacuum cleaner, to the Kings for about 7000. The Winkinses were moving to California and didn't want to be burdened with their furnishings. The Kings needed everything to set up a home.

But Mrs. Wilkins was sad to leave her home. It was neat and convenient. It had an attached garage that had double doors thru both the front and back. It had a very pretty front screened porch and large awning. There was an attic fan.

"Listen," said Mrs. King, "I hate to take your home from you. When the war is over, come back to Memphis and maybe we can sell your home back to you."

Stanton Abernathy of George Abernathy & Co. handled the sale.

Joe and Babe bought a fully- furnished house on Spring Street in Memphis from Mr. and Mrs. Oscar Wilkins for about $7000. Babe lived here with her children until Joe returned from the war. The Memphis Press-Scimitar, 1943.

time to get all of this stuff done. Too, if we find a place to move the dispensary, that's going to take just that much more work. I am getting in new combat crews all the time. When each one comes in, it means a few days' work within itself. Maybe I'll catch up some of these days.

Maybe you shouldn't write me about Kay's antics concerning picking up my picture and saying "My da-dee". I don't know whether I can take it or not. Sorta hard to. However, you know I want to hear about everything she does and says — gosh, I'd like to hold her — or is that still an impossibility? Honey, I haven't got time to write more now. I'll do better tomorrow. Take care of yourselves and remember I love you with all I got and will continue to do so.

Your man,
Joe.

August 1, 1943
Sunday, 10:30 p.m.

Dearest,

I'm getting up at 5 a.m. in the morning to go into Salt Lake. I'm going for two reasons — firstly, I'm going to buy my wife and daughter a little something, so they won't think that I have forgotten them. Secondly, I'm going to get away from this hole for half a day.

I certainly have been thinking of you a lot today. For some reason, it has been even worse today than usual. You know, when I'm in this present mood, the army had just as well let me go home for I'm not worth a damn to the army, myself, or anyone else. I really believe you would be terribly surprised to know how many hours of the 24 that I'm thinking of and wondering about you and Kay. You mentioned me in one of your last letters as a good soldier. Not to make you lose confidence in me, honey, but I expect I'm probably one of the worst soldiers there is. I believe I probably miss and hate being away from my family as much or more than any of these guys. Really, I'm a sissy. God only knows I wish I were more a real man — but I ain't. I wouldn't confess all of such to just anyone — but when you mentioned that you would try to be as good a soldier as I, it just made me feel so terribly small. I hate for reality to come to the front and me realize just how small and little I am. You know there is no comparison of character — true character — between you and me. I'm not trying to do anything in the world but confess, honey, and try to tell you in a very meek and humble sort of way that I know that you have got it and I haven't.

Regardless of my shortcomings, I will try to make it up to you in our married life in faith, fidelity, thoughtfulness and a few other ways, so you won't be getting gypped too much. I wish that the Creator had seen fit to put more guts in me, so you could have had more to be proud of. Some of these days you may be so fed up with me that you may ask yourself "just what did I see in this guy anyway" — that day will be the most unhappy day of my life. We have adjusted ourselves to one another so beautifully once before that I'm sure with even less effort, it can be done again even if I should return from this affair with my brain twisted and cynical and disproportioned. Asking such as that is just another good example of my smallness — but I love you. Looks like I can't write about anything else tonight doesn't it.

I went to the picture show tonight and saw an all negro cast in "Stormy Weather." It had Fats Waller playing the piano in it on Beale Street. It brought back rather pleasant memories of Beale St. Palace. The only thing is that we weren't side by side then as we should have been.

How are Kay's teeth and disposition? I hope, for your sake, both are doing nicely. Also hope it has gotten cooler. By the way, it rained a few drops here today, and everyone was amazed. All the guys in the barrack are trying to go to sleep, and I'm keeping the light on. Will write tomorrow. Take care of you and Kay for me. Remember I love you both,

Joe.

August 2, 1943
Monday, 8:30 p.m.

Dearest Babe,

Well you came through today in the true Babe fashion. Didn't get any letters yesterday, but I got two today. It surely is swell. It's funny how our minds run alike on almost the same days. Trace the letters back and see for yourself sometime. When you write in a given manner within 24 hours I will write you in the same mood.

By the way honey, you and Kay got left out again. This morning I got up at five to go into Salt Lake to get you and Kay a little something — just so you'd realize that I still thought of both of you frequently, but I didn't have time to do a thing. We had to get back here by 1 p.m. — naturally we had a blowout and had to change tires. I went with Arnold to buy his wedding ring, and as soon as he got it, we turned around and drove the 127 miles back. Anyhow, I thought of you and at least had good intentions.

Next night — Tuesday, 9:30 p.m.

Honey, I'm terribly sorry I didn't get this letter off as I should have — but I didn't. Tonight I'm O.D. We are O.D. out here every 4 days, and it really is a job. I forgot to go down on the line tonight to make a pre-flight check, and one of the Majors came in and really did "rack me back." I guess the Colonel will get on me in the morning.

We have been doing overseas exams for the past two days, and today we were informed that the rest of the overseas exams are going to have to be done at night. I have about 150 men to examine, and I have examined 40 of them in two days, so you can imagine how long it's going to take to do them at night.

I'm looking forward to getting the pictures of Kay. I just wonder whether

I'm going to be surprised that she has changed so much or so little. It wouldn't hurt if you threw one in of yourself — if you are so self-conscious you could make it from the waist up. However, you should know I love you just as much one way as another — and how else am I going to know how you look or how the "little one" looks without your whole body. This is the most rambling letter. But, honey, I'm being constantly interrupted. You should try to write a letter when there are 29 guys around just trying to do something to upset you. That's what's happening to me right now.

Oh, honey, there is no use sending me an itemized account of what you have spent and so forth. I only want to know whether you have enough to live on or not, and that you aren't wanting for anything. Soon as I get some money, I will send it to you. You know I love you two, and I don't want you to have to go without anything. By the way, honey, do you still think that you are going to deliver in September or do you think it will be October? I just want to know when I'm going to be daddy again.

I'm signing off honey, so I can drop this in the box tonight. I'm fixing to go over now for a coke — wish it was with you. Remember how much I love you and will always.

Joe.

August 5, 1943
Thursday, 10 a.m.

Dearest Babe,

Here I am sitting around doing nothing at the dispensary for the first time in I don't know how long, so I will write my honey and get some practice on the typewriter at the same time.

Before I go further, in your next letter, tell me whether you ever sent my trunk or not. I am in no particular rush for it, so don't put yourself out too much. I just want to be on the lookout for it.

I am going to go to Ely, Nevada tomorrow with Arnold to help get himself wed. We are going to drive over and right back after the ceremony. Kremers and I are both going. We are starting our teaching schedule now, and it looked as if I wasn't going to get to go over with them, but the schedule has been put off for a few days.

I wish it was put off for good, cause I really hate to get up there and talk to a bunch of guys about a lot of things that they aren't particularly interested in and I can't blame them for feeling that way about it. We start off by teaching the ground personnel medicine first and then work on the flying crews. The army expects you to make doctors out of them apparently in short order.

Honey, you asked about some of the boys wives coming out here. Yes, quite a few of them have come out. As far as I know, they are all living over at Wells, Nevada which is sixty miles from here. Three of the boy's wives came in yesterday. I played with one of the fellows' baby girl in the officers club last night for about 30 minutes. The baby was 7 months old. Born exactly one year after Kay to the day. It was nice to just hold one. I certainly would have liked to have been holding my own instead.

Apparently, it isn't too hard to get a place to live over in Wells. The boys however don't get much of a chance to go over there. Most of the time, their wives are over here instead. It has been hard to keep from writing and telling you to come on out here, but this really isn't the place for you and Kay. If you went into labor in Wells, Nevada, the good Lord only knows what your outcome would be. The nearest hospital is 70 miles away, there wouldn't be anyone to take care of Kay, we may be leaving here at any time, and I might have to go right off and leave you, etc.

Honey, I am going to sign off now so I can get this in today's mail.

I love you,
Joe.

August 7, 1943
Saturday morning, 11 a.m.

My dearest,

I was terribly sorry to hear about Kay's accident. I surely wish I were there to help you with all of the troubles. Sounds as if Kay is being put through the mill. It doesn't seem quite fair. I have been in a dither for the past 48 hours. Day before yesterday, I got a letter from you just saying that Kay was restless and telling me about her bandage and sutures. Later in the letter, you mentioned that your mother and Dad had been by and you acted as if nothing had happened. Last night when I got back from Elko, two letters were on my bed telling me what had happened. I'm terribly sorry about the little family troubles that come up, honey, maybe it wasn't such a good idea to get you a house in Memphis after all. I wish that I were there to help out some.

We went over and got Arnold married off yesterday — didn't get back in until late last night. I'm paying the price of staying out late today. I'm tired and irritable and just don't feel up to par. Arnold's wife seems to be very nice, as was her mother.

Honey, I've got to go get a bite to eat so I can get down and do pre-flight checks at noon so I must hurry. Keep a stiff upper lip and remember that I love you more than I could possibly tell you. Take care of yourself and ours.

All my love,
Joe.

August 7, 1943
Saturday night, 11:30 p.m

Dearest Honey,

Got a letter from you this afternoon. One of the very few times that I have gotten a letter in the p.m. Most of the time, if it doesn't get here in the morning mail, I just don't get one. It's tough that our mail gets so messed up en route for I know it makes the sequence of thought all cockeyed.

You mentioned Kay talking. That is simply impossible for me to even imagine. I can't imagine her even saying "Daddy" now, much less putting a whole sentence together.

I'm O.D. tonight and have to get up at 3 a.m. to go check out the crews. I'm pretty tired, and I have only 3 hours to sleep, so I'm going to close short on you tonight. I love you most dearly and think of you and Kay and the little one almost all the time. You have never told me what we are going to name the new baby. Remember that I love you.

Your hubby,
Joe.

August 9, 1943
Monday, 9:30 a.m.

Dearest Honey,

We had a severe crash on the field last night — second one we have had in a week. Thank goodness, neither of them were in our group. However, I had met some of the fellows that cracked up last night.

Enclosed is some money, use it as you see fit. We had better start saving for next month. I may have to take a six day furlough, and I'll have to draw the money out of the bank to come home. As paradoxical as it may sound, I hope I don't get one then. I had rather wait until the little one came along. If I do get it, however, it will be that furlough that comes before departing. It is rumored that they are going to start them next month. Don't pay that too much mind either for there are more rumors around here than there is anything else. You can hear almost anything that you want to believe.

I have been awfully smart this morning. I was up eating breakfast at 5:30 then went back to the barracks and read the paper (first one I have read in days). Then I went over and had sick call. Then came back to the barracks and shaved — took some cleaning over to the cleaners — went to the officers club and paid my dues. (Dues here for a little over a week were $6.00 — guess they charged for the whole month). Now I'm sitting down writing and it's only 9:30. When I finish this, I'm going to go get a haircut (the C.O. insists on at least 1 per week), and then I'm going over to the dentist and see if I can get some teeth filled. Tonight, we have got to do overseas physicals again. We should finish those about the middle of the week, and I certainly will be glad when we do. I don't like this night work worth a damn, but, after all, it helps to pass the time away. Went over and bought me some low-quarter G.I. shoes yesterday. Today, I gave those good shoes with the strap on them to McFarland. They got where they hurt my feet too much to wear. My new ones are plenty big — 7 1/2's E. First time I ever wore larger than C in my life. Now my feet can swell all they want to and I'll still be comfortable.

I'll probably write to you again tonight after I get your letter. Love yourself and Kay for me for I love you both tremendously.

Your loving husband,
Joe.

Army Bomber Crashes on Utah Desert, Wrecking Freight Train

WENDOVER, Utah, Aug. 9 (AP) —Crashing in the night on western Utah's dreary salt desert, a four-engined Army bomber killed one flier and caused the wreck of a freight train, leaving 26 box-cars stacked up like splintered toys on the Western Pacific railroad's main line today.

The big ship, groping for emergency space on the flats, smashed down on highway U. S. 40-50, slithered at terrific speed across the salt crust before hitting the rails and winding up 100 feet on the opposite side.

The westbound freight, powered by a double Diesel locomotive, roared along 10 to 15 minutes later and plowed into scattered wreckage and a spread rail.

The engine stuck to the rails, three freight cars were derailed but stayed intact, then 26 more crashed together in a dizzy pyramid of destruction.

Second Lt. Richard L. Blue of Rantoul, Ill., the plane's co-pilot, died today at the hospital at Wendover field, where the plane was based. Ten other fliers were dragged injured from the wreckage and some were critical.

One rail official estimated damage to train and freight at $200,-000.

Front page of the The San Bernardino Sun reporting the plane and train crash. August 10, 1943. Courtesy of the UCR Center for Bibliographical Studies and Research, California Digital Newspaper Collection.

August 10, 1943
Tuesday, 11:30 p.m.

Dearest,

The crash I mentioned to you yesterday really was a catastrophe. Perhaps you have seen something in the newspaper about it. It hit the railroad when it crashed and a fast freight came by 10 minutes later and 29 box cars are just piled up on one another. It really is a mess — I went out and saw it yesterday.

One of the squadrons started giving overseas furloughs today. I'm going to wait as long as I can to take mine. Don't you think that would be best? I really don't know when our squadron is going to start giving theirs. Even when we start, I have got a lot to do before I can leave. I believe I will finish my overseas physicals in a couple of nights, and then I have got to finish immunizations, which I am starting now. Tomorrow we are supposed to start teaching again — don't know how long this teaching program will be going on.

Honey, I've got to go get to work. Hope I hear from you this afternoon.

All my love,
Joe.

August 11, 1943
11:30 p.m.

Dearest Honey,

Have you ever received 4 letters in one day from the one person in the world that you wanted to hear from most? If you haven't, it certainly is a wonderful feeling. For six days, I had gotten no word at all from you — today I got 4 letters. I know it wasn't your fault, honey. It was because of that wrecked train only seven miles from here that mail couldn't get through.

I'm keeping all the guys in the barracks awake writing this letter — and probably getting more unpopular by the minute. All the lights were out and I came in and turned on a light and started writing here at midnight but I have been pretty busy and just haven't had a chance in the past 24 hours and for all I know may not have a chance tomorrow — and I have gotta talk to you.

Got up at 7 this morning and went over and did preflight checks. Then I went over and held sick call. At 10 a.m. I had to go over and give a lecture on sex hygiene and personal hygiene to a number of officers. At 12, I had to do preflight exams again. I had lunch, gave my officers shots from 2 until 4, lectured again from 4:30 until 5:30. Then I had dinner, preflights again from 7 to 8, then overseas exams from 8:15 until 11. At 11 I went over and made the day just barely existable by having a nice cool double scotch and soda. Anyhow by quoting that schedule, you can see why I say I sometimes just don't have time. Even worse at 4 a.m. I've got to get up and go over to give preflight exams again. All the flying personnel are examined before each mission.

Again, I think it best to put off my furlough as long as I possibly can. I suppose I will get 10 days. If I can catch a plane both to and from Memphis, I should be able to stay about a week. Naturally, I would rather wait until the little one comes, but I'm afraid I may not be able to put it off that long.

Honey, everyone is looking over here with scowls on their face wishing that I would put out the light, so I guess I'd better. I will try to add another note on it in the morning before I have to send it in order to get in the mail tomorrow. At present, for protection alone, I had better say goodnight, honey, I love you most dearly — I think of you and Kay continuously.

All my love,
Joe.

August 13, 1943
Friday, noon

Honey,

Gotta write you fast for I have to be over to the dispensary at 1 p.m. and it's 12:37 at present. Gotta letter from my honey this morning, so now I'm feeling a whole hellova lot better.

Kay and I are having our troubles. I'm an invalid now with a pretty severe care of Vincent's (trench mouth). Have been to the dentist for the last three mornings with it and I think it is doing some better now. Poor little Kay — glad the stitches are out. Today is Friday 13th — 52 days since I have seen Kay and exactly half that many since seeing you — two eternities I assure you.

I got some more men a couple of days ago — making 8 to my squadron medical staff. Believe it or not, all with Intelligence tests of over a hundred (very unusual for medical detachment men) and one of them with 3 years of college behind him who majored in biology. Two of them are high school graduates. It is up to me to train them now while doing millions of other things.

The names you have picked out for the little one are O.K. by me. I feel honored that you want it to be Junior if it is a boy, and Carol sounds good with Kay. Don't think it is going to be a boy, however, for if I felt that Kay was going to be a girl I feel doubly sure that Carol will be — why, I don't know, but I'll believe I'm right until proven otherwise.

Honey, did you ever send my foot locker? I gotta go, honey, I love you so much it hurts. Keep a stiff upper lip and remember to hug Kay for me sometime when you can stop her long enough. I love you both.

Joe.

August 14, 1943
Saturday afternoon, 4:30 p.m.

Dearest Honey,

Didn't get a letter from you again today. When it becomes time to vote for a new postmaster general, I'm not going to vote for the so-and-so that is up there now (nor his boss). Can't ever tell after Roosevelt puts on his propaganda campaign, I might even vote for him.

Things are not going so well. I'm having a hellova lot of trouble with my squadron C.O. I don't know what is going to be the final outcome of our little private feud. Some of it is supposed to come up before the Colonel today. I'll probably get put in my proper place, but at least I will know what they want in this outfit. If they want a doctor OK — if they want a guy that will make a diagnosis to suit the whims of a 24 year old kid that hasn't even got the manhood about him, so he can look you in the face when he talks to you then I can degenerate a little bit farther and do that too I guess — but not until I know damn well that that is what is expected of me. This outfit is really a MESS. If all combat organizations are as sorry as this one, I can assure you that we might as well lay down our arms this minute and quit. Until we are on the boat going overseas, I will never believe that we are going to combat as a unit, somehow that just can't happen unless some fellows learn a few principles that are fundamental for all mankind to have to get along as a unit. I could rave on about this subject, but I know it is boring to you and to me. Sometimes I wonder if it is just me — I really don't think it is though.

Hey, you promised me some pictures of Kay, and I have asked you for one of yourself. Sometime send those out for I certainly would like to see what both of you look like. I wonder whether I will be surprised at how Kay looks now. You know you can't imagine her changing any even though it has been quite awhile since I have seen her. Really, I can't imagine her talking so much. Something tells me that I am in for a shock.

Later on (9:30 p.m.)

Well for some reason (Colonel was flying), I didn't get to see him this afternoon. So that's that. Guess I will have to sleep on everything and see what tomorrow brings. I'm sorry I brought such petty stuff up. After all, this is war and everything is topsy-turvy.

You know, I remember distinctly those long six weeks that we went together before we were married, that I thought that medicine was my prime life and my married life would be secondary to it. That's the only way I know that I have progressed any in the past 3 years — medicine is so far behind in importance to my married life and my family life that it barely deserves a thought. I wonder if Mr. Ellis, the preacher, knew just what a nice thing he was doing to us when he married us — probably not.

I was supposed to go flying tonight but was just too good for nothing to do it. My boys were going to take me up and down over the Grand Canyon to Phoenix and back. I would have gone, but I didn't want to make the mission — particularly at night. After you have been up for six or eight hours it becomes tiresome. In fact, I get pretty tired of it in four hours.

Monday, we are starting to teach our Medical Enlisted Men. The four of us together will have a full time teaching job. Seven lectures a day for about a month — besides having to give 22 lectures to the flying personnel and 22 to the plain enlisted men. For a couple of months, we are going to be more lecturers than we are doctors. It's gonna be rough too for during the same time we are going to be taking our furloughs.

Honey, I know I've griped, griped and griped tonight — please forgive me. I'm just not being very nice. The answer to all of my problems is you. If I were with you this post would probably seem O.K. with me. This is really an epistle, isn't it? Remember I love you and Kay with all I've got. You are on my mind most continually as you will be until we are together again.

I love you,
Joe.

* * *

The B-24, more commonly known as the Liberator, was the most produced heavy bomber during World War II. The large four-engine plane gave the B-24 advantages in long distant bombing missions, high altitude flying, and the ability to carry heavy bomb loads. The B-24's twin vertical tail made it easily identifiable.

The bombers were slower and more difficult to maneuver, making them easy targets for the enemy. Fighter squadrons accompanied the heavy bombers on missions. Used in all theaters during the war, the B-24 was the most widely flown U.S. aircraft.

While the B-24 was a larger aircraft, the interior was compact, built for efficiency rather than comfort. The crew's positions were spread throughout the plane, accessed by crawl spaces or catwalks. There was little to no space for additional passengers to sit or stand. The Liberator crew was made up of 10 or 11 men, including a pilot, co-pilot, navigator, and bombardier, who were all commissioned officers; and a nose gunner, radio/radar operator, flight engineer/top turret gunner, left and right waist gunners, tail turret gunner, and ball turret gunner, who were all enlisted men. Sometimes the crews included a photographer.

The crew entered the plane through the large bombay doors on the belly of the plane or by climbing around the front wheel. The bombs were held in racks on either side of the interior frame. The bombay doors rolled up into the wall bulkheads when the bombs were released.

The temperature at high altitudes could drop to 70 degrees below zero. Many of the men suffered frostbite on their fingers, toes, and faces. Crew members wore bulky, uncomfortable clothing to keep warm, including electric flight suits that plugged into the plane. However, the suits often overheated and caught fire while in flight. At high altitudes, the crew wore oxygen masks, which often failed or were inadequately connected, depriving the men of oxygen. Uninsulated fuselage brought deafening noise constantly and isolated the men from each other, despite their headsets. On returning to base, crews fired a red flare to notify ground personnel that there were wounded on board or the plane was damaged.

Nicknamed "The Flying Boxcar" by the crews, the Liberator was prone to accidents, gas leaks, and crashes on take off and landing. While faster than the B-17, the Liberator had no power steering and relied on throttles and brakes to navigate. Accidents also stemmed from the fact that most pilots had less than 500 flying hours, the equivalent of three and a half months of training. After missions, crews often bailed out of planes that were too damaged to land safely.

* * *

Letter from Joe on a B-24 to Babe in Memphis

August 15, 1943
Sunday, 11:30 a.m.

Dearest,

Just flew over the old Deluxe Tourist Courts in Boise and saw where we used to live. We were too high or perhaps I could have made out whether the Koenigs were staying there or not. We have turned around now and are high-tailing it for Wendover. I have been sitting in the Bomber's compartment (the glassed-in portion of the nose) all morning — can get a better view from there. At present, I am at the radio operator's desk and there is so much vibration that it almost puts your hand to sleep.

We were supposed to take off at 6:30 a.m. for the Grand Canyon, but the plane wasn't ready to fly then; we waited around until 9:30 and then hopped to Boise instead. It's rough as so and so today. The weather is closing in from the west and it is pretty gusty. Too it's cold. Boise looks the same except there is no snow on the mountains now.

Next day about same time —

Didn't write anymore "upstairs." Hit a rainstorm and it got too rough. I was O.D. last night and had to spend much time up on the line. So this morning, I slept. Didn't even go near the dispensary.

Was called in at an Evaluation Board meeting before the Colonel. They were trying to "bust" the boy that we have been having trouble about. After much arguing to and fro, they aren't going to bust the man so I guess I can say I won a moral victory. I would tell all the particulars but it is very nasty and uninteresting and unfair.

I am enclosing a check. All of it is not profit for you because I cashed a check for $50. However, this will cover the check with some left over.

We are going to move the whole group area tonight, dispensary, tents and everything, for they finished building our new facilities on base. We are overcrowded where we are and have got to have more room. That's a lot of trouble, and it is going to take time to get straightened out again.

Honey I wanna go get a cashier's check and get this in the mail, so I'm leaving you. Take care of yourself and I love you — remember?

Joe.

From Joe in Wendover, Utah to Babe in Memphis

August 16, 1943
Monday night, 10:15 p.m.

Darling,

Today has been another day of about the same variety. There isn't enough difference in one day or another out here to mention — except Tuesdays. Tuesdays are bad for you have to wear a gas mask all day, which is certainly a bore. If you are caught without it, you are fined $25 on the spot. There is talk of making us wear our tin hats and leggins all the time too. That will be interesting, won't it?

Honey, I sent you a $100 cashier check today. I wouldn't always remind you of such things, but I just want to know if you get them. By the way, I got a $2 a day raise. My flight orders came through and beginning next month it will mean $60 more a month. I went to see finance about raising your allotment, but it seems that they would have to cut the $150 allotment out completely before I could raise the allotment. I haven't decided whether it would be best to do that yet or not. I don't think it will make much difference for I'll send you all I can get my hands on anyway — I had rather do it by allotment, however, if I possibly could.

Sorry the heat is so terribly bad there. No, it isn't so hot out here. I don't know what the mean temperature has been, but you don't notice the heat much. More importantly, the nights are usually quite comfortable. If the heat gets too bad, we can always go up to 10 or 20 thousand feet and freeze.

You mentioned in your last letter that you and Kay would try to be worthy of me. Golly, honey, that's a joke. I have tried to impress on you that I know our respective worths. Regardless,

I love you,
Joe.

52

August 19, 1943
Thursday night, 10:30 p.m.

Dearest,

Gotta go over to the line and do pre-flight checks on the flying person-nel at midnight, at 4 a.m., and 6 a.m. Then sick call at 8 a.m. Don't have but one lecture tomorrow so that won't be so tough. Have done all of my overseas examinations except for 22 now. The trouble is that we keep on getting new men, so you can't ever catch up. So goes the life in the Army.

I have been thinking about calling you every night for the past week, but by the time that I can get to it, it is 10 o'clock and that means it's 12 midnight there. By the time a line got through, it would probably be 2 or 3 in the morning. Don't think that I don't think of you, honey though, for I do — almost constantly.

Being busy has helped out a lot. Damned if I didn't damn near go crazy for a while. Even though I think I'm making myself do better now, I'll never be able to really be happy, or content, or even merry without living with you. It certainly is nice to love someone as I love you, and it is nicer to know that that love is fully returned — but it certainly does make it tough to be separated. I don't have the same conflicts that the other guys have away from their homes. It isn't difficult for me to be true to my love, honey, it's a cinch. Don't ever worry about that — just have half the faith in me that I have in you.

Honey, it is now 11:35 p.m., and I have got to bathe before I go down to the line, so I gotta hurry. Remember how much I love you and my family and how I'm looking forward to being with you again. I shall never forget what you are going through for us.

I love you,
Joe.

August 20, 1943
Friday, 11:30 p.m.

My Honey,

Got three letters from you this morning. You certainly have been nice about your writing, honey, and you don't know how much I appreciate them. They certainly help an awful lot.

Honey, don't be afraid to have this baby. I'll grant you that I would prob-ably be scared to death if it were me, and it's a hellova note for a man to tell any woman who has born children about bearing them. However, there are a few logical, I think, points for you to take into consideration about having this baby. Firstly, your second baby is usually much easier to have than the first one. Secondly, you won't remember anything about having this one except snatch-es of your labor and maybe nothing at all. Thirdly, honey, you're going to be alright. I know you will for everything is in your favor. I know you are dreading it, honey, that's human nature, but don't be scared you're gonna be O.K.

You asked me to pray, remember? To begin with, I can't see why my prayers would be answered. But, if they are, your worries are certainly over — for you and Kay are my feeble prayers, every night that comes — maybe a few times in the day. You're gonna be O.K., honey. If anything about it is bothering you in particular, write and tell me about it. You just need someone to talk to and confide your anxiety to and you will be O.K. I'll bet 10 to 1 that you felt considerably better when you just wrote it down on a piece of paper. Having this baby is going to be so much easier for you than Kay that you will probably be wanting to have one every nine months. I'd give anything if I were there to help you somehow, however, husbands are no good at such parties — but it ain't right for you to take the knocks and me out here taking things easy. Remember though, honey, that I know, as much as a man can, that labor is no joke, and I appreciate what's up. I'll try to make everything up to you sometime — don't worry, honey, I am at least with you in spirit. I may even be with you in body for I don't know whether I can get a leave as late as October or not. I love you, honey, very very much.

I'm sorry it's so terribly hot there. It's hotter in Memphis at 97, than it is out here at 130. Our nights are cool. I always have to use a sheet before morning and many times a blanket. A pity we couldn't trade.

Had another hard day today. Pretty soon though I'm going to have things in pretty good shape I hope. If I could only get really good and caught up for a while! One nice thing is how much more I've felt like working lately.

Today, I drew out most of my flying equipment. You should see my fur lined leather coat, trousers, shoes, gloves and helmet. They should really be warm. Also got me one of those suitcases (valet-pak) that I have been wanting, and I will get my parachute tomorrow if I possibly have time. It is certainly nice looking equipment, but is going to be hell to lug around all over the world.

Did overseas physicals again tonight. I shouldn't have but about 10 more to do, but as soon as I get caught up, we will get a batch of new men in and I will have to start from scratch on them. There is no rest for the wicked, is there?

Honey, it is past midnight, so I'm going to stop on you. I got the cookies yesterday, and they are delicious — I can prove it by everybody in this barrack. It was certainly nice of you to send them to me — I'm not used to getting nice things through the mail, you know. It's a good feeling — so thanks.

Take care of yourself honey and don't worry too much. Again, maybe I will be there when you "shoot the works." I love you so terribly much.

Goodnight honey,
Joe.

August 23, 1943
Monday, noon

Dearest Honey,

Got two letters from you today. Like you, I rarely get a letter a day any more, but more frequently skip a day or so and then get all of them at once.

All this morning I have been doing 64 Exams. Yesterday I grounded all my pilots, even the C.O., until I could get them done. My records are in pretty good shape. In fact, my squadron is in better shape now than any of the others, by far. Too, now I'm doing much more work than they are. But, as I have said before for a while I wasn't worth a damn.

Honey, I have tried to call you for the past three nights, but so far haven't been able to get a line through Chicago. Lots of rumors are making the rounds — one that we may move out of here before October 1. There was one yesterday that we may even move back to Dyersburg.

Honey I am going to have to take my furlough between September 20 and September 30. Ten days all together including travel time. I'm hoping that we will have moved closer to Memphis by then, but doubt it very much. Maybe something will come up and it can be postponed but, as it is now, that's when I am slated for a furlough and that is as late as anyone can get one.

Honey, the next time I am in a town, I will certainly get you the prettiest blue gown I can find. However, remember I have been off this post only 2 times since we have been here. Once to Salt Lake (for 1 hour or so) and with Arnold to get married. I'm getting pretty fed up with it now, however, so I may take some time off pretty soon.

Sounds like Kay may have whooping cough — and I wouldn't care too much if she has it if it is that light. It would be impossible to say whether it is or not. However, if she is coughing 1 month from today, perhaps I can tell you whether it is whooping cough or not.

Thanks for sending my foot locker. I haven't gotten it yet, however. The cookies were swell, honey, all of us enjoyed them. Honey, these guys in the barrack are hollering "Doc, how about this and that, etc" every 2 seconds. This letter is probably even more disconnected than usual. These guys are like a bunch of little boys. They are pretty swell guys. One just came in and brought me a couple of sandwiches and ice cream since I haven't had time to go eat.

Honey I gotta go in order to get this letter in the mail. I'm going to try to call you again this afternoon. I love you — take care of yourself.

Joe.

August 25, 1943
Wednesday, 11 a.m.

Dearest,

Looks as though I'm getting in the habit of only writing every other day — it isn't premeditated, honey, so forgive me.

Things are happening fast. We have orders to have everything arranged, so we can pull out of here in a moment's notice. The Colonel has said that when we moved, it would be under secret orders and apparently he means that they are going to be secret. Don't be surprised if you get a telegram some day saying that we are leaving and that I don't know where to or when, and if that happens, that's going to be how it is. Probably we will be moving eastward — that is nearer home. As far as I am concerned, that will be swell. I have been tired of this place for many a day and hour.

Trying to practice medicine in a tent and doing the million other things that we are supposed to do is certainly no snap. Where we live, there is just about 6 inches of the finest, powdery, dust you have ever seen. The wind blows continually out here, so you can imagine how everything is. Everything is covered with this fine white powder.

Don't worry about my immediate C.O. and me. We are getting along somewhat better. He has finally decided to let me run the medical end of my outfit. He even asked me today when I was going to take a 48 hour pass — saying that I hadn't had one since I have been here. He also asked that I get another haircut — but in a fairly nice way — I think it is because he damn near drank a bottle of scotch off me the other night. And whiskey, of any variety, is plenty tough to get out here.

Honey, you asked about how you could call me. I live in barracks #904. The telephone number is 189 for this barrack. I would like to hear your voice anytime. I have tried to call you a number of times but can never get a line through. The other night (2 nights ago) the operator told me I could get a line through in 45 hours. I asked her if she meant 45 minutes and she said "no — 45 hours." I think Chicago is what holds the wires up. I have been so busy lately that I really haven't had time to sit around waiting for a call to go through. But if you can get me, I would really love to hear your voice.

Now, honey, please, please, please, don't be trying to do a hellova lot of things cause I will only be in Memphis for 6 or 8 days. Don't worry, I want to see you and Kay — not the house. If you were to happen to go into labor while I was there it would be so much the better — then I could sorta look after you. Don't worry about anything else. Not only don't go to trouble fixing up the place, but don't be disappointed if some changes are made and I don't get home then. We aren't told anything in this outfit, and things are changed considerably on a moment's notice. So don't count on me being there at a given time until you are looking at me.

I truly love you and my little family — bless her heart. I'd really love to see you both right now. My trunk hasn't arrived yet. Guess it should be here in a couple of days anyway. I love you.

Joe.

P.S. Had that tooth filled yesterday — quite a job — sat with my mouth open for 1 hour and 15 minutes — but didn't say a thing. Don't worry about Kay — if she has whooping cough, she will start whooping pretty soon or following her coughing with vomiting. If she coughs in paroxysms (spells) that last for 2 or 3 minutes — it is a good chance for it to be whooping cough — Hope Mrs. so and so across the street didn't upset you.

I still love you.

Stateside Training

From Joe in Salt Lake City, Utah to Babe in Memphis

August 27, 1943
5 p.m.

Dearest,

Just arrived in Salt Lake City for a 48 hour pass with a friend of mine from Maryland, Lt. Young, who is a swell fellow, a gentleman of the Virginia variety and our squadron's operations officer. We are staying here at The Utah. They had only one vacancy and it is a suite costing $15.00 per night. In other words, for two nights I'm going to live very comfortably — from double-decker beds, 30 beds and 60 men to a room to a suit of 2 rooms with a bath — single beds air conditioned and all the trimmings. This is, by far, the swankiest room I have ever seen. The furniture is white, the upholstery is a slate blue color. The walls are light blue, and the carpet is dark blue. It all goes together nicely — believe it or not.

Do you want to know something that is beautiful? Grass and trees. I had almost forgotten what they look like. They are really pretty when you haven't seen anything green in over a month.

Honey, I'm terribly sorry that the mail hasn't been coming through so good. I'll admit that I haven't written you every day lately, but I have never missed writing every other day since I have been here. I received two letters from you today — One had Kay's pictures in it. You can't imagine how she has grown. I'll bet I've looked at them 15 times already. She must have been mad that day by the looks of her mouth — not a smile in the batch.

I have heard from some of the base officers, honey, that we will probably leave sometime around September 1, and we are going to either Kearney or Grand Island, Nebraska. I know you remember both places, and if I remember correctly, both places have Wendover, Utah beat 2 to 1. In fact, I can't think of a single place that hasn't got Wendover beat.

Next morning — 10 a.m.

Honey, I talked to the sweetest voice in the whole wide world about 8 hours ago. It surely did sound good and familiar. I don't know what it cost yet, but it was certainly worth every penny of it. Surely hope by now Kay is better, and I wish I was there to help you — and hold up my end somewhat. I've thought of quite a few things that I wanted to talk to you about since our conversation but you know how it is.

Well honey I've gotta do some shopping. I'm going out to buy my best and only honey a nightie. A blue one with all the trimmin's. Certainly hope you like what I end up with. 'Cause I love you, and I don't want you to think that my taste was bad. If I see anything else that strikes my fancy, I'm going to buy that too. Hope our baby is alright. Take care of yourself and Kay. I love you both.

Joe.

Dearest Babe,

* * *

In August of 1943, Wendover Field experienced an abnormally strong wind storm. The wind blew down tents and covered everything with salt and dust. The men's belongings were blown all over the field. The next day, men repaired over 100 tents and continued with their training.

From Joe in Wendover, Utah to Babe in Memphis

August 30, 1943
Monday, 11 a.m.

Dearest,

Well I wish you were here for the benefit of the weather. It is colder than the proverbial "well digger's — ." We had a wind storm last night, and it blew about 40 of our tents away. All three of our dispensary tents were down this morning. I couldn't find my records for awhile, and I thought I was going to have to do everything over again — overseas exams, flyers exams, immunizations and everything, but, thank goodness I finally found my box stored in one of the ambulances. I certainly was glad. The dust was so thick yesterday that you couldn't see a patient in front of you. All of us were white when we got through processing 20 new crews that came in. We have really been going through it. If the Sahara Desert is worse than this, I won't believe it until I'm there for this is right rugged. It is really pretty cold — everyone is wearing leather jackets or field jackets. Sleeping with a couple of blankets. In fact, we have been sleeping under blankets for quite awhile. Hope your hot weather has passed off somewhat, but I am afraid that it hasn't.

How is Kay? Is she still sick and what was wrong with her? Maybe if I get a letter from you this afternoon I will find out. So far I haven't gotten one, and the morning mail is in.

We still hear many rumors every day. The most often repeated one is that we probably will be pulling out of here about the 8th for somewhere in Nebraska (Kearney or Grand Island). Lord only knows what is really up. We have reason to believe, however, that we are a pretty "hot" outfit and will probably be heading for distances before too long. I will see you before then however, and I'd rather you wouldn't mention these things to anyone. It is also rumored that Lewis' outfit has been converted into an R.T.U. outfit, which means that they will be made static personnel at some field and will just train replacements. All of these things are rumors, however, nothing definite being settled yet by action.

The nearer it comes to September 20 the more anxious I am. By the time it gets here I won't be able to sleep at night. I am so damn anxious to see you. After I see you, I think everything will be alright. Wish to hell I could be with you through your next ordeal, and maybe I will be.

Take care of yourself and our little ones. Don't be working now regardless of what or how the house looks. I love you so terribly much.

Joe.

August 31, 1943
Tuesday morning, 7:00 a.m.

Dearest Honey,

We had a parade this morning at 6 a.m. — I thought we were over that stuff but apparently we aren't. Some lousy stuff. Anyway, we got up at 5:30 this morning. Put on our leggings, helmets, etc and went out and had the parade. I got out, but I wouldn't parade. I contended that someone needed to be in the dispensary during a parade, so I just sat there and watched. The boys looked pretty good too.

By the way, honey, I haven't received the package from you yet. Guess it will come in today or tomorrow. We should get paid today. When we do, I intend to send you as much as I can 'cause it's going to take a lot for me to come home to see my honeys. Personally, I don't give a continental how much it costs — I'm so eager.

Honey, I've got to eat and then get to sick call. The windstorm night before last left our dispensary in such a hellova mess that we didn't have much of a sick call yesterday — so we will probably be rushed today. Everybody is gone today but McFarland and me, and he has to be a juror on a court-martial at nine, so it looks as though I'm going to be the only one around. Quinn and Wagner are both on pass, and Kremers hasn't returned from furlough yet. Wagner leaves on his furlough tomorrow. See you pretty soon, honey.

I love you,
Joe.

September 1, 1943
Wednesday night, 10 p.m.

My dearest Honey,

Wagner came in yesterday and told me that he was leaving on the 31st (yesterday) instead of today. So he took off — all of our reports go into the Surgeon General's Office, the Air Surgeon's Office, the 2nd Air Surgeon's Bomber Command and Wing Surgeon on the first of the month. None of these reports were up, so you can understand what I have been doing since Wagner left. I was appointed Acting Group Surgeon, so it has fallen to me to get those reports out. I got most of them out today but still have a couple more to do in the morning. Since I had never attempted to do it before, it has been a hellish job. To cap the climax, I was O.D. last night and we were running double missions, so I had to get up to preflight the boys every 2 hours instead of the usual every four hours. To make it a little worse, I have had the G.I.'s (Army slang for diarrhea) all day long and am weak as a kitten. That in short is one of the reasons, or some of the reasons you didn't get a letter yesterday.

The Colonel called me up to his office (tent) twice today. Once to show me the order which arrived two days ago putting us on the "Alert." In the order it specifically instructed that all men be inoculated against Typhus and mentioned Typhus only. To me, that means we are heading for the Mediterranean Basin I'm quite sure. For if it was the Southwest Pacific, it would have mentioned Plague, Cholera, Yellow Fever, etc. I am now fairly firmly convinced that we are "hotter" than a firecracker. When it will all happen, I don't know but probably in early November. You can tell the family that I expect to go overseas almost anytime after I'm home. I'm sure I will be home first though.

The other thing that the Colonel wanted with me was to tell me that I am on the Evaluation Board tomorrow where a poor scared guy is going to be kicked out of the Army to a civilian and have his wings taken. Tough luck, but I guess this is war. He isn't my man, thank goodness, I'm just a member of the board because Capt. Wagner is not here. This Army stuff isn't very interesting, honey, but it is all that is happening about and around me, so it is all that I can write about.

Don't know yet whether we are going to leave here or not. Nor when, nor where. Nothing definite at all, just rumors and more rumors. Today was laundry day, however, and I didn't send any laundry 'cause it takes 10 days to get it back. I'm just going to be mighty conservative with my clothes for a while.

Now that I've told everything that I know about the Army, I would like to tell you that I love you and Kay and even Carol very much. I will tell you again and again in about 3 weeks. That almost sounds fantastic — I haven't seen you in so long. Looks as though I'm going to be pretty busy for at least 10 days doing Wagner's work and mine, so that will make the time pass more quickly. I can't possibly put into words how much you mean to me.

Tell all of the folks, honey, that I know that I owe them letters, but I'm not going to be writing soon so they will just have to get second-hand information from you.

Love,
Joe.

September 3, 1943
Friday night, 10:15 p.m.

Dearest Honey,

I'm so tired, and I'm going to bed very shortly. We had another wind and dust storm in our area last night, and I was down there practically all last night. Today Colonel Kinard, the 2nd Bomber Command Surgeon, was on the base, so I had to stick close to the dispensary and get the place in ship-shape until he came around. He finally got around about 3 p.m., and we talked down there until 6 about everything. He is still my favorite — really a swell fellow, a gentleman and scholar. Has all the poise that is possible to have — and still an army man.

About the time to leave there, I found out that one of our ships was circling the field — and couldn't get his landing gear down — so I went out there in my ambulance. Then I found out that it was not only one of our group's planes, but one of my squadron's and that Young was flying it. I listened to his conversation with the tower telling what he was going to do — just as calmly as if he were talking to me — told his crew that any man could bail out — none did, but all wanted to ride the ship down — which they did. They could get the landing gear down, but it wouldn't lock. When he landed, he did it in such a manner that he made the thing lock, and it was really inspiring to see what some fellows can do under extreme stress and strain. The fire trucks and the ambulances were lined up on the long runway, and the runway at a B-24 outfit is plenty long. But there is much suspense when a plane is coming in — whether the wheels are going to fold under or not. General Eubanks (C.O. of 2nd Air Force) was on the field watching and when it was all over, he went out and told Young what a swell job he did — it really was beautiful.

We had another evaluation board yesterday, and I convinced the Colonel again that we medics know a little something. I think we made more headway with him than ever before. Pretty soon we may start getting a little cooperation — Hope so.

Colonel Kinard says tonight that we will probably go to England. He also told us that all the squadron surgeons would fly across rather than ride a damn transport. You may rest assured that all of us were glad to hear that.

Enclosed, honey, is a check for $150.00. That will cover that $50 check and leave a hundred dollars over. You take some of it and get you some bedroom slippers to match your nighties and also buy Kay something to play with from her Daddy.

Yes, honey, I received my trunk and don't bother about sending the key. I already have it open. My bedroom slippers arrived yesterday. They are a little large, but I can probably swap them with someone, or if not, I will bring them with me on furlough. They are real nice and it was swell of you — feeling as you probably do — to go to the trouble of sending them. When I married you, I surely got the best there was and there has never been the slightest doubt of that in my mind. Honey, I'm going to bed for I am really tired. Don't forget how much I love you and how terribly much I want to see and be with you.

> All my love,
> Joe.

> September 6, 1943
> Monday, noon

Dearest,

I'm still pretty busy — working like everything and am enjoying it. As you know, I have been doing Wagner's work and mine too and that's plenty. In fact, some of mine has just been riding.

Dearest Babe,

We are fixing to move, so that is even more work. Figuring out the health schedule and precautions for a group (1600 men) on a three day trip is quite a job particularly when they are on 4 different trains. The following are military secrets — 2 trains are pulling out Wednesday. The other two trains will pull out Saturday. I will probably be on the last train to pull out. I wish they would consolidate one of those trains, so I would get to fly rather than ride a damn troop train, but each train will have to have a doctor, so I suppose I will have to go on a train. We are going to Fairmont, Nebraska, I'm quite sure, but continue to write to me here until I can give you my address.

Another thing, honey, don't be planning too much on me coming home the 20th. That may be changed. However, I hope it won't because I am eager to see you and Kay. I just don't want you to be disappointed. Things in the Army change so damn rapidly. The General suggested the other day that we are bound for England.

I sent you $150 yesterday. I only mention this, so we will know whether you receive it or not. I think you are doing swell with running the business — better than me, in fact. Better watch out, in future years you might have another job besides raising the 12 kids (6 adopted).

Honey, I'm going to send you my suitcases within the next couple of days. I will put your black one inside the large one. I don't need them now — I don't think.

Don't be too surprised if I don't wire you when I leave here because I may not be able to. However, I will if I can.

I'm glad you liked the negligee. If the outfit is not going to fit tell me, honey. Or if you would like some other color or model its O.K. I gotta go work, honey. I love you an awful lot, and I am continually wishing that I was with you. Take care of yourself and I will be with you soon.

All my love,
Joe.

September 7, 1943
Tuesday, noon

Dearest Honey,

Just a note — not much more. Haven't really got time at the moment to write but I want this to get in this afternoon's mail.

We are still packing. In the morning, two of the squadrons pull out with McFarland and Quinn. Kremers and I then will have to be up every other night until we pull out Saturday. I will certainly be glad when this is all over for I'm already pretty tired of moving, and we haven't gone a step. I certainly do dread that ride on the troop train. Rumor has it that we are going to hold ground school, lectures, talks, etc. while traveling. Certainly hope that's a false one.

Honey, when I get home with you for those few days we are going to rest. I'm so damn tired. All I want to do is sit and talk and play with Kay. Get out some of my clothes for I want to get out of this uniform for a while. Bought me some undershirts this morning. They are the first bit of wearing apparel that I have bought since I have last seen you. I gotta buy some shorts now for I need them worse than I did the undershirts. Ought to be able to get them from the Quartermaster cheap — got the undershirts from the PX however.

I gotta go honey, I'll write more later. I love you and Kay so terribly much.

See you soon,
Joe.

September 8, 1943
Wednesday night, 9:30 p.m.

Dearest,

Honey, looks as though I layed down on the job and didn't get a letter out to you. I'm sorry. You have been so good about writing. I hate to not do my share.

Last night I quit writing because the crash trucks and ambulances went out and that always causes a bedlam around aviators. Nothing serious happened however. There has been an awful lot of crashes lately around here, and everyone has gotten sorta jumpy about them. Thank goodness none of it so far has been in our bunch, but we are living on borrowed time.

Two of the squadrons moved out today — leaving my and Kremers squadron. I have been running around like a chicken with his head cut off. Trying to get typing done when I have no typewriters, trying to doctor without medicines, trying to inoculate without any vaccines, etc. It's been hellish. Kremers decided today apparently that it wasn't a working day — for I haven't seen him since 9 this morning until just now. I have been O.D. today and he heard I was sick, I guess, and came down on the line to relieve me. Thank goodness for I have been sick all day, and I certainly didn't refuse him. I'll certainly be glad when we get out of this dump for, damn it, something has been wrong with me ever since I have been here. This time it's my belly with nausea and vomiting — nothing bad and I'll be O.K. tomorrow I'm sure.

As soon as I get this outfit moved and Wagner is back and I get my 8 boys where they can work while I'm gone, I'm going to take off and come on home and let the rest of these guys do the work or let it go undone. I think it griped me more today than usual for I have just been in a bad humor to begin with. Maybe I still am, honey, I certainly shouldn't be writing such as this to you. You have more to gripe than I have and under far worse conditions and more strain. Forgive me for being so nasty. It certainly isn't my intention to make you feel badly.

I hear that Italy capitulated today. Thank goodness! The sooner they all capitulate the better I like it. Again, it is rumored that we are slated for England. I don't care if they would have the Germans licked by the time we are supposed to pull out, and we go to Japan. That would suit me O.K. If we are really slated for England, however, it is my firm bet that we will get there in plenty of time to get into the scrap — and that is the big-leagues of this man's aerial war without a doubt.

Honey, if I fail to write you as often as I should until we get to Nebraska, don't hold it against me. I'm thinking of you as usual and wishing for you. Take care of yourself and Kay for me — I'll be seeing you pretty soon I hope.

All my love,
Joe.

September 9, 1943
Thursday night, 8:10 p.m.

My dearest Honey,

I received two letters from you today. I'll admit that I didn't get any yesterday but that wasn't your fault — the old Uncle Sammy again, no doubt.

Tonight again I'm O.D. due to the fact that Kremers worked for me after 9 p.m. last night. It's O.K. by me for I can assure you that I feel much more like working tonight than I did last night. Last night I felt terrible.

Got everything in pretty good shape now and am just about ready to move. The sooner the better, I'm pretty fed up with the sand and dust, tents, etc. I hear the next place is supposed to be fairly nice — certainly hope so.

Yesterday, the Colonel stopped me and wanted us to give shots and lectures while travelling. I was able to talk him out of the shots, but we are still going to have to give lectures — dammit. We will have to give the same lecture 4 or 5 times in order to get it to the whole train. I'm sure about the fifth time I lecture on the same subject in one day it's going to be damn boresome, both to me and the fellows I'm lecturing to.

Tomorrow should be an interesting day. I have got to go down and direct building a kitchen in a baggage car to accommodate feeding all the men in our outfit. That's something. When I get home, you will be better suited to take care of the finances, and I probably will be suited to inspect the kitchen to see whether it is sanitary or not. Maybe, honey, then I'll make you a good wife. For example today I supervised building an ice-box — a pretty damn good one too. You really get pretty good at improvising in this man's army.

Kremers ran out on me again today, and we got in 20 new Navigators, so I had to process them. He came in at 5 p.m. when I had only two more to go and wanted to know if I wanted him to do them, so I was nasty and refused. I don't know what's in him, but he has certainly been worthless lately. If I was the group surgeon permanently I'd certainly put a stop to a lot of this loafing — I might make myself unpopular in doing so, but I would do it anyway. He informed me today that he thought he would fly tomorrow, so I guess I'm stuck again. So far, I haven't even had time to get my flying time in this month. I really think the whole answer is that Quinn and Kremers both out rank me in seniority and during Wagner's absence I was appointed by someone, either the Colonel or Wagner, to take his place. It certainly doesn't matter to me even though I have learned a lot and have learned our Colonel a lot better. I believe, as much as I dislike the guy, by the time we go overseas, we will probably have a pretty fair outfit due to his efforts.

Honey you had better start writing me in Nebraska about now. I'm sure I will get it if you address the letter to:

        451st Bomb Group Dispensary
        Army Air Base
        Fairmont, Nebraska

By the way, it's only 580 air miles from Fairmont directly to Memphis, so if you see a B-24 buzzing Memphis within the next couple of weeks, I may be in it even though we can't land. See you soon.

All my love,
Joe.

* * *

Joe traveled to Fairmont, Nebraska with his squadron and stayed for a few days before continuing home to Memphis for his furlough and the birth of his second child. Joe's ten day leave, including travel days, lasted from September 15 to September 25. According to family story, Babe wanted Joe to see the new baby before he had to leave for overseas duty. By the end of Joe's furlough, Babe still had not delivered Carol. The night before Carol was born, Babe took caster oil hoping to bring on labor. It worked. Carol was born on September 25 at 6 a.m. Joe stayed through the morning but returned to Fairmont that afternoon.

Fairmont Army Air Field, located southwest of Lincoln, Nebraska, was the final phase of the unit's training before moving to the staging area. This time, the 451st was the only group on base, and living conditions were much better with barracks instead of tents. While at Fairmont, there were a few deadly training accidents. The last training mission was one of the worst accidents they experienced, when two B-24 planes collided midair. Only one man survived.

From Joe in Fairmont, Nebraska to Babe at Baptist Memorial Hospital

September 27, 1943

Dearest,

Sorry I didn't get to write to you yesterday, but I didn't get up until 10 a.m. and I left here at 1 p.m. and got grounded in Sioux City and didn't get back here until midnight. Consequently, I didn't have too much time — that however is a sorry excuse.

Had some harrowing experiences getting back to report for duty. Everything turned out O.K. finally, and I got in at daylight yesterday (Sunday). After leaving Memphis at 4:00 p.m., I got to St. Louis to find out that the plane I was to catch there was 4 hours late. If I waited to catch that, I would have missed my plane from Kansas City to Omaha — consequently, I caught the next plane out that was going just anywhere near where I might make some kind of a connection. It happened that that was Des Moines, so I caught a plane for there at 7:15 p.m. Got in Des Moines at 8:30 p.m. and there was a plane coming through for Omaha at 11:15, but it was all filled up with guys with priorities. There was a guy that had a priority for 5 days that was supposed to show up. I waited around the airport, and he never did appear, so I got his seat to Omaha. Got in Omaha at 12:40 a.m. and my bus pulled out of Omaha for Fairmont at 12:45. The airport is about 15 miles from the bus station, so I called up the bus station to find out if they could tell me where I might meet the bus in time to get on it. Finally, they consented to hold the bus for 10 minutes for me. Luckily, they held it a little longer 'cause I made it. So it all worked out that I arrived in Fairmont at 4:40 a.m. which was the time that I had planned on. I finally got to the Base about 6 a.m. got in bed about 7 a.m. — got up at 10 and came to the dispensary to find there was plenty of work to be done. So I got my men busy. At 1 p.m. I took off for Sioux City with one of my crews to get another airplane — got up there and found out that the plane was not fit for flying and waited around until about 11 p.m. for them to get the 4th engine where it would "putt-putt." Then took off for here even though we had no landing lights, instruments, or anything that's good to have in an airplane. Got in here about midnight, went to bed and got up at 7:30 and find that I'm O.D. So I have been O.D.'ing all day. That's my life history since I have last seen my 3 honeys.

To get at something much more important to me, what about your life history since then? I hope you have been doing O.K. and are not in any pain. Let me know when you are going home, etc. whether you moved rooms or not, and anything that is of interest — that's everything about you and ours. How is Carol doing and is she getting any prettier? Certainly hope that neither of you are having any trouble of any kind.

Honey, I want you to know that I had the best time I ever had while I was home with you and Kay and our house, and yard and garden and grass and weeds and doorknobs and so forth. I really did enjoy it, for you don't know how much you appreciate home until you have been in the army for awhile. Home to me of course, is where you are. It sure was good being with you. I shall remember those few days for a long, long, long time.

Honey, I've got to do pre-flight exams. Will write again tomorrow.

I love you,
Joe.

September 28, 1943
Tuesday, 4:30 p.m.

Dearest Honey,

How are you doing today? I wish I were there to see for myself but I presume I will have to sit tight and wait until I hear from you to find out. I can't wait to hear from you but don't you write until you feel like it. Get Sis to drop me a line if you don't feel up to it yet. I'm still hoping and praying that you are progressing satisfactorily. How is Carol doing? Lastly, how's Kay? Is this too many questions?

Things here are about the same. Still worrying about immunizations, examinations and such rot. Can't say that there's never a dull moment for I'm afraid they are all pretty damn dull.

I have some good news. I hear that my papers are going to be sent on October 1 for my captaincy. Certainly hope so for it will mean a few more extra dollars for us. Of course it sometimes takes 2 or 3 months for the damn things to go through. It will make me very happy to get a $33 increase per month.

I have been loafing today. Came to the barracks and shaved and bathed at 11 a.m., and this afternoon I just took off and have been getting my pay voucher fixed up. Finally finished it and hope it goes through in a hurry for pretty soon I'm going to be running short for eats.

Got two sore arms plus a sore tail now. Took my last cholera shot yesterday and my yellow fever in my right arm. So now I'm practically sore all over. That damn cholera is really rugged. It really makes your arm sore and doesn't improve your disposition.

How is Mother Black's cold, Honey? and did Janie get over there to work for her? Certainly hope Mrs. Black feels better. I think she felt about as low as I did Saturday and that was pretty low — coming home is swell. I enjoyed being there a whole hellova lot. Leaving you, Kay, and now Carol is a tough job for me. Gosh, it was awful.

While I was O.D. last night, I went to the picture show by myself which is an unusual thing for me to do and saw Jimmie Cagney in "Johnny Come Lately." It was pretty good — one of the better pictures I've seen of him. By the way, while at Sioux City the other day I went in to see Jimmy Stewart. Sat in his office and had quite a conversation with him. He made like he had met me before, but he hadn't. He's a swell guy apparently as everyone that knows him says he is. You would never know he was anything but a "commoner" by his actions or personality. His voice is quite natural and he's very tall, but you never see the sheepish look when you meet him that he often exhibits on the screen. He is an operations officer for a squadron and seems quite capable.

Honey, I love you tremendously. Hope you feel well.

All my love,
Joe.

September 29, 1943
10:30 p.m.

Dearest Honey,

For the first time since I returned to the base from my too short visit home, I'm human. Why? Because I received two letters from my honey today. These letters telling me how my two little honeys are but never mentioning herself. I wanna know how you feel and are doing truthfully! Even though I am kept in the dark about the aforementioned matters, I can and will truthfully say that with all of my heart (and mind), I hope to the fullest that you feel as well as it is possible for one to feel after going through everything you have gone through in the past few weeks. But, please, let me know honestly how you are getting along.

I felt rather mad at myself after leaving you Saturday. I wanted to tell you how much I appreciated you. However, Saturday it wasn't in me. In fact, about all that was in me were tears, and I couldn't have talked to you about anything that was on my mind without shedding them. Saturday, I was definitely weak and I don't like weakness in myself. Anyhow, though late, I want you to know how terribly proud I am of you. Honey, I love you, and the more I'm around you the more I love you.

There is no place in the barracks to write. In our room, Zraick and myself have exactly two cots and one chair. There isn't room for the chair, so it is folded up and placed behind the door. Consequently, I am writing on my knee and it isn't so good. So far I haven't gotten any laundry back since I've been here nor any cleaning — you can imagine how I smell.

Again, I surely did enjoy the letters today, but don't you write if you don't feel like it.

I love you,
Joe.

October 1, 1943
Friday, 8:30 p.m.

Dearest Honey,

I did you bad again today, didn't I? I'm sorry but last night I didn't get through until midnight and then I was just too tired so I put writing off until this morning — again, this morning it was the same old story. I was too busy so I'm writing tonight. Yesterday was payday for all the enlisted men so I took my needles, vaccines, sterilizers, etc. over and got the poor guys while they were lined up for their paychecks and let them have it there. I certainly was effective for I got a lot of guys that I have been hunting down for quite a while. Certainly wish I could finish up but this is something that never stops apparently.

How is my best Honey today? Don't worry about Carol losing weight — every baby loses weight for the first 4 days after birth. Which pediatrician is seeing Carol, Dr. Tom Mitchell? If so that's swell. Even though I don't know him very well and he probably doesn't even remember me, I have more confidence in him than any other pediatrician in Memphis. However, I'm certainly not worrying about Carol — new babies rarely have much trouble except feeding problems as far as I'm concerned. How is Kay? — guess she is into everything as usual. Tell her that her Daddy says hello even though it may not mean anything to her.

I'm O.D. again tonight. They are throwing a dance over at the Officer's Club. Here the dances apparently come on Friday night rather than Saturday. The last one they had must have been a drunken brawl by the way that the boys were talking about it when I returned from leave.

We got in four brand new airplanes today. Those four will go overseas with us. They are pretty heavily gunned and heavily armored. They should deal someone a lot of misery.

Really, to me it looks like I won't have to be in England too long. The way the Russians have progressed in the past week is remarkable. Not that I think we can whip the German Army in a matter of months for we can't — but I'm hoping that the same upheaval occurs in Germany as happened in Italy — and it may happen in the not too distant future. However, we will get to Britain before that happens easily. So looks as though I may get in the European and the Southwestern theaters before this war is won.

I had another pilot decide today that he doesn't want to fly anymore. That means another Evaluation Board. This time I'm going to let them throw the book at this guy for there is nothing wrong with this fellow physically or mentally. He just has decided he doesn't want to fly after getting $20,000 worth of education. In other words, he is minus a few things called "guts".

Honey, take care of yourself — I wonder when you are going home. Let me know so I will start writing you at home rather than at Baptist Memorial Hospital. Remember that I love you and the two chillun' more than anything else and you most of all.

With all my love,
Joe.

October 1, 1943
Again Friday night, 11:30 pm.

Honey,

Maybe if I write you again tonight and send my first letter by air mail special, you may get it a day early. In that way, perhaps you won't go a day without hearing from me. Not that I am so doggone important, but laying in a hospital isn't so damn cheery even at best. Hope it works out that way anyhow. I really was terribly busy yesterday. I even forgot to eat, and even worse I forgot to send any laundry or cleaning off. You can imagine what a shape that leaves me in. I haven't had any laundry done since two weeks before I left Wendover, and I haven't had any cleaning done since you and I were in Boise, except that little bit that Janie did for me while I was home. Anyhow it will be next week before I can send out any more, and I just don't know how I am going to get along that long without losing a roommate and a lot of so-called friends.

I have a crew that is flying down to Oklahoma City tomorrow morning, and I think that I will fly down with them. I don't know of a better way to spend the day under the conditions. I can wear flying clothes down there, so I won't need any clean clothes. By the way, we went into winter uniforms today.

Imagine, in the morning our new baby will be one week old. I wonder what she will be like. I'll bet you ten to one that she won't have the same personality and ways that Kay has at all, but I bet that she will be just as cute in her own little way. Reckon what she will look like when she is one and two years old, and how Kay and Carol will look together. It's going to be fun watching them. They can't both have the same personalities or there is going to be hell to pay. I don't know about Carol yet, but two Kays can't exist under the same roof without bars and chains. I can't begin to tell you how much I enjoyed being with you and Kay for those ten short days. Living with my three girls is going to be so damn much fun when this war is over.

I went over to the dance for a little while but it was as boring as they usually are without you. I didn't see anything that was the least bit exciting around there but a crap game, and I didn't feel in the mood to indulge in that, so I came back and am doing something far more useful and entertaining. I also wrote to Sis and Royce tonight.

We got paid today, so as soon as I can get to the post office or to a bank I will send you some money for you to do with as you see fit. Don't think that I expect you to put it all in the bank for I know that you will want some new clothes to fit your lovely new figure. You well deserve them so go ahead and get you some. I just wish that I was there to see you in them.

Honey, it's getting late and I have got to get up at four to preflight these boys. Too, we will be pulling out for Oklahoma City about seven. Remember I love you with all my heart.

Your Honey.

From Joe in Oklahoma City, Oklahoma to Babe in Memphis

October 3, 1943
Sunday morning, 9:30 a.m.

Dearest Honey,

Here we are in Oklahoma City grounded. Have been here since yesterday noon. Had plenty of excitement on the way down but everything turned out O.K. The plane caught on fire about 200 miles from here. Consequently, when we got here, there were plenty of repairs to be done, so we had to stay overnight. We expect to pull out for Fairmont Army Air Field about noon today.

This is the driest town you ever saw. Sorry whiskey costs $7.50 per pint — consequently, there has been very little drinking since we have been here. One of the other officers and myself went to the Oklahoma - Oklahoma A. & M. game last night and it was a pretty fair game. At least I will get to see one inter-collegiate game this year.

Honey, the fellows are here from upstairs to get me to go eat so I gotta quit writing. If I get a chance to finish before leaving Oklahoma City, I will. If not, I will send just this. I love you so take good care of yourself.

All my love,
Joe.

12:20 p.m. A continuation —

Here we sit out at Tinker Field where our ship is — trying to get ready to take off. Apparently, that's going to be a little while even yet. I hope today will be better than yesterday. We had just about all the bad luck you could have yesterday and still made a safe landing plus a four-hundred-foot ceiling to land here at Oklahoma City, and that's a pretty low ceiling. Today it is clear and pretty, however, and we should have smooth sailing back to Fairmont.

I'm here in the Red Cross canteen at the airport where army transient personnel can get all the coffee, tea, cookies and sandwiches that they can eat free. The surroundings are pleasant, and it is very nice.

I don't know whether to send this to the hospital or to home. I presume you are staying at the hospital for at least 10 days, so I will send it there. I will send my letter from Fairmont to Spring St. tomorrow. Hope you, Kay and Carol are doing well.

All my love,
Joe.

From Joe in Fairmont, Nebraska to Babe at Baptist Memorial Hospital

October 4, 1943
Monday night, 9:15 p.m.

Dearest Honey,

Was I glad to hear from you today? I got three letters from you, and I'll have to admit that I had grown rather anxious about you. When I arrived on the field yesterday, I was sure I would have at least one letter from you for I hadn't gotten one the day before — well, there wasn't any. I knew you had not skipped writing more than one day unless you either felt badly or Uncle Sam was messing me up — today all three letters came.

I'm certainly glad that you are at home. Let's see, you went home either yesterday or today. I have been writing you right along at the Baptist Memorial Hospital, so if you haven't been getting any mail since you have been home it is probably at the hospital. Hope it has followed you O.K.

The trip back from Oklahoma City was uneventful — just another airplane ride. We had enough excitement the day before to last each of us for quite awhile I'm sure. Left Oklahoma City at about 2:45 p.m. and arrived here at 4:45. Went to the dispensary and worked until 6 and have been working fairly hard today too.

Went out late this afternoon for P.T. That was the first time since I have been in this organization — and, boy, I'm plenty sore already. There's no telling what I will be like by tomorrow.

Had an Evaluation Board on one of my pilots today. He not only got removed from flying status, got his wings and commission taken from him, but also his uniform. He is busted to a civilian and his draft board will have him back into the Army in 5 days as a private. He got just about what he deserved — he had nothing wrong with him, just got "overseas jitters" without ever going overseas. I'm pretty sure that I didn't help him out either with my testimony. It's sorta nasty, but you certainly don't want to be risking your neck with a guy like that.

Certainly was glad to hear that Ma Black was up and going again. Tell her not to do too much taking care of you now. Janie can do all the work. Also tell Inez that I certainly do appreciate her coming up to see you so often and being so nice. She is one of the nicest people I know, and a hellova good friend of yours — I like her.

All the boys send their regards — Zraick, in particular, just got through asking to be remembered to you. He sends his congratulations, etc. He has been standing up here over me giving an extemporaneous lecture on "harmony," which he is pronouncing "hominy" since he is giving this lecture in what he calls "Memfis" dialect.

Glad to hear Dr. Stabnick says you can come up so early. However you can't come up at all without some help. I will begin looking for a house and you try to get in touch with Mattie. I will try to find a place large enough for all of us.

If and when you come up, don't expect it to be like it has in the past. I will have to stay on the post at least 2 nights out of four, etc. This outfit is pretty much real army as you will see when you get here.

How's Kay and Carol? Certainly would like to see your slim figure and maybe I will pretty soon. I love you so much that it hurts.

Your Joe.

HQ AAB
WF 158

LAST WILL AND TESTAMENT

I, *Joe W. King*, a legal resident of *Helena,*
(City, town or county)
ARKANSAS, United States of America, now in the activ mili-
(State or district)

tary service as a *1st Lt. M.C.*, (Army serial No. *O-491213*), in the Army
(Grade)
of the United States, do hereby make, publish and declare this instrument as my

last WILL and TESTAMENT, in manner following, that is to say:

1. I hereby cancel, annul, and revoke all wills and codicils by me at any
time heretofore made;
2. I hereby give, devise, and bequeath to *Olive B. King*
(Name of person or persons who are

my wife, now residing in *804 Spring St. Memphis*
to inherit,with relationship,if any, (City, town, or county)
Tennessee U.S.A. all my estate and all of the property of
(State or district) (Country
which I may die seized and possessed, and to which I may be entitled at the time
of my decease, of whatsoever kind and nature, and wheresoever it may be situated
be it real, personal, or mixed, absolutely and forever;
3. I hereby nominate, constitute, and appoint *Olive B. King*
(Name of executor or executrix

my wife, of *804 Spring St. Memphis Tennessee*
with relationship, if any) (City, town, or county) (State or district)
United States of America, as my executrix and request that she be permitted to
serve without official bond or without surety thereon, except as required by law;
4. I hereby authorize and empower my executrix within her absolute
discretion to sell, exchange, convey, transfer, assign, mortgage, pledge, invest,
or reinvest the whole or any part of my real or personal estate.

IN WITNESS WHEREOF, I have hereunto set my hand and seal to this my last
WILL and TESTAMENT, at *F.A.B.F. Geneva Nebraska*, this *4th*
(Place of execution)
day of *October*, 194*3*.

Joe W King, 1st Lt MC.
(Signature of Testator)
Signed, sealed, published, and declared by the above-named testator,

Joe W. King, to be his last WILL and TESTAMENT in the pre-
(name of testator)
sence of all of us at one time, and at the same time we, at his request and in his
presence and in the presence of each other, have hereunto subscribed our names as
witnesses, and do hereby attest to the sound and disposing mind of said testator
and to the performance of the aforesaid acts of execution at *Fairmont*

Nebraska, this *4th* day of *October*, 1943.
(Place of execution)

(Name) (Address)
339 Bratt Street
Inglewood, California
(Name) (Address)
6531-B Stafford ave
Huntington Park, Calif
(Name) (Address)

*On October 4, 1943, Joe filed his last will and testament, leaving all property to his
wife Olive B. King and naming her the executrix of his estate.*

From Joe in Fairmont, Nebraska to Babe in Memphis

October 5, 1943
Tuesday, midnight

Dearest,

How is home and how does it look to you? I'll bet it's nice to have Kay and Carol both around. You must be careful not to try to do too much at first with the little rascals. Please take care of yourself. You know I can't see you, so I worry about what you do much more than if I were there. Hope all of my girls are doing nicely.

They tell me (the Colonel) that the officers that inspect everything before an outfit goes to the Port of Embarkation are coming in tonight. I really do dread for them to come around this dispensary. I have been in such turmoil the past 48 hours that I don't know what to do first — there are so very many things that need to be done. I wonder if I will ever finish.

I am sending you a money order for $100. That should help for when you are able to start northward. Use it for whatever you see fit.

I don't know about you coming up in two weeks — to me that is a bit early. However, if you have plenty of help, and Dr. Stabnick says you can, I guess it is O.K. I will try to find a place to live just as soon as I possibly can. I don't know how soon that will be with the inspectors here. I will do the best I can. But in two weeks if you don't feel like making the trip, don't do it.

I wish I could really tell you what to do. But I can't for I haven't the slightest idea how long we will be around. I have even heard that our ground echelon will pull out in 2 weeks. I rather doubt that. We, the echelon that flies across, will pull out sometime after that — when, I don't know. It may be that I will have to leave in a very short time but don't let that stop you.

Honey, it's 1 a.m. and I've got to get up at 4. I am O.D. and have to preflight crews. Take care of yourself. I love you so very much.

All my love,
Joe.

October 6, 1943
Wednesday night, 9:50 p.m.

Dearest Honey,

Am I an old fool? The old axiom "there is no fool like an old fool" is certainly correct. The reason I know is because I am both old and a fool. Old because I certainly wouldn't be as sore as I am right now from playing a little basketball if I weren't, and certainly nothing but a fool would go out and get himself as sore as I am. I can't walk, I can't sleep, I can't sit. I was supposed to go out for P.T. again this afternoon, but I couldn't have made it to save my life. I told them that they could kick me out if they wanted to. I still wasn't going out for P.T.

Honey, I'm inquiring about a house. I'm trying to get a whole house, but I don't know how successful I will be. At first I believe it was easy for the fellows, however there have been 2000 men move here, so it is harder to get a place now. I have some people looking for us though, and we should find something. I suggest that you contact Mattie for we are going to need some help. I certainly won't be able to come home half as much as I used to, and many times you will have to come out to the base to be with me. So far, you've never been in army life like it is in an O.T.U. outfit. So I believe I would try to bring someone along with you, or we may not get to spend too much time together.

As I explained last night, I can't tell how long we will be here nor where we will go when we leave. I do know we are really getting on the ball, so I imagine time is drawing near. Too, I'm pretty sure the Colonel asked for a two week extension on our departing time, and I'm also pretty sure it was refused — but I just don't believe they will send us across so very soon with some of our crews as green as they are. If they do, it certainly is going to be mass murder if we go to England — unless they give them considerably more training over there — which may be the case. I wish I could tell you something that was somewhat definite, but in such an outfit as this that is quite impossible.

I am glad to know that you got your flowers and am glad that you enjoyed them so much. Wish I could have brought them to you personally. You are a good brave girl, honey, and I appreciate you. Take care of yourself, Kay and Carol and remember how much I love you all.

I love you best,
Joe.

October 7, 1943
Thursday night, 10:30 p.m.

My dearest,

Got a swell letter from you today. It was one that was long enough for me to really enjoy, and it built up my ego tremendously. In the army, one needs his ego built up occasionally — it is torn down so frequently. It was a swell letter. I have already read and reread it many times — as I do most of your letters. It was written last Saturday, and five days is certainly a long time for an airmail letter to get here but that is just about what it takes all the time. It certainly does put me way behind in the news.

I have some news for you. I rented a house today. Nothing fine but something we can manage in very nicely. It is an old house — two stories but there is no stairway from the inside to the top story, and the top story is only used to store stuff by the lady that owns it. It has a garage and an awfully large yard which is not fenced in. There is a living room, kitchen, 3 bedrooms (one not furnished). The front of the house is somewhat separated from the other two bedrooms, kitchen, bath, cellar, front and small back porch. The furniture is everything but good — but I'm sure we can manage. The rooms are all moderate in size. In the kitchen there is a Roper Gas Range and a Frigidaire. There is also a washing machine. In the living room, there is a couch which is apparently made to be converted into a baby bed. It could easily accommodate either Kay or Carol — even both if there were some way to keep them separated. As I remember there are certainly no surplus chairs. There is a furnace in the basement and a stoker. So far this sounds O.K. but there are some disadvantages. We will have to buy coal, furnish our own gas and lights. That in itself is not too bad, but we have to also furnish pots and pans, dishes, and linens. I thought perhaps you could bring a minimum of each, and we could just eke out. I'll get the coal and gas and lights when I find out definitely when you are coming.

The house is in York which is 20 miles from the base but where most of the officers have their wives. It is not a very large town — a couple of thousand I guess, but the largest that there is around here.

When you come you will have to bring money with you for at the present time I'm practically penniless. I think I have the total sum of about $8.00, however, I think I have $41.00 out that the boys owe me — that won't last too long however.

Be sure to have the heater fixed before you start, and be sure it puts out heat. I'm quite aware that it runs now but don't let anyone talk you into believing it's putting out heat 'cause it ain't. Remember, I love you with all of me.

Joe.

October 8, 1943
9:45 p.m.

Dearest Honey,

I have really been busy the last few days. I have three typewriters going all day long — trying to catch up on records. We are making some progress — I think. Certainly hope so. There is so much to do and so little time to do it. We will keep at it however, and by the time we pull out of here, I hope to be in damn good shape.

This afternoon I had to go out on the range and fire the pistol, which I am and always have been very poor at. Tonight we had to go to a lecture which we have to do 2 or 3 times per week. We are really making an honest effort lately. I found out today that I'm going to have to give 28 more hours of lectures before we leave here, which within itself is going to be a real job.

You say my grass is coming up already? Boy, would I like to see it. Hope we really have a bumper crop. How uneven is it? I'll bet I skipped so many places that it looks like a pieced quilt.

I'm glad to know you are already sitting up. I certainly hope your strength returns rapidly. As far as I'm concerned, the more rapid the better. Of course that is a very selfish way to look at it. I really don't want you to leave Memphis until you feel like it and are able to travel. Please above all, don't do that. I really feel that I should probably suggest that you stay there, but I am just too weak to say it. So therefore, just please use your better judgment. You know how you feel better than I do or anyone else. I can't say anything but that to see you would make me extremely happy.

Honey, I have much work to do before I can go over to the barracks and go to bed, so I had better say goodnight for today. The only other thing that I have to say tonight is that I love you more than anything. Kiss my babies goodnight.

All my love,
Joe.

October 9, 1943
Saturday night, 11:30 p.m.

Dearest Honey,

Another day, another dollar! I am a tired little boy. I have really put in a hellova day today. A few more like this one and I am going to retire from this man's army. Tonight, for the first time in a long time, my legs ache like I had played about 36 holes of golf. I am not the only one that is tired either. These boys that are working for me won't have any trouble going to sleep tonight, I betcha. We have a few more days like this, and we ought to be in pretty good shape. I certainly hope so, and we will be if we can get the proper cooperation from the squadron headquarters. Enough said about business.

Got your letter that you wrote last Monday today. You asked me about this place. It is not bad at all. However, one would be sacrilegious to gripe about anything after living at Wendover for a couple of months. In fact, I sincerely believe that almost anything would look pretty good to all of us. I am not the only one that feels that way about it. Our dispensary is very nice. It is every-thing but elaborate, but the dust doesn't blow all over you continuously. The Officer's Club could easily be improved upon, but it is passable, and we don't have time to spend there anyway. The food, like most army food, is good before it is cooked, but once gotten in the hands of an army cook anything can happen and usually does. The barracks are only livable, but at least one doesn't have to go outside to go to the latrine, and that is something that we all appreciate very much. So all in all, I don't think that any of us are griping about this field, however we are all griping about something.

The weather is definitely cold early in the morning, but by ten it has warmed up considerably, and it does not get chilly again until about nine at night. The morale of the outfit is definitely sub-par as it has been ever since we activat-ed. I certainly hope that it gets in better shape before long. However, I hardly see how it can. It is not the Colonel's fault entirely, although he is responsible for his share, but the training schedule is so terribly rigid that the boys never get any time off for anything. One can't blame them for being tired of doing nothing but flying a B-24 and sleeping, and that is all that they are doing. I know all of this is terribly dry to you. I don't mean to bore you with it.

I don't think that there's a possible doubt that we will get away from here before November 7. When we do leave here, we will be going to a staging area. That is the same thing to the flying echelon that the Port of Embarkation is to the ground forces. We won't be able to get off the post, call, write or anything until we get away. Well we will talk more about that later.

Honey, if by chance you want to call me just call this base as you address your letters. Then I will either be at No. 6, 25 or the Officer's Club. No. 6 and 25 are the dispensary and my barracks, respectively. I have got to get up at 4 a.m. So I gotta quit. This is certainly too long anyway. Remember that I love you and always will and the kids too. Certainly hope that you are all doing fine.

> All my love,
> Your honey.

> October 11, 1943
> Monday night, 9:30 p.m.

My dearest Honey,

I'm sorry, but I just didn't have even a wee minute to spare to write. Yesterday was a real tough day, as has been today, and tomorrow has all the markings of being of the same variety. We are really getting "on the ball." Night before last I got only a couple of hours sleep while I was O.D. Then yesterday we worked hard, and yesterday afternoon the Major informed me that I was to give a 4 hour lecture today beginning at 1 p.m. So last night I got my references together to try to prepare my lecture, and I just was too tired to work on it. Besides I had a headache, so finally I went to bed about midnight. This morning I slept through the alarm, so someone else had to hold my sick-call for me. Then later in the morning, I started doing overseas physicals and did those until noon. At 1, I started lecturing and lectured all afternoon without preparing a lecture, which for me is a tough job. After eating, I went back to the dispensary and have been working until just now. As you see, I have been pretty busy. Not only that, but I have to give another 4 hour lecture tomorrow, and 4 hours is just too damn long to have to talk or listen. Don't think I am the only one that is working for, on the contrary, some of the fellows are working much harder than I am.

I am wondering whether you have been able to arrange to come up and, if so, when? Certainly hope you feel strong enough to tackle it. I think I will call you in a couple of nights or so to see what you think about it and also to see how you feel. I certainly am glad to hear that Carol is getting along so nicely and that Kay's cold is getting better.

I finally got some scotch. One of my Bombardiers brought me a quart from Omaha yesterday. Looks like I may have to write a check. The boys owe me $40.00, but they haven't paid me yet, and I have 45¢ left and breakfast costs 35¢ — Almost broke. I love you and am eagerly looking for your arrival if possible.

> All my love,
> Joe.

October 12, 1943
Tuesday night, 11:15 p.m.

Dearest Honey,

Well I just finished lecturing to the boys. From 8 until now is a long time but I had a good audience, so I didn't mind so much. I had to cut it short an hour, so the boys could get on the line to get to work, and I certainly wasn't sorry. That's over for a few days anyway.

Today has been almost wintery. In fact, I guess it is a winter day. The wind has really blown hard. This morning it was coming from the south, but tonight it's directly out of the north, by morning it's really going to be cold. It has rained, snowed, and sleeted practically all day — not a very good day for flying, and we haven't been doing too much today.

The real overseas inspectors arrived this afternoon. We have to stand formation in the morning for them to check us personally and our personal effects — whether we have dog-tags, etc. I think I have everything but my "pay-data card," and I am going to try to get that in the morning before the formation.

Pretty soon things should lighten up somewhat for I'm almost up to date — or at least I think I am. I believe, medically speaking, my squadron is in as good shape as any of them. I will certainly find out tomorrow. After then, things should slack up some for me, and I am ready for it to do so.

The C.O. is back, and I'm quite sure he didn't get any extension on our training time, however, I'm not positive. I believe if he had, it would have leaked out before now. I do know that in the past extensions on training time have not been granted by the 2nd Air Force.

I am certainly eager to know whether you are going to be able to come up or not. Perhaps such a long trip for such a short time is expensive and impractical, but that doesn't matter if you are in shape to make it. It would be worth the cost just to see you.

Honey, I must go to bed. I'm pretty tired, and I need sleep, so I'll cut this short tonight. However not really happy until I can again live with you. Our marriage is certainly the most perfect one that I have ever seen. Remember how much I love you.

Joe.

Parading at Fairmont Army Air Field, 1943.

Babe, along with Kay, Carol, and Janie, moved to York, Nebraska for three weeks to be with Joe while he was in Fairmont.

At the beginning of November, rumors spread through the camp about deployment. The unit was complete with 3700 men and 62 B-24 Liberators. The squadrons began their overseas movement mid-November with a stop at Lincoln Army Air Base as the final staging area. Lincoln Air Base was established in 1942 as an aircraft mechanics school. The staging area readied the planes with mechanical checks and any modifications or improvements. All of the men's belongings and equipment were labeled and counted, and the men had their final medical approvals.

KING MADE CAPTAIN

Joe W. King of Memphis and Helena, Ark., has been promoted to captain in the Medical Corps attached to the Army Air Forces. He is a flight surgeon with a bomber group, now stationed at Geneva, Neb.

Capt. King received his medical degree from U - T Medical School and interned at John Gaston Hospital.

CAPT. KING

He was practicing in Helena when he entered active service in October, 1942, as first lieutenant.

His wife and their two daughters, Kay, 2, and Carol, four months, are with him. He is the son of Mr. and Mrs. A. C. King, 3422 Summer.

On October 26, 1943, Joe was promoted from 1st Lieutenant to Captain. Memphis Press-Scimitar, November 18, 1943.

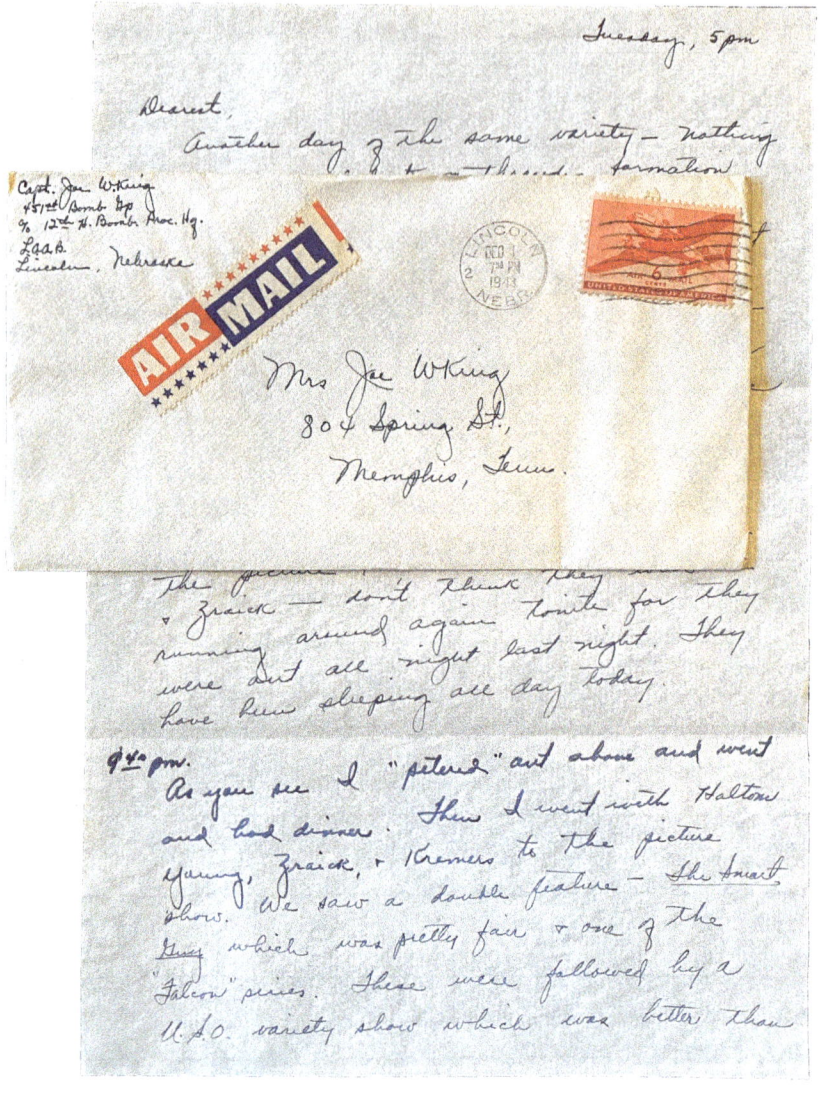

From Joe at Lincoln Army Air Base, Nebraska to Babe in Memphis

November 23, 1943
Tuesday night, 10:45 p.m.

Dearest,

I wonder where you are tonight. Possibly somewhere around Kansas City, since I know you probably didn't get off too early with the two girls, packing, etc. I felt for you and have been thinking about you most all day.

This post is O.K. Everything is run according to a prearranged schedule, and everything is well organized. It so happens that you aren't busy but about 4 hours per day, but that is two hours in the middle of the morning and again in the middle of the afternoon. It is apparently a very good post.

We aren't restricted. We were last night, but that was lifted today. I started to call you last night and tell you, but with the indecision of when we are leaving and all, I thought perhaps it was best to leave things as they were. If I have to stay here for a couple or three weeks I will be mighty mad at myself.

I certainly hope you, Kay, and Carol made the trip alright, and you had no bad luck or any mishaps. Surely am glad to know that you had three good tires on there and wish you had a fourth one. Hope Kay nor Carol got a cold.

Speaking of colds, Young has the worst one I have seen in quite a spell. He has really suffered with it. I am trying to get it from him apparently. However, so far, mine hasn't been anything to talk about.

We all damn near froze to death last night — sleeping under 1 army blanket with no fire in these damn cardboard barracks. I don't think I slept a wink after about 4 this morning — gosh, it was cold. I'm going to be warmer tonight for I have already confiscated an extra blanket off of some bed farther back in this barracks. Young, Zraick and I have turned out as roommates here. The three of us are in a room with two "double-deckers" in it.

Honey, I want to take this opportunity to again tell you how much I hated to say goodbye to you yesterday morning and how I wish I could tell you what is inside of me, so you would better know how much I love and care for you. However, on being with you the last few hours, it isn't in me to be able to talk with you and tell you how I feel. I suppose I am feeling too sorry for myself knowing that I'm going to have to try to exist for a period of time before I can get back and really start living again with you. You are a swell person — absolutely the best one I know or have ever known. You have gone through a hellova lot with me and for me, and I appreciate every bit of it. Too, I want you to know how much I enjoyed you and our family coming up to F.A.A.F. It's just a pity that the time passed so rapidly while you were there. Had you been in Memphis, that would certainly have been a long, drawn-out affair. Honey, I will try to get into town to get the pictures made as soon as I can. However I don't see how I'm going to get there in the daytime, and I'll bet 10 to 1 I won't be able to find anything open after 5 p.m. However, I haven't forgotten and will do the best I can.

Next morning —

It looks as though there isn't anything for me to do today except make two formations of about 30 minutes each. It certainly does feel funny to have nothing to do but sleep, carry clothes to the cleaners and "piddle" around. It certainly is boring to do nothing when you are used to keeping busy. At the moment, I'm about to go to the PX and carry my blouse to get it cleaned (which you can do in 36 hours here) and come back here and shave. I bought you a little pin at the PX yesterday — will send it to you the first chance I get.

I believe I will be transferred to Blackie's plane before leaving here. At least, he thinks so. Don't know when we will leave but probably pretty soon. Pretty sure we will fly over Tennessee somewhere on the way down for we will probably leave from Morrison Field, West Palm Beach Florida then to South America. Then across to somewhere — don't know where. Also pretty sure we will be fighting Germans instead of Japs. Other than that "I don't know nuthin".

Tell all the folks hello for me — you can write to me here. Don't mention where we may go however to me nor anyone else — please. Tell Kay hello for me and kiss Carol. Hope you are all doing fine.

All my love,
Joe.

From Joe in Omaha, Nebraska to Babe in Memphis

November 25, 1943
Thanksgiving night, 7:00 p.m.

My dearest Honey,

Here I am in Omaha with a bunch of the boys — Young, Zraick, Cue and Roach. I don't think you know the latter two. This afternoon they let us off for some reason — maybe this post doesn't know that there is a war going on. All the rest of the boys are down stairs at the bar, and I suppose I will join them as soon as I finish writing you a note.

We had the first five crews of our bunch pull out today. I'm quite sure for West Palm Beach. It may be quite awhile before I pull out however. I'm going to leave on the last ship, so someone can be with Major Blackmon. So I won't be flying across with McAllister, Young, and company. However, it is Threadgill's crew and that is a perfectly good crew. There is nothing to worry about. We aren't through being processed yet, and I don't know when we will be. However, we will leave here as soon as we are through and our ships are ready.

Today is Thanksgiving. I am truly thankful, honey. Thankful for you primarily — the kids almost as much. All of our health and happiness. We have been extremely lucky honey — or to put it another way the Lord has been extremely good to us. It won't be too long until we are back together again and even happier than ever, if that can be possible.

How did the house seem to you? I'll bet it looked pretty good. How is my rye crop coming along? Is Janie in better shape now that she is back with her husband? Are Kay and Carol O.K. after their trip and how are you? I know I have to ask specifically, or I will never find out. I suppose our families are alright or we would have heard.

Jimmy Dorsey is playing at one of the theaters tonight, and I think I will go over to see and hear his show. Went to see Guadalcanal Diary at the Post Theatre at L.A.A.B. last night, and it was worth seeing. I personally thought it was pretty good.

Honey, I think I will go join the boys before they run off and leave me. I love you more than I will ever be able to put across in my failing vocabulary. Tell the folks and children hello.

All my love,
Joe.

From Joe in Lincoln, Nebraska to Babe in Memphis

November 28, 1943
Sunday, 3 p.m.

Dearest Babe,

Well, honey, I was finally able to get into Lincoln and get my picture snapped for you. The lady took 4 poses all with my hat on, and I don't imagine any of them are good, and she is going to send you the proofs. You get 4 pictures, all 8 x 10. They are already paid for — it will be up to you to either have them tinted or not — I had that in the bargain. You should hear from them in 7 to 10 days. I will send the receipt along, so if there is any trouble, you will know who to write, etc. By the way, I had my picture taken with my wings on. Just fudged on the government a little bit. A TWX (wire) was sent in yesterday for all of our flight surgeons' ratings. Don't know for sure whether they will go through or not — it may be that we shouldn't have put in for it until we got overseas.

While in town yesterday, I went shopping and bought you and Kay a little something. I will send the tickets along with this letter, so if they don't fit or you would like something else you can return them. Honey, I looked hard to try to find something for Carol, but I couldn't find anything that struck my eye that she didn't already have — tell the little rascal that her Daddy loves her too and that even though I didn't send her anything that I thought of her just as much as I did Kay.

They have been issuing us more and more equipment — when they get through there won't be room for any personal clothing. The Army is really trying to prepare us for any eventuality. Still haven't an idea when I will be pulling out of here — probably quite a spell since I have been transferred to the last ship pulling out.

Things are pretty dull around here. There isn't enough to do to really keep you busy. We are all certainly getting enough sleep. Some of the boys are getting too much time off to run around. The peculiar thing is that I believe I have more guys with colds now than I did have when they were getting no rest and were exhausted all the time.

All the boys have just congregated in here — shooting the bull about

85

everything — women in particular. Too, they have borrowed all the money I have except $20.00, and I'm not going to lend them that 'cause I have gotta eat for the rest of the month. Honey, I'm going to get Young to mail this for me uptown, so I'll close until tomorrow.

Take care of yourself and love the babies for me.

I love you,
Joe.

November 29, 1943
Monday, 9:30 p.m.

Dearest Honey,

Last night Zraick and I went down to Lincoln to see a picture show. Saw Dangerous Blondes, which was almost fair. There was another one which was sorry, but I don't remember the name. I do remember that there was no comedy and no newsreel. After the show, we tried to get a taxi to come back to the camp but couldn't because it wasn't midnight, so we went over and caught the bus with all the enlisted men and finally got here about 10:15. Hit the sack and slept well.

This morning, the boys had a difficult time getting me out of bed early enough to make a 9 o'clock formation down at the hangar, however I finally made it. After that formation, I went to PT for 2 hours and then went and had "brunch." After that "half-breed" meal, I took Young over to the dispensary and dressed the boil that is located on where he sits and then held sick call for our men. Then, I came over to our barracks and slept until 5 p.m. — then to the Officer's Club to dinner. Finished dinner and went to the picture show on the post with Marco and saw Riding High with Dotty Lamour (the shapely lass who cannot sing nor act), which was a little better than average. Cass Daley was in it, and as usual she was a scream. I like to see that homely looking creature act — she is pretty good. I left the show and came back to the barracks to write to — guess who. So there's all the happenings since I last wrote to you.

The post is very big. You walk and walk and walk when you have something to do. Right now my legs are hurting from walking back from the picture show. Guess it is good for me, but I still don't like it.

Honey, my surgical experience has finally become of use to me. At least I learned in my few years of practice a few stitches. Last night before going to the show, Zraick and I had a "sewing circle" in full swing. I sewed up a pair of pants for him — my blouse lining was coming loose, which I fixed. I altered a shirt collar, which I never liked before, and sewed up something else, I've forgotten what it was now. Anyhow, I hadn't done anything like that since I was in college, thanks to a good wife. I rather enjoyed it, however, for it gave me something to do to help pass the time away.

It's mighty lonesome without you up here. I might be here well into December. No one can tell — we aren't moving out as fast as we thought we would. I wish we would hurry up and get on our way. The sooner we get over, the sooner I will get back to live with my three honeys — the three best in the world. It certainly is nice to have you all to think about.

Zraick and Young are in Omaha tonight. I think they had dates in Lincoln and were either going over to Omaha by rent-a-car or train. Guess they will be back in before the night is over. Everybody is making use of the lull apparently and are in Lincoln or Omaha practically all night.

Honey, I received your telegram this morning. I'm certainly glad you reached Memphis safely. By the way, Pfau's crew, the boys that I went to get when they parachuted out in Northern Nebraska, brought me a quart of scotch in appreciation of my efforts Saturday night. It was a nice gesture, which I appreciated.

Hope I hear from you tomorrow. Goodnight honey.

All my love,
Joe.

November 30, 1943
Tuesday, 5 p.m.

Dearest,

Another day of the same variety. Answered roll call this morning and have been playing poker at the Officer's Club all day long — and I mean all day. Won $20 but lent $45 out, so that leaves me with a meager $5. However, that is plenty for I have about $125 lent out and pay day is tomorrow. Then, I will be in the chips again. Think I will go to the picture show tonight with Young and Zraick — don't think they will be running around again tonight for they were out all night last night. They have been sleeping all day today.

9:40 p.m.

As you see I "petered" out above and went and had dinner. Then I went with Haltom, Young, Zraick, and Kremers to the picture show. We saw a double feature — "The Smart Guy," which was pretty fair and one of the "Falcon" series. These were followed by a U.S.O. variety show, which was better than the average and far less corny.

Since receiving your telegram, I sorta expected to get a letter from you today, but it didn't get here. I'm pretty sure that I will hear from you tomorrow. Until then take care of yourself, tell Kay and the family hello, kiss Carol for me and remember that I love you very very much.

All of my love,
Joe.

Photo that Joe sent to Babe before he left the States, 1943.

December 1, 1943
Wednesday, 10 p.m.

Dearest Honey,

Looks as though Uncle Sam has gotten his mail all mixed up again. No letter again today when I was so damn sure it was going to come. I know it wasn't your fault — just the traffic of air mail.

Nothing to do again today. Played poker all afternoon with Blackmon, Young, Zraick, Marco, Leeser, and Stinwinter — won about $40.00; wish to hell you were here to spend it on. As it is, I think I will send you a money order for it tomorrow for I haven't use for it.

Tonight, I have been real extravagant. I went up town and went to a barber shop and sat down and ordered the works. Got a haircut (short because you like it that way), oil, shampoo, and tonic. Cost me $1.50 but it surely was nice. After getting the haircut and the trimmings, I went out with Young and Zraick to have a few drinks and dinner. Then they stayed in town, and I returned to the base.

I think it is safe to tell you honey that we are leaving here for West Palm Beach, Florida. From there probably to Port Said — then to Trinidad — then to Natal, Brazil. Then to Ascension Island (Gr. Britain), which is in the middle of the Atlantic Ocean. Then to the Gold Coast Africa — then up through Africa to Sfax, Tunisia probably. We will be there for a spell perhaps and then go on up into Italy — just where, I don't know but probably around Foggia, since that's the only large base we have there.

Thursday, 10 a.m.

Don't know how I will pass the remainder of the day away. Guess I will hunt me up a nice comfortable poker game or sleep, then go over to the dispensary about 1 p.m. and see what few of my men that come in — and I should go over to the hospital and check on a couple of boys that are in the hospital. Tonight, it will probably be another picture show. I've seen so many picture shows lately that pretty soon I'm going to know one actor from another.

Well honey, I'll close this time and get this in the mail. I haven't gotten a check or money order yet, so I guess I'll just start by sending you a bill by mail. Let me know when you receive it. Surely hope all of you are OK and none of the babies are sick from their trip.

All my love,
Joe.

December 2, 1943
Thursday night, 10:30 p.m.

My dearest Honey,

I held sick call at 1 p.m. and saw two of my boys. Then I "piddled" until about 4 p.m. at which time, I found myself in the Officer's Club with nothing to do except play poker. Threadgill, Blackie and I were playing with four other fellows from the base. Peculiar as it may seem, Threadgill, Blackie and I all won — don't know how the other guys came out but someone had to lose, so I imagine it was them. Threadgill won over $100; I won about $70, and Blackie won about $20. As I said yesterday, I have no use for so much money, so I will send you some. Don't think I'm sending you money I won. I'll keep it and send you the money I already had — just to keep your hands clean.

Dearest Babe,

Tonight, I went to the picture show with Marco and saw Bette Davis and Miriam Hopkins in "Old Acquaintances" — one of the most unique movies I have ever seen. Silly, but it still has something about it that makes you think. Both actresses play their parts very well, but it certainly is a screwy one.

I think everyone has sent their wives home now except Blackie. Mrs. Blackmon is leaving this coming Monday, I think. Mrs. Wagner left today, and Wagner is pulling out in the morning — which leaves only Kremers and myself. Marco leaves in the morning too. By the way, Wagner told me today that his wife is about two months pregnant. They are very happy about it and rather amazed since she has been trying to get that way for quite awhile. In fact, I'm sure that he just signed some papers not too long ago concerning adopting a child. Honey, this is all the gossip I know. Hope I hear from you tomorrow so I'll at least know you are still living.

All my love to the best woman there is,
Joe.

December 3, 1943
Friday, 7:30 p.m.

Dearest Honey,

Zraick and Young are again in town tonight. They would have saved money to just buy the hotel, I think. They haven't been out here for so many nights that I doubt they could find their room now if they were looking for it. Sorta like two high school kids after their first date — sorta letting it run them wild. No — they're OK — still just having their last fling I guess. Apparently tonight is their last night for they are to be "briefed" tomorrow, so tomorrow night they will be restricted to the post, and they will probably leave here early Sunday morning.

I wouldn't be too surprised if we weren't "briefed" Sunday, which would mean that we would leave Monday. There's just no way one can tell — particularly since we are supposed to bring up the tail of everything. At least, leaving is within sight now for the first time — I really need to do something. I have never been so restless in my life.

It is true that I have received no letters from you, however, there is a slip here on my desk saying that I have an insured package at the post office. So at least I will have received that before I pull out of here. Just thought I would tell you that, so you wouldn't be wondering as to whether I received it or not — I imagine it is something from you.

Believe it or not, I'm not going to the picture show tonight. The first night I have missed one in quite awhile I think. Instead I am going out as soon as I finish writing to you and shave and bathe. Then I'm going to come back here and get in the "sack" and read. I have a Saturday Evening Post, a Look and a Yank to read tonight. I will either read every line in them or read myself to sleep. I'm not really concerned as to which happens. I wish you were here. I would try to beat you on the Photoquiz in Look. As it is I will just see by myself whether I'm average or not.

Honey, I suppose I will say goodnight to you now. It's been pretty cold up here — in the 20's all day long and night too. In fact, if you were up here, we would both be getting in the bed with cold feet I guess. As it is we can't even get one another's feet warm. I love you — always will — and count myself extremely lucky to have the chance.

All my love,
Joe.

P.S. By the way honey, they took $250 out of my pay for the first of this month, so you should be getting $250 per month starting December 1, 1943. That other $100 might be a little late, so if you didn't get it don't let it worry you, but let me know when it starts to come through. Love.

December 4, 1943
Saturday, 1 p.m.

Dearest Babe,

As Li'l Abner says so frequently "O happy day." Received two letters from you this morning. You can't imagine how much it picks you up to get a letter when you haven't heard in so long. I'm happy to know for sure that all the kids and you and our families are doing O.K. Received the package from the folks today — it certainly was nice of them to get all those things up here. Tell them that I think I'm all fixed now. I have 150 razor blades — enough to last 1½ years at least, and I certainly hope I won't need more than that. I have two tubes of shaving cream and two small tubes of toothpaste — personally think I am fixed.

This afternoon, I am going down to the hangar and exchanging my B-4 bag. (That's the val-a-pak). I have busted it by over-packing. Still don't know how in the devil I am going to be able to get all this stuff in such a small amount of luggage. I guess if I use the old adage of "there's always room for 1 more," I can get it in. What's worrying me is whether said old adage is true. Let you know more about that later.

I may take off from the base this afternoon and go uptown to mail a couple of things to you. While I'm up there, I probably will go to another picture show since my one day's rest from them.

By the way, we got a wire back from the air surgeon saying that we couldn't put in for our flight surgeon's ratings until we leave the Port of Embarkation. So therefore, we will put in for it then. There is one nice thing about it, and that is the rating will be retroactive, and we will get the raise in pay from the moment we leave the U.S. Probably I will be able to send you some money from overseas yet.

Honey, I just received the third letter from you today. This is swell. I got your picture which is very good but not nearly as pretty as you are. However I shall buy a frame for it this afternoon just the same. Also, as usual, I received the itemized bank account, which I'm sure you know is something that is unnecessary for you to do. I'm not in the least worried about how the money is spent, just as long as I can have it there for you when you need it — and I shall make every attempt for it to be there.

In case it gets real cold before Christmas, you might open up the package you got from Miller and Paine's. It is the one that has Kay's little suit in it. And she might as well get all the use out of it this year as she can for by next year she will probably have outgrown it. It should be nice and warm.

Well, honey, thanks for the letters. I feel much better already. I will send you my A.P.O. address as soon as possible, and you can spread it around to the family. I love you so much. Tell Kay hello and kiss Carol for me.

All my love,
Joe.

P.S. Sorta watch the skies Monday or Tuesday. If we come close to Memphis, I will tell Threadgill to "buzz" our house. You can tell Mom and Dad.

December 5, 1943
Sunday afternoon, 5:00 p.m.

Dear Honey,

I just put in a call for you, and now I'm in the process of waiting for it to go through. Hope I don't have the same luck that I had a couple of times while I was at F.A.A.F.

Sorry that Mother and Dad Black haven't heard from Hugh in so long. Tell them not to worry too much — Hugh is probably on a troop movement some-where. If so, he could not have told them he was going to make it probably. Be sure to let me know when you hear from him next.

Honey, I don't think that picture of you is bad. In fact, I believe it's the best that I have seen of you. Speaking of pictures, I will enclose this snapshot of myself that Prewitt took of me one day while standing in the dispensary.

By the way, I finally got the other little package off to you today. If you polish the little pin, it won't look so bad. Hope you enjoy it.

Well, honey, apparently that time has come. We begin in the morning at 6:30 weighing our equipment in and loading our ship. Tomorrow afternoon, we get briefed on our route and early Tuesday morning, we will take off

Photo of Joe in the dispensary, 1943.

southeastern-ward. It will probably be two weeks before we reach our final destination. I will let you know where as soon as I can — probably be on the North African coast — Sfax, Bengasi, Tobruk or something of that sort.

At the time being, Lt. Tucker and I just came up from the base and rented a hotel room precisely for the purpose of him calling his family in Alabama and for me to call you. We happened to meet up with the Blackmon's, and they came up to our room, and we all had a drink of brandy together, which is all you can get up here with the exception of gin, and you know my opinion of gin.

As soon as our calls go through, we are going to go with the Blackmon's and have dinner and then all of us are going to the picture show to see Cary Grant in "Mr. Lucky." I don't know whether I have seen it or not but something tells me that perhaps I have — maybe you can tell me. Last night, I came to town to get this package wrapped and sent to you and couldn't for it was too late. It seems that the stores don't stay open late here on Saturday. (Tucker's call just came through, and he is talking to his Mother now) I hope that mine goes through pretty soon.

Zraick and Young left this morning. By the way, all those B-24's that have been coming over have probably been ours. Some of our outfit have already reached their final destination. We know that Manoogian is already there.

If we pass over Memphis, which I am going to try to do, we are really going to buzz the house. Hope it doesn't scare the babies or your neighbors. I will try to throw you a "streamer."

Honey, don't be sitting in Memphis worrying now for two reasons. (1) Worrying won't do any good, and (2) I'm going to be OK. Nothing can possibly happen to a person who is as anxious to come home to his wife as I am, and if something did happen, I would get out of it and get to come on back to you — so please don't worry. I want you to think of me, yes; but no worrying. Tell the folks the same thing. I'm going to be O.K.

Now, honey, you probably won't hear from me as often in the future. Don't let that bother you for it will all depend upon the mail and the exigencies of the situation. Don't think for one second that I have forgotten you nor cease to care so much. That will not happen ever. Too, from now on my letters will not be quite so informative for this is probably the last letter you will get from me until we see one another the next time that won't be censored.

Regardless of that, I want you to remember for always that I love you, Kay and Carol more than anything else there is on the face of this earth. That one thing is the most important thing to me. No other thing that has ever happened in my life has been so completely satisfying and nice as is my marriage to you. Take care of yourself and remember how much I care.

All my love,
Joe.

P.S. Your call just came through at 10:50 p.m. Hasn't Kay's voice changed so much since I last saw her?

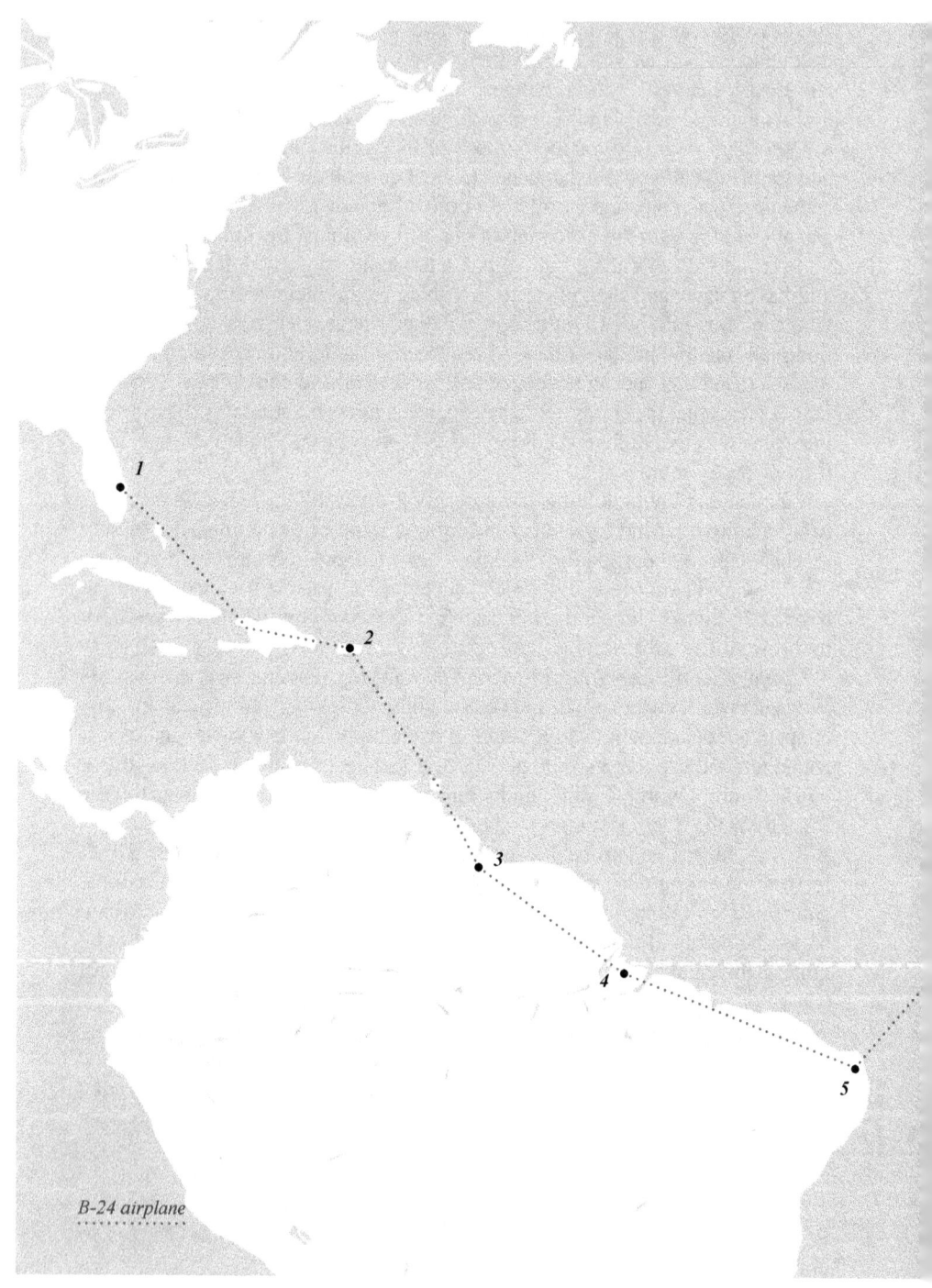

B-24 airplane

Part II

En Route to Italy

The 451st left for Italy in two groups. The ground echelon traveled by train to Camp Patrick Henry in Virginia and departed on December 3, 1943. They sailed in two ships across the Atlantic through U-boat territory to Naples, arriving before the end of the year.

The air echelon flew in four phases, and Joe's group was the last to leave. The B-24's were packed so tightly that the few passengers they carried had little to no space to sit or lie down. To get to Italy, the 451st flew from Florida, through the Caribbean, to South America. After a stop-over in Brazil, they headed to North Africa. On January 20, 1944, five weeks after they left the United States, all 62 B-24 air planes arrived safely and in formation in Gioia del Colle, southern Italy.

Joe's journal, 1943-1944.

Dearest Babe,

Joe received no mail while he was traveling; however, he kept a journal documenting his travel and the censored events that he could not share in his letters. The following section aligns Joe's journal with the letters he sent home to Babe. Some of the letters have missing words where the censor cut out what Joe was saying.

Joe's Journal

December 8, 1943
Today I left Lincoln, Nebraska for Morrison Field, West Palm Beach, Florida. Flew with Lt. Threadgill in "Supermoose" B-24-H #4442. Intended to "buzz" home but Memphis "closed in tight." Left Lincoln at about 6 a.m. Arrived at Morrison Field about 4 p.m. in woolens, which are much too hot for this weather. Co-piloted for about 1 ½ hours. Buzzed Birmingham. Morrison is a nice field. Nice officers club.

December 9, 1943
Spent day on field getting lectures and some additional equipment ready for hop tomorrow. Was briefed on how to ditch, mosquitoes, how to beat the jungle and sea. Briefed to Trinidad or Atkinson Field, British Guiana.

From Joe to Babe

December 9, 1943

Dearest,
We are terribly busy, so I will just make this a short note. I will do better later. By the way, I may cash a check for I may want more money. However if I do, I will send it back to you if I don't use it. Just thought I would tell you.

I am now rooming with Major Blackmon, who is a damn good roommate — wish I could keep him as such but realize that is quite impossible.

Hope you have heard from Hugh by now. I know that you are worrying about him, but keep your worrying down to a minimum.

Capt. Joe W. King, O-496213
727 Bombardment Squadron
451st Bombardment Group
A.P.O. #9200 c/o Postmaster
New York, N.Y.

Henceforth write me at this address and tell the folks the same thing. Kiss the kids for me and remember that I love you all very much.

All my love,
Joe.

December 10, 1943
Was awakened at 12:30 a.m. and carried to the plane. Plane wasn't ready to fly — hydraulic system out. However, operations pushed us out anyway. Take off time about 4 a.m. Flew out over the Bahamas. Nassau looks beautiful, like a perfect circle of lights from the air. Then over nothing but water until we sight Haiti and the Dominican Republic. Overcast is about 8,000 ft but spotted and can see through to the ground. Country looks as though well taken care of, beautiful and green. High mountain peaks sticking up through the overcast. Flying almost over the center of the island which is to the right of our course. Thought we sighted submarine this morning, but not too sure. Was briefed not to get close to them anyhow. Flew over mountain top projecting up out of strait between Dominican Republic and Puerto Rico. Landed at Borinquen Field P.R. at about 11 a.m. The field is beautiful. Very picturesque, just on the north west tip of P.R. Beautiful buildings, hospital and officers club. Bar, dining room, waiters and all. Plenty of rum and whiskey. Cigarettes 5¢ per pack. Taking Atabrine to prevent malaria. Is warmer.

December 11, 1943
Laying over at Borinquen today. Too nice a spot to leave so quickly. Slept this morning. Hired a "publico" this afternoon and went down to Aguadilla, the nearest town. Town very dirty. Amazing — can get all Kodak films, any size you want but no Kodaks. Went down coast to Mayagüez, which is said to be more modern in the wildest taxi ride I ever had. Scared to death but don't know how to tell the driver to slow down. Mayagüez is a little cleaner than Aguadilla. Kids bother you just as in Mexico. Don't know how to rid myself of them. Hungry but scared to eat in any place due to disease. Returned to Borinquen with the same wild ride. A dance at officers club — went over for a few free drinks of scotch and then to bed preparing for tomorrow.

December 12, 1943
Awakened about 4 a.m. Take off from P.R. about 6. Flew along the northern coast of P.R. past San Juan, which from a distance looks like a nice town. Flew over water only until we hit Saint Lucia on which Bean Field is located, just south of Martinique. Turned south and headed toward Trinidad. Scattered cumulus clouds at about 2,000 ft and up. So flying at an altitude of about 700 ft. Thought we saw submarine this morning but not positive — anyhow briefed to leave them strictly alone. Hit Trinidad (north eastern tip) about 11 a.m. at which time I was piloting. Raining like the devil over the island. Instructed to continue to Atkinson Field, British Guiana. The northern coast of South America is just flush with the ocean. The water is really muddy for a mile or so out in the ocean.

Dearest Babe,

The land is thickly covered with trees. Water is standing over all the land appearing to be about knee-deep — no houses, no cultivation, no habitation, just jungles and rivers. First fly over small amount of Venezuela and into British Guiana. Atkinson Field down inland about 20 or 30 miles on a rather large river, which empties into the ocean at Georgetown. No road from camp to town. Restricted to camp anyway. Runway sorry, camp sorry. Has an Officer's Club, which serves plenty of fruit juice (canned) at 20¢ per glass but nothing else.

December 13, 1943
Woke up late. Nothing to do but swelter and sweat. Thank goodness for summer uniforms. Went to a show and saw one of Dr. Gillespie series. Not too good. Atabrine upsetting my stomach.

December 13, 1943
Monday, noon

My dearest Honey,

Well a lot of miles have passed under me since I have last written to you. We are slightly weary, too, even though we have been stopping about every 1000 miles for Major (Blackie) to rest. The trip is really getting him down — apparently, it's the constant vibration that one experiences in an airplane. I have enjoyed the trip so far. I have co-piloted quite a bit of the way, which certainly makes the trip pass a great deal faster. So far, we have had no trouble at all.

It is very hot here. You are damp all the time. It is very very humid. All the houses are built far off the ground, apparently to make it cooler so living can be more bearable. Too, I guess it helps as far as the insect problem is concerned. It rains many times per day — no one seems to mind. In fact, it is raining now, which is about the third rain so far today.

There are millions of birds — all chirping, and all colors. The foliage is very heavy. The trees are all tall and skinny — no short thick trunks like you see in the States. Somehow they remind me of a kid in adolescence that just starts to grow when he hits the "bean-stalk" age. No foliage on them until the very top.

Where we spent the first night out was really beautiful. The people reminded me very much of the people we met in Mexico City. The kids hounded you continually for money just like Mexico. In fact, everything reminded me of old Mexico.

Here it is completely different. This is far more jungle in nature. The people speak better English of course than I do, and it is almost too sultry to breathe. I think that I can tell you that I am now in ▮▮▮▮▮. But I don't think it is permissible to tell you where.

By the way, the water down here on the coast is really muddy. I suppose it is muddy 500 yards off the coast. On leaving the U.S., the water is first blue. Then as you near here, it gets green. Then it gets as muddy as the Mississippi. In the first part of South America that we flew over, all of the land was under water — looks as though it was just continuous with the ocean. I certainly would hate to have to bail out in that mess — you never would get out of it. I wouldn't know whether to wear my life preserver or not.

I bought you some ear-bobs at our last stop and some here too. I know you don't wear them much, but they are small and take up practically no room, and that's what I need to get. Also bought 4 small Caducei. You can keep one and distribute the others to Sis, Katy and Mother if they want one.

The paper is so damn wet, I can hardly write on it. You can bend a cigarette double without it breaking. Paper matches are no good at all.

Honey, I will write you again the first chance I get. I'm O.K. and will be O.K., so don't worry. Hope you, the children, and our families are doing just as well. Wonder whether you have heard from Hugh yet and certainly hope that you have — not only for your sake but for Mother and Dad Black's too. Tell Mother and Dad that I will write to them as soon as I get to my destination or before.

Remember that I love you more than anything and kiss the kids for me.

All my love,
Joe.

December 14, 1943
Took off from Atkinson about 6 a.m. Going to fly the semi-inland route to Belem, Brazil — not straight over the jungle and not around the coast — more emergency fields this way and more radio. Plenty of jungle but some spots look almost decent (palmettos). Flew over the Amazon River just as it enters the ocean. Mouth of River about 60 miles wide with a large island in the center. Southern branch of the river is known as Porto Rio (I think). Belem just on the southern bank. Plenty of shipping in the River. Belem looks busy and is definitely large. Landed at Belem about 2 p.m. and, boy, is it hot. All the difference in the world. There is a nice runway, field and Pan American Airport. Carried to barracks and then to mess — primarily of green bananas. Six miles to town, but we are again restricted.

December 14, 1943
Tuesday

Dearest Honey,

Well when your mail is censored somehow or another it cuts down considerably on what you can write about. At present, I am sitting in a very fair Officer's Club. Just finished my second scotch and soda, so I don't feel the heat so terribly. When I say heat, I really mean heat. I think it was around 97 degrees when we landed here today. I imagine that is quite a bit different than what it is at home.

We are restricted to the post I think, but if we can get off, we are going to go into town tomorrow. If so, I will try to buy you some nylon hose, which I hear might be possible. I started to buy you a purse today made of alligator hide, which cost about $12.00 but would last you a lifetime. However, I doubt seriously you would use it. They were $22.00 at another place I have been. You can get them with a complete alligator (small) on them if you want to.

Practically all of my boys are sick. I had to give Blackie some codeine today to make the trip even though we rested over 1 whole day at our last stop. We are going to rest over tomorrow too I'm quite sure. I'll bet the Colonel is really raising the devil wherever he is, but he is going to be raising more than that probably by the time we get there — which is O.K. by me. As I understand it, that was why they wanted me in this ship. Lt. Threadgill is also sick with diarrhea tonight. Me, I'm perfectly healthy. Feeling fine except hot and sticky. Also somewhat weary. It is a tiresome trip getting to the main event at least.

Funny how the prices change down here. Cigarettes are 5¢ per pack. Scotch ranges from 20¢ to 50¢ per drink. Could have gotten all the film I wanted for 40¢ per roll but can't possibly get a camera. All the prices seem to be peculiar, and one never knows just how much money his dollar bill is worth — I have probably been gypped many times but don't have sense enough to know it.

Have a good mess here. All the fruit you want. I ate 4 bananas at lunch — all were as green as grass on the outside but ripe on the inside. Oranges are the same way. We always have a heck of a time making the natives who work in the mess halls understand what we want to eat. So far no one has gone hungry however. Mahogany is used for everything. The floors are made of it, the boardwalks, the beds — they use mahogany just as we use pine in the states. The jungles are unbelievably dense and vast. More rivers — real rivers — than the states can boast of all put together. You can hardly go 15 or 20 miles without crossing a nice river. Insects are terrible — last stop, it was ants and I mean ants of all sizes and varieties, here it is mosquitoes.

Honey I've run out of things that I can tell about other than I love you still and always very very much. There is no change and will never be. I think of you many many times per day. Sometimes wishing you were with me — other times glad that you are not. Tell the children how much I love them but more important remember how much I love you.

All my love,
Joe.

P.S. Somewhere in Brazil.

December 15, 1943
Laying over here today. Blackmon not feeling too good. Is hot and sultry. Field (living quarters) just cut out of the jungle. Playing bridge and poker and drinking scotch and bourbon just to pass time. Leaving tomorrow.

December 16, 1943
Briefed to either Fortaleza or Natal. Practically nothing but jungle all
the way. Took off about 6:30 a.m. Landed about 11 a.m. Flew over over-
cast practically all the way. The land is everything but beautiful, unless
you like the jungle. Field at Natal loaded down with planes of every
variety — B-24's, B-17's, B-25's, B-26's and plenty of them. 6 groups
of heavy bombardment going through here that I know of. Most of our
fellows are still here. Only a few crews have flown over. Blackie having
more trouble with his back.

December 17, 1943
Met LaVigne, an old San Antonio buddy and Santa Ana classmate who
is a Flight Surgeon for an Air Service Squadron here. Went over to PX
and bought me a pair of Brazilian Boots — very nice but hurt my feet. La
Vigne is going to get me a pass to go into town tomorrow. Played poker,
bridge and sticks to pass time. Quite warm.

December 18, 1943
Got up late and ate lunch. Then Blackie, McFarland, and I went into
Natal shopping. These boots hurt my feet terribly so I'm going to buy
another pair — which I did and much nicer at the same price — $5.00.
See much pretty silk, all one could want. Bought 13 pairs of stockings
for Babe, perfume, and alligator bag at Casa Rio. Think she will like it
all except the perfume, which is probably imitation. Stopped at the Hotel
Grande and had a few scotch and sodas. Got to talking to a kid 14 years
old who was the smartest kid I think I ever saw — spoke 5 languages
Portuguese, Spanish, English, French and German. Knew plenty about
the States though he had never been there. Told me about Jesse James
and his family — where he was from etc — but insisted that New Mexico
was in the "South." Of course, being literally true. Smart as a whip.
Went back to camp to do nothing.

December 18, 1943
Saturday

My dearest Honey,

How I wished for you today. We surely could have had a heck of a lot of
fun together. Blackie and I went to town. Blackie could get off the post because
he is of field grade. I got off because I know the base Flight Surgeon, and he got
me a pass through pull. None of the other boys could get in — consequently, I
had to shop for everybody and did I shop. By the way, the base Flight Surgeon
is LaVigne. If you remember, he is one of the fellows who was in that Buick that
stopped and lent us a tire tool when we had that flat in the middle of the desert
on our way to California from Randolph.

Well anyway, Blackie and I went into town and we bought you and his wife all we could afford which was pretty good. We also bought ourselves a pair of leather boots, which are quite the vogue in our outfit now and in everybody's outfit who passes through here. The boots look like a cross between an English riding boot and a cowboy boot. They have low flat heels but are a little more pointed than an English boot — however they only come about 3/8 up your leg. They cost $5.00 all day long, and you can stand there and watch them make them for you. I could make a fortune on them in the States.

Now to tell you what I bought you. Firstly, I got you one dozen pairs of silk hose. They sorta stick you for them, but they are supposed to be the best there is down here. I do know that they are pure silk 'cause I bought them at the most reputable place here, which may not be any too reputable at that. I got about 3 different shades — all as light as I could get. I hope you like them. Now to keep them from rotting — put them in a fruit jar and seal them air tight — that is with a rubber ring under the cap and store them in the refrigerator. You probably already knew about this, but I thought I would tell you just in case you didn't. Next, I bought you an alligator pocketbook which looks O.K. to me. I don't know whether you will like it or not, but I can assure you it is a genuine alligator hide. Next, I got you some Chanel #5 perfume — no Cassandra here. One of these bottles may be, and probably is, imitation. I think the darker perfume is the real McCoy. It's very difficult to tell what is good and what isn't. You can get "Tabu" all day long, but all of it is imitation here. Perfume is something you just have to take a chance on. The other little trinkets I bought at other places. I will try to send all of this stuff as soon as possible. Hope you get the stockings in time for Christmas.

This place is somewhat different from any place I have ever been before. Not exactly what I expected. Here, like in Mexico, they play the American for a sucker (as we are). But they do it in a little more genteel way. You see so many things that are so much cheaper than in the States that it almost takes your breath away. All the silk stockings you want — all the rubber tires — no cameras and no film though. In fact, a soldier in the U.S. Army is not allowed in any shop where they have such. But bolts and bolts of beautiful silk and velvet for approximately what you paid for it in the States prior to the war. I didn't buy anything for the children 'cause I didn't see anything small enough to send. I saw a rocking horse that I would have loved to see Kay on. You will have to get Carol and Kay something for me. There I go again always throwing things on you. I can't help it, honey. Again I wish you had been with me today. We would have had a wonderful time together.

Telling you how much I love you — how much I think of you and how much you mean to me every day of my life means a lot and helps a lot to talk about. And I'll start to feel much better when I get to wherever I'm going and hear how you and the children are faring. Here's praying that you all are in the best of health and no worries at all.

I expect this will get to you around Christmas. Anyhow have a swell Christmas and remember that I hope to be with you for the next one.

All my love,
Joe.

P.S. Give my love to all the folks. Tell them all I will write when I get to my destination. Wish them a Merry Christmas for me.

December 19, 1943
Up late and to mess hall for lunch. Pillow terrible and bed hard with no springs. Sleeping in double deckers. Blackie having plenty of trouble with his back. Carry stockings for Babe to Censor's office and sent to her air mail. Censor won't let me send undeveloped film or purse by air. Mess here is terrible and remains so. Primarily green bananas. Play bridge, poker, and sticks to pass time.

December 19, 1943

Dearest Honey,

Today has been just like the previous ones down here with nothing to do but just sit around and wait. We are still in Brazil and have been in this one place for about four or five days now. None of us feel too good. We have to take Atabrine prophylactically to prevent malaria, and it doesn't induce pleasant symptoms. In fact, it is causing a bellyache in me right now. I've firmly made up my mind to skip the dose tomorrow — malaria or not. I'd cut off my nose to spite my face, wouldn't I?

Next morning —

Here it is 10:30 a.m. and I'm just barely getting up. Living the life of Reilly, I am. Better hurry or I'll miss lunch. The beds here are horrible — no springs at all, and I believe the cotton in the mattresses has never been seeded. The pillows are harder than the floor being made of cotton, too. The food is terrible. I have been doing alright, however, living off of bananas, Brazilian oranges, and pineapple, which are plentiful every meal.

I got some laundry done yesterday — the first since Fairmont. The bill for all was $1.12. Had 5 shirts and 5 pairs of pants too — many towels, etc. And it is done beautifully. Instead of putting starch in your trousers, they put tallow and it does a beautiful job. They iron with an iron that has a red-hot fire of charcoal going inside of it — I'm becoming educated. By the time I've knocked around for about five more years, I will be about half as smart as my wife but not nearly so good looking or nice to be around.

Honey, here comes Blackie with his laundry, and we are now ready to go mail this and eat. I love you so much that trying to tell you about it almost seems a waste of words. Hope you and the children are doing swell.

Joe.

Dearest Babe,

December 20-23, 1943
Doing nothing. Get up late, from nightmares on this sorry bunk with no springs and this mattress which is made of cotton not seeded. Continue to play bridge, which I am getting somewhat better at — poker, sticks and whatnot to pass the time. Looks as though I might fight the war here. Gets more boresome every minute.

December 22, 1943
2 a.m.

Dearest,

Guess I'm just plain dumb. You sorta lose track of time around here — I am doing nothing except sitting around and waiting to go on. At present, I am still "somewhere in Brazil." That's all I can legally say. I think it is December 22 but I'm not positive. I think I have been here one week today.

I wonder how Kay is taking Christmas. Is she old enough to get "worked-up" about it? I guess not, but I'll bet Christmas morning when she is opening packages, her eyes will be something like saucers. It would be fun to see her. However, it will be even better to see her next year or the following at the latest. Surely wish that I had sent her something.

Certainly hope by now that you have heard from Hugh for I know if you haven't your Christmas is, or was, a flop. If you haven't, don't worry too much — he is probably O.K. There hasn't been any action in Sicily in quite a spell, so my bet is that he was just moved somewhere and couldn't write while en route. If I had had to ride a boat, the same story would have gone for me. These boats seem to take forever to get almost nowhere apparently.

Hope by now you have received your silk stockings. Maybe they got there by Christmas. How did the red corduroy house coat fit — and was it too loud? That red should look good on you. Did Kay's little snow suit fit or was it too large? Isn't it silly asking all of these questions? I know you have already told me in your letters, but of course I haven't received your letters and won't for awhile. So if you don't mind I wish you would send me just one letter, your next, with a brief recap of everything that has happened since you last wrote me in Lincoln. Such as about 1. Hugh, 2. whether any of you are sick, 3. how the folks are, 4. and whether you need anything and lastly, but not least just reassure me that you love me, which I know you do but just like to hear it. Send this letter to A.P.O. 9101 but this letter alone; send all others to 9200 just like you have been doing.

There's nothing to say about this place except the beds get harder by the night and the pillows also. By the way, the mattresses really are made with cotton that still has the seeds in it. I said it more as a joke in my last letter, but it really is the truth — so you can imagine how hard and lumpy they are. I wouldn't be surprised if pretty soon this mattress didn't seem like a Beautyrest to me. This place is still as hot as always, and it rains frequently. However, lots of the time, there is a fairly good breeze blowing. It really is boresome to just sit and wait. After doing just that for one solid month — more or less, it really is going to be hard to get me working again. Hope you had a lovely Christmas.

I love you,
Joe.

December 24, 1943
Everything just the same. Hot, sorry food which costs 50¢ per meal, green bananas, sorry cot, nothing to do. Went to picture show tonight. Had to go and sit on those hard wooden benches for 1½ hours before the show started so I could get a seat. Rained during the show, but I just sat there. Show over at about 8. Went down and heard Nelson Eddy sing, which was very good. I enjoyed him more in person than on screen. He tried to tell a few nasty jokes which didn't go well with his personality — so were a flop to me. A little more blue perhaps than usual. Christmas Eve night and all. Go home and to bed and lie awake quite a spell.

December 25, 1943
Christmas Day with nothing to do. Sent Babe's purse to her yesterday by air. Up about 11:30 and to mess of sauerkraut and Vienna sausage and green bananas. Rumored that we might leave tonight but probably won't. Play bridge. Go to supper early. To everyone's surprise have turkey with dressing and cranberry sauce — very good. First decent meal since the States. Serve cigars, cigarettes and candy too. Then to barracks to pack just in case we leave. To bed about 11:30 p.m.

December 25, 1943
Christmas Day, 1:15 p.m.

Dearest Honey,

Well here it is Christmas and I'm still in Brazil — don't know when we will leave but I hope soon. We are all pretty tired of this place. I went to the theatre (open air job) and saw Nelson Eddy and his troupe. It was pretty good. That guy can sing too, but I don't care for his nasal twang to the singing. I had to stand up for the whole performance, but it at least passed two hours. We then came over to our barracks and drank up a quart of scotch and a pint of Canadian Club that we had been saving for the occasion. There were six of us, so none of us were affected by it except our bombardier, Oleen, who was pretty drunk. Finally about 11:30, we got him home, so the rest of the fellows could go to sleep. We were afraid that if we didn't the M.P.'s would be around for the guy in the next room was raising plenty of Cain on account of so much noise. The fellows over in an adjacent barracks must have really thrown one — when I finally went to sleep about midnight they were singing, and I woke about 6 a.m. this morning, and they were still singing. Just a bunch of homesick kids trying their best to put up a front — it was really evident last night.

We had Christmas dinner today — but not quite like the one that I'm sure was described in the newspaper. Beans with onions was the main dish with a dab of sauerkraut on the side, tepid cocoa, the drink and desert. Beautiful and tasty. Not quite like the one that was served at March Field last year, but I didn't expect so much. I would give anything for a coke or a glass of milk. The food really is terrible though — I'll bet I've lost a good 5 pounds in the past week, and so has everyone else. We really will be glad to get a change.

I'm still mighty thankful, honey. I know I have you waiting back there for me, and two beautiful little girls. As long as I have that, and you three have your health and I have mine and hopes of getting back to you, I won't really gripe seriously.

By the way, honey, I was finally able to get your pocketbook out of here by air. I guess you should have gotten it by now. I found out there was a light shipment going out and they let me send it. I noticed that one of the "scales" of the alligator hide was scuffed-up. If you will take it over to Dad, I think he will be able to glue it down for you where it will hardly show. I hope it was suitable. Honey, I also tried to send you 6 rolls of film that I bought along the way, but the censors won't let any film be sent, undeveloped or not — consequently, I have 6 rolls of 620 film.

Well I hope you all had a very lovely Christmas this morning. If I had been there, I would have seen to it that Santa would have been good to you. All of you deserve it whether you got much or not. I'll personally see to it from now on. I love you so very much and wish I could hear from you.

All my love,
Joe.

December 26, 1943

Awakened at 1:30 a.m. and carried out on line. Try to get K-rations or sandwiches for long hop — no soap. Pay lodging bill and go out to "Supermoose." Ship hasn't been serviced. Have to wait until they fill her with 2700 gallons of gas, which they won't do until we have already run up our engines — conserving every drop. Finally take off for Dakar about 4 a.m. Trip across almost uneventful, pass over rock jutting up about 200 miles out to sea, which is on course. Flying at about 9000 feet until we hit bad weather about half way. Weather too high to go over without oxygen, so we decide to go under; dive down to 1,000 ft and still have to fly on instruments. Raining so hard you can't see nose of ship. Rain lasts about 1 hour. I sleep about 2 hours. Then pilot about 2 hours. Sight land at about 3 p.m. which is Dakar — Navigator Adler really "on the ball." Circle field and start to land. Co-pilot "Scotty Scarlata" too busy reading Ellery Queen's Murder Stories didn't let down flaps until too late, and we damn near missed runway but made it O.K. Runway made out of steel mats, short and rough. Field shows effects of much rain. Ship and us get sprayed for mosquitoes and then carried to sorry wooden barracks — sorry mess barracks and latrine. Blackie and I go up to Officer's Club for drink, and we decide to pull out in morning — Go to barracks and to bed under mosquito nets. Seems too damn cold for mosquitoes to me. About 10 p.m. I have to get up and go. Atabrine has got me down. Vomit and have g.i.'s all night.

December 27, 1943
Leave Dakar at about 6. Country different than anything I have ever seen. Land first very rough and craggy but no vegetation. Then the Sahara Desert with miles and miles of sand and sand dunes and nothing else. Pass over the landing strip at Tindouf. The place of the Foreign Legion and where Beau Geste was written. Very desolate place, so we raise cruising speed up to about 195 miles per hour. Radioed that there is bad weather in the Pass through the Atlas Mountains, so we decide to turn west below (south of the mountains) and fly to the ocean and go around the mountains as you have to fly at 17,000 feet to clear the Atlas Mts. Make 90 degree turn and head for ocean completely going outside of Spanish Morocco, which is neutral. Hit the Atlantic coast at Agadir just south of the Mountains and head north over the ocean. Hit a storm, which we can't get through and have to turn back because gas is getting low. Radio out so can't request landing at Navy field at Agadir but decide to land anyway. Make approach for landing and horn starts to blow warning us that our landing gear is not locked down, so Threadgill gives her the gun and we circle the field again. This time the hydraulic system works and landing gear is locked so we make uneventful landing. Report to operations, and he instructs to go to the hotel in town to spend the night. Ride 6 or 8 miles to town in G.I. truck and look or stare at camels, Arabs, cactus, etc. along a rather pretty paved highway. Pass nice resort hotel but positive we aren't going there for the Army wouldn't have such a nice place for transient officers. Correct — we don't stay there. Instead go down the road 1 more mile and drive up to one much nicer — Bewildered. Beautiful resort hotel Terminus at Agadir, French Morocco. Here we are given very nice rooms with private sun porches to each room. Beds with inner spring mattresses and sheets. Private shower (cold water though) very modern, thick glass doors, chromium plate everywhere, douche bowl in toilet but no private commode. Funny people. This hotel run by Frenchman, was prewar resort hotel patronized by English primarily. Had been rented by U.S. Army for rest home for soldiers until 3 days previously. Beautiful place — all rooms overlooking the ocean and what little beach there is.

Dearest Babe,

December 28, 1943
This place too nice to leave, so I ground Blackmon for 24 hours saying
he needs the rest! Go up into the town, which is very dirty. Stinks. Sewage
flowing down center of street. Nothing but Arabs — who defecate along
the sides of the street, while you watch them. Shopkeepers try to get us in
but we see nothing worth looking at. Town on side of hill at top of hill is
a "place of prostitution" known as the bull pen. How many live up there
is not known. They grab men's caps and run inside to try to get the men
in. The price is 2 francs or 4¢. They will give you 40 francs if they "don't
satisfy." While having intercourse, anyone can stand around and watch
that cares to — no modesty. I returned to the hotel. Some of the boys go to
watch — no takers, so they tell me. Hang around the hotel — play bridge
and poker. To bed to sleep on those wonderful beds.

December 28, 1943

My dearest Honey,

Well here I am in Africa. I am not to my destination yet but should be soon. Much water has flown under the bridge since I last wrote to you three days ago.

I couldn't possibly describe what I have seen in the past three days — if I were adept in the art of writing, it would take volumes. As it is, I will merely say it is much different than anything I have ever seen before, and I am surprised to a certain extent. This part of Africa looks a good deal like the Middle East, which I did not expect it to.

At present, I am not at a scheduled stopover. We were forced down yesterday here, which is a Navy base, because of weather. We are living in a hotel which during peace time is a resort hotel. It is very nice, glass doors, private tile sun porch, private bath, beds with inner-spring mattresses, no mosquitoes and fair eats. Wine but no scotch. All very nice — much better than we are accustomed.

French is the native tongue and not only have I been wishing for you personally but for the use of your French also. I can get what I want in Portuguese, but in French they don't understand me and I don't understand them.

I walked up in the town today, and it is much different from anything I have ever seen. The main thing is that it is just much dirtier. To begin with the town just smells. I tried to buy you something, but I didn't. It's pretty crummy. However, I'll bet it was very nice during peacetime. It looks rather queer to see camels walking down the middle of the street with someone holding its tail. Too, it is common for the sewage to flow down the middle of the street. It looks funny to see somebody "depositing" their "sewage" just off the sidewalk. All in all, their customs are very different from ours. The Arab women turn their faces when we walk by. Of course most of their faces are already covered, but you don't even see their eyes if they can possibly help it. Very peculiar people.

I guess I can say that yesterday we flew over the Sahara desert. And, lady, that is a real desert. Nothing like the Arizona kind — just sand dunes for miles and miles and more miles.

The place I am now reminds me somewhat of the Edgewater Gulf Hotel — smaller but nicer and nearer the bay. I am staying on the top story and the breeze is wonderful. Don't need much breeze, however, for it is quite cold. I got my woolens out when we arrived and my blouse feels good at night.

Well, honey, I will write to you again probably the day after tomorrow. When I get to my destination I will cable you. Honey, if you can find a nice pair of large Flight Surgeon's wings in town send them to me. I have only been able to get some 85¢ ones which aren't very nice.

I love you and think of you continually. Wish you were right here with me. You would certainly enjoy this spot. From now on, I'm glad you are where you are however.

I love you,
Joe.

December 29, 1943
Arise about noon. Ground Threadgill. Place is definitely too nice to leave. Go down to an old fishing pier and watch two Frenchmen and two Arabs fish. Wish I could rent a pole from them, but see none around and couldn't make either understand me anyhow — so just watch. Neither couple seem to be having much luck. Huge waves coming in. Sea rougher than usual. Stand by pier and just let salt spray hit me. Feels good. Return to the Hotel. Play bridge and poker. Decide to leave the following day.

December 30, 1943
Arise about 8. Eat and go out to Field. Our ship is sitting away over in mud. Threadgill gets in and runs up the four engines and literally flies it to the runway to keep it from bogging down. Almost overshot the runway. Taxi down to end of runway to load on our fuel. While sitting there, the French Fliers in their peculiar looking ships are taking off and landing in all directions irrespective of instructions of the tower or wind direction. Damn Fools. A French General walks over in all his pomp and uniform and stands under our wing tanks smoking a pipe while we are loading her up with 2300 gallons of 100 octane. Of course, he was told to move and quick by us, which he did. No wonder the Jerries ran over them so quickly. We take off about 11 a.m. and head again to the ocean. At 500 ft, it sounded as if big hunks of gravel was hitting our ship and plenty of it. I looked out and the sky was black with grasshoppers. Millions of them over an area of about 25 square miles. I wonder how many those 4 props killed making 26 hundred R.P.M.'s. We hit the ocean and flew up the ocean coast line to the next large town. Then turned westward for Marrakesh, which is north of the mountains and a

very large native city. The Atlas mountains are beautiful. They are high (about 16,000 ft). They are a continuous chain rather than a peak here and there like the Rockies. The whole range is covered with snow. They are pretty. Wonder why I haven't heard more of them. We fly over rather smooth land at a low altitude. This land has occasional Arab palaces on it and orange groves. The land is irrigated. We land at Marrakesh at about 12:30 p.m. Were assigned to tents out in olive grove. No lights, dirt floor, canvas cots, 4 blankets. Went into town to look around and to have wine at Hotel Mamounia. Also a few glasses in some joint downtown. Arab part of the city in the very center is off limits — plague and typhus rampant. Contains many people, probably 500,000. Known as Medina. Marrakesh is almost modern and built around Medina. Orange trees full of fruit growing all along side of all streets. Oranges on them look O.K. but are too sour to be edible. Back to camp by bus and to bed.

December 31, 1943
What a night!! No sleep, froze to death. Freezing cold with 2 blankets under me and two over. No one slept; decide to get out of here before another night. Get crew together and briefed to Oran. Take off at about 10 — only short hop. Land Oran about noon. Assigned to French barracks, which are cold but much better than last night. Officer's mess just across the street. Go to club about 6 p.m. sit around with Blackie and drink vino and watch everyone have good time. To bed about 10 p.m.

December 31, 1943
New Years Eve, 7 p.m.

My dearest Honey,

Haven't written you since stop before last. However I sent you a wire this morning at which time I was farther in the interior.

I don't know what your opinion of Africa is, but if it's anything other than a very, very, cold place you are mistaken. 'Cause, honey, it's cold. Last night, I spent the most miserable night ever just trying to keep warm. I gave up trying to sleep early in the evening. They gave us 4 blankets apiece, but since we were on canvas cots, and you had to cover the bottom as well as the top. It wasn't half enough. We got out of there early this morning, and I haven't warmed up yet. Tonight, we are staying at a permanent camp that was at one time French, and though we have dormitory type barracks instead of tents, it is nevertheless still cold. Somehow or another, I am going to confiscate me another blanket or so from somewhere. We will start out in the morning or the next day for our final destination. Maybe there I may receive a letter from "the one and only," which if I do will boost my morale to no end, I can assure you.

Talking about being cold, I am sitting in my room now writing with my flying jacket and my short coat on, and I am still far from the warm stage. I have seen some beautiful mountains since I last wrote you — tall, stately, and snow-capped. They seem taller than the Rockies, but you know more than I whether they are or not.

I still could starve to death for my lack of French if the French were not brighter than I. I just don't pick that stuff up at all. I wish you could see the French "commode." There just isn't any; I will describe it to you at a later date. Too, the French seem especially dirty to me, perhaps I am getting the wrong impression, but so far that is it.

Yesterday, we flew through a scourge of grasshoppers. The 'hoppers were about as long as your finger, and the sky was black with them. I'll bet we killed billions. It sounded as if the plane was going to get knocked to pieces. When you hit something the size of a large grasshopper going at the rate of speed that we travel, it really makes a loud noise and just juices the grasshoppers.

Well it's New Year's Eve. I hope and pray that next New Year's this horrible nightmare is over. If that is not so, I hope it will be over enough for me to be with my wife just somewhere.

Honey, you know it is impossible for us to realize just how rich the poorest of the people in the States are. I have heard it all my life, I have known it, but again it is impossible to absolutely realize until you see the ways that so many of the other people have to live. Boy, if we don't have anything to fight for, no one has.

Well before I freeze, I'm going to close and either hunt some heat or go to bed. May God bless you through the coming year and take care of you and ours. I have never even heard of a finer, sweeter and better woman than you, nor any who is half so deserving. Telling you how much I am looking forward to returning to live with you seems to me to only be wasting words — for, you already know it. Too, you know how much I love you.

I shall love you forever,
Joe.

P.S. Tell Kay and Carol that Dad sends his love and wishes them a Happy New Year. Also tell Kay that I expect her to be a "good girl" and not worry her mother too much. I love you.

January 1, 1944
What a New Year's. Play bridge all morning. Catch ride into town in afternoon. Try to go to Army-Navy Touch Football Game and Camel Race in Oran. Couldn't find Stadium, so go on into town. Go to the Red Cross and Officer's club. Streets very crowded. Go back to camp. To bed.

January 2 and 3, 1944
Nothing to do. Play bridge. Get haircut. Try to keep warm. Stay on post. Field at Telergma is wet and can't get in.

January 4, 1944
Weather good in Telergma today. So we take off at about 10 a.m. Only about 1½ hour hop. Over pretty rugged mountains which are covered with snow. Finally see landing strip, which is practically mud. Land and plane almost bogs down as soon as we get off runway. See "Craven Raven," Prewitt's Ship, bog down to its belly on taxi strip. Met by Lt. Bartlett in jeep and carried to tent area. Mud, mud, mud, sloppy mud everywhere. Just up past your ankles. Assigned to tent with Threadgill's crew. No stove and can't get one. Water ankle deep under my cot. Mud everywhere. To bed to try to sleep and succeed in freezing.

January 5 and 6, 1944
Nothing to do but fight mud and try to keep warm, which is impossible. Am filthy. No way to wash and too cold even if there was a way. Trying to keep warm by various contraptions using 100 octane gas, which has been captured from Germans. Little heat but plenty of soot. So get dirtier.

January 7, 1944
Take ambulance loaded down (19 of us) into Constantine. Most of us to get baths. Go to hospital to take in 2 patients and eat there feeling very conspicuous for my filth and beard. Go to public bath house on the 2000 ft gorge that runs right through middle of town. Bathe standing up in bathtub cause mud is all over floor of tub and it's too dirty to sit in anyhow. Then to Red Cross to get a shave and check in excess clothing. Walk around town and back to camp.

January 7, 1944

Dearest Honey,

Well I'm at my destination in North Africa. Safely and soundly. No trouble. Everything went off nicely all the way. I arrived here two days ago but have not been able to get to a place where I could write.

You should see our camp — it's a beauty. To begin with, it's icy cold. Tents — no heat — no water, no nothing but mud and gobs of that — oceans and oceans of it. It snowed all yesterday and today, quite a snow too. I have made us a stove out of a wrecked airplane, a half of an oil-drum and some number 10 tin cans. The cans are wired together to make a chimney. You'd be surprised at how well it works, and boy it really puts out heat — even so we all have colds and will have until we get out of here. Sick call looks like the chow line.

I am 39 miles from camp now. Drove in so I could get a bath and shave. Hadn't bathed for about 10 days until a few minutes ago — nor shaved for about 5. Boy, you should have seen me. You wouldn't have known me.

I haven't heard from you yet, nor will I while I am here. Honey our A.P.O. is changed again. It is now A.P.O. #520. The rest is the same. I'm quite sure I will get the mail that you have already written however. The town I am in is quite large and looks cleaner than the others I have been in; it has a 2,000 foot deep canyon running right through the middle of the city — quite impressive.

Boy, it surely feels good to be half clean once more. I have on clean under-clothes, but it would be silly to change the outer garments. The mud is real sloppy and about 8 inches deep. Kay certainly would enjoy it — until she got cold.

By the way, I have a little 5 by 8 foot tent I use for a dispensary. 1 litter — no heat. Two boxes for shelves. The hospital is here in town. I brought Zraick into the hospital today — he damn near has pneumonia.

Honey, I will write again as soon as I can. I think of you all the time quite naturally and look forward, oh so much, to being back with you. Hope you and the children have missed the usual colds and earaches and sore throats this winter.

All of my love,
Joe.

January 8 and 9, 1944
Mud and more mud still. Heavy snow — cold, but no wind. Just fight mud and try to keep warm.

January 9, 1944
Sunday, 2:50 p.m.

Dearest Honey,

Well, I'm in town again — I've lights here at the Red Cross, so I'll attempt to write again. I had to come into town today to bring in a couple of boys to the hospital. Zraick is still in the hospital but is doing O.K. with his cold. The boys seem to be standing up physically regardless of the rain, mud, and cold. I'm rather surprised myself.

I'm going back out as soon as I write to you. I haven't even had time to go to the public bath and have one myself today. Guess I will have to get by on "those kinds of baths" until I come into town again. Had one this morning and felt pretty good after it. You can get used to it, I guess.

Honey, I cashed a check yesterday for $200.00. Blackie got paid and was going to send that much home. I owe everybody and have been broke so long that I wrote him a check, so he could send it. As soon as I can manage to get paid I will send it back to you. As yet I haven't been paid and probably won't for a little while and that's the only way I could cash a check. Just thought I would tell you so when you saw a check made out to Mrs. Blackmon, you wouldn't think that I had taken it upon myself to keep up some other woman — of course, you would know better than that — I already have the only real one in the world, so why have more than one? I've known that all along.

My squadron has gone and gotten itself restricted again. We were flying the other day and a couple of the crews didn't get up early enough to be briefed, so the C.O. restricted us to the post and made us hold a formation at 7 a.m. every day, which doesn't bother me a lot for I can get into town anyhow with patients.

Honey I will try to write a little more frequently if and when we ever get slightly settled down. Don't worry about me — I'm O.K. Tell the folks hello.

All my love,
Joe.

January 10, 1944
Back to Constantine to hospital. See Zraick. To Red Cross to write. To the casino — which is the only place in town clean enough to buy vino. Back to camp.

January 11, 1944
Getting gasoline and new pipe for our improvised stove all morning. Told at 11:45 that a B-24 had been sent down for Wagner and me and was waiting for us at operations to take us to Italy. Told that our ground echelon had landed in Italy, all sick, lousy, etc. without medical aid. Told to be ready to take off in 15 minutes. Much confusion, no time to try to clean up. Just grabbed all of my stuff I could see and set off for operations. Took off in B-24 named "Doodlebug" for Algiers. "Doodlebug" had been on 52 bombing missions had shot down about 10 or 12 Jerry fighters, was all shot up herself with "flak" holes all over it, and machine gun holes in plexi glass all about co-pilot's seat but could really travel. Cruised about 195 mph indicated to Algiers and landed at Maison Blanche Airport about 19 miles south of Algiers. Pilots name was Watson — who was a kid from 376th Group stationed in southern Italy. Don't remember co-pilots name but he was brand-new at the game. I know as much about co-piloting as he did. Watson 21 yrs old had been on 13 missions. Young but definitely stable. On landing at Maison Blanche no.1 engine went out. We were on ground so was O.K. To operations and caught G.I. truck to Algiers. Passed Harbor. About 3 beached ships laying in harbor on their sides. Had been sunk in battle. Streets of Algiers near harbor stacked 20 ft high with supplies. Looks like enough stuff here to win war by itself. More planes on airfield than I have ever seen at one time before — all varieties, mostly fighters but plenty of medium and heavy bombers too. Go to the transient officer's office and after much persuasion (I have no orders) am allowed to stay at Argo Hotel which is rented by the Army for transient officers. Wagner and I sleep in large bed with only 3 blankets. No sheets. No window panes. Cardboard covering windows. Cold water. Wash up a bit and walk up to Hotel Aletti, which is fine hotel where you have to be Colonel or better to stay there. Also rented by the Army. Have 1 glass vino and go to transient

officers' mess — then to Red Cross. Find out that there is a G.I. play at the Opera house and catch street car to there. Opera house beautiful 5 or 6 tiers of balconies — very nice and modern. Bars (not open) on every tier. Play fair. To hotel and bed.

January 11, 1944
Tuesday morning, 9:30 a.m.

Dearest Honey,

Well I am out of the mud temporarily anyway. I am now en route somewhere away from our outfit. Yesterday a big ▨▨▨▨▨ bird swooped down out of the skies onto the mud and plucked Wagner and I up with about 10 minutes warning and was taking us away. It was sent for us because we are needed elsewhere.

I am now in ▨▨▨▨▨▨. It is very pretty. It is more like an American city than any of the others I have seen here. Electric lights and show windows in the front of the stores. The first I have seen in quite a while.

Boy, was I glad to get out of that mud. Even if it is for only a short time. I'm here at the Red Cross now and in a very few minutes I'm going to be taking a nice hot bath. In the morning, as soon as we can get some medical supplies, we will move on.

Darling, I will write to you as soon as I can after I get there. I love you so much — maybe I'll be hearing from you in a couple of days or so, which will please me no end. Much of my love to Kay and Carol and the folks.

All of my love to you,
Your Joe.

January 12, 1944
Wagner and I up early trying to obtain medical supplies without orders. Finally after walking all over city (impossible cause it's a big city and most modern one we have hit) we are instructed to go out to Maison Blanche and try to get supplies out there. Traffic in Algiers very fast. All sorts of vehicles from burros to trucks, limousines and what not. French drive like damn fools. Finally catch truck out to airport — wild ride. Finally find medical supply battalion and get plenty of supplies up to delousing outfit — very lucky — Come back to Algiers and go to Red Cross. Get hot bath. Shave. Eat at transient mess and go to our quarters.

January 13, 1944
Up at 5:30 to catch truck out to field. Arrive at field at 7:30 and try to get engines of poor old "Doodlebug" started. Engines cold and won't. Finally at 10:30 get all four going at once and take off for Italy via Bizerte and Tunis. Low overcast and mountains high. Finally see Bizerte on our left at which time we are flying in rain and bad weather. I'm

silently hoping Watson will land at Tunis but it apparently doesn't enter his mind. Turn north over Tunis and head out over Mediterranean flying low due to weather. Weather clears and only a couple of hours hop to Sicily. Hit Sicily just east of the western tip and continue over Sicily to the northern coast. Pass just to the right of Trapani, which is just on southwestern tip of Sicily. Cross range of mountains to coast. Land all cultivated even up on mountains. Peculiar — no trees. Hit northern coast and turn east flying just off coast. Many rather large towns very close together. No beaches. Just sheer cliffs to water. Towns much closer together than in U.S. Pass Palermo which is very very large or apparently so, lying just at foot of mountain, nice size ships anchored there. Chain of mountain tops apparently jutting out of the water to our left and Sicily to the right. Fly on down coast to Messina and Messina Strait. And keep following western shoreline of Italy. Can see Mt. Etna in distance on Sicily and many mountains in lower Italy. Fly on up to the "instep" of Italy and alter course across a narrow low pass through mountains to Taranto Sea. Fly across Taranto Sea to Manduria which is in "heel" of Italy. Land at Manduria on roughest runway I have ever seen B-24's land on — mud runway. Taxi to operations and report to Col. Cogland 47th Wing Surgeon, who instructed us to get a few bites to eat as it was suppertime and then report to him, which we did. About dark — pulled out from Manduria to Taranto in an old ambulance. Highways seem very good between towns but terrible in towns. Get to Taranto in 30-45 min. Taranto large. Rubble in streets showing evidence of semi-recent bombing. Finally found way through Taranto heading toward Bari to Gioia Del Colle where our group was at present and where we were going to be stationed. Good highway all way. Too dark to tell anything else. Arrive at Gioia about 10 and find all executive officers playing poker. Borrowed blankets and to bed — in a barracks.

Part III

Gioia del Colle

The four ground echelon squadrons reunited on January 2, 1944. By January 4, they were in Gioia Del Colle, their first base in Italy, which was nothing but a runway and an olive grove. The men had to build an air base from scratch, working and living in tents that included mess halls, workshops, dispensaries, operations offices, and living quarters.

The ground echelon needed medical aid, so Joe and Wagner were picked up in Telergma and taken to the new base on January 13, arriving before the B-24's and their crews. The air echelon arrived on January 20, 1944. By the end of the month, the 451st started flying missions into Germany.

The 15th Air Force was based in the nearby city of Bari and serviced multiple Air Force units in the southern part of Italy. Bari had a hospital, movie theater, Red Cross, Officers and Enlisted men's clubs, and military operations office that included postal service.

Mail usually came in two forms — Airmail and V-mail. Airmail was regular 1st class mail. The V-mail system, Victory mail, was established by the Army as a way to expedite the mail for servicemen overseas. A typical V-mail letter was a single sheet of paper that folded to create an envelope after the message was written. The Army post office would then scan each letter onto microfilm, which was sent to the combat area, reprinted into letters and delivered. The reprinted V-mails were miniature, less than half the size of the original. This saved cargo space for much needed supplies. Joe frequently referred to V-mail as "shorties."

Winter 1944

Winter in Italy was wet and cold, and staying warm became one of the great challenges for the 451st. The men built wood or gasoline fires in stoves made of metal oil drum barrels. It was a common problem to set the tent on fire while trying to light the stove. Men would end up without clothing or shelter after one of these fires, and many men were injured or burned.

During this time, the men flew missions over Italy and Germany. The weather played a big part in the success or failure of those missions and the ones that followed. With snow and rain came mud, causing problems for the crews. The runway was constructed with strips of perforated steel planking connected end to end. These strips provided a way for the planes to take off and land, but they began to sink in the mud with continual use. The planes would get bogged down and at times would have to land at neighboring bases.

The last mission flown from Gioia del Colle was February 25, 1944 to Regensburg, Germany. This mission merited the Distinguished Unit Citation for the 451st, the first of three such awards. However, on returning from this mission, the airstrip was ruined and the planes, damaged and carrying wounded crewmen, had to land at nearby airfields, scattering planes and crews across Italy.

* * *

January 14, 1944
Arise early and to chow. Find my area about 1½ miles from headquarters on rock knoll. Sanitary devices are entirely inadequate. Back to sick call in cold dirty barracks. Spend remainder of day trying to arrange decent dispensary. Many colds but apparently not as bad as I expected. Few lousy G.I.s.

January 14, 1944
10:30 p.m.

My dearest Honey,

This is the first letter I have written to you from Italy. The country is quite pretty in a peculiar sort of way. It is rough in places but the mountains are not high. Every inch of the ground has been cultivated. I think the thing that strikes you about the place is the nearness of fair sized little towns to one another and the lack of any trees except olive trees, which are quite bountiful. It is chilly here but not as cold as Africa. The place is muddy but not like the mud where I was. Another peculiar thing about Italy — the roads are very good in between towns but in the towns themselves it is just like driving over cotton rows. The roads are straight and well marked — not like Mexico a bit. We are near a small town, but as yet I haven't had time to go in. Now that I have given you a brief description of what it looks like I will go on with my letter.

When I arrived here, there wasn't any mail for me at all, and I was very much disappointed. However, the following day I got a V-mail letter from you which was written December 13. That as yet has been the only mail I have received but, oh, how welcome. Even for its shortness, I found out that you and the children were O.K. at that time and that Hugh was O.K. I certainly was tickled to hear of both things. I had sorta been worried about Hugh, and I am glad to hear he is in England. Maybe he won't have it quite so rugged.

As yet Wagner and I are the only medics here. It has been keeping us on the run trying to tend to the sick and get everybody to put in the right kind of latrines, drainage pits, etc. It's quite a job. We have already had to move our makeshift dispensary once. After the others get here and we only have to look out for our squadrons, it should be far easier and much more interesting.

By the way I am getting a Kapok mattress and pillow. So pretty soon I'm going to be sleeping in style. I'll bet I won't be able to sleep on it at all at first. I'm eating more than ever I guess, but I expect I've lost a little weight — don't think I've lost much since Brazil however. My hardest job by far is trying to keep clean which is impossible.

Honey, they just turned out the light on me and I'm writing by flashlight, so I'll draw this to an end. I love you as much as ever.

Joe.

January 15, 1944
About same. Spend day telling details how I want 4 latrines built and looking over kitchen. One latrine, largest one, is going to have to be dynamited in order to make deep enough.

January 16, 1944
Take ambulance and two drivers down to Manduria (80 miles) to get 2
ambulances, which we need badly. Same road traveled when coming up
to Gioia. Thousands upon thousands of olive trees along way and many
fig trees. Thick rock wall chest high along road on both sides all way.
Drive through Taranto. Probably best seaport I have ever seen. Loaded
down with ships. Must be 100 in harbor. Busy town. Get ambulances —
both dilapidated, 1 with about 30 machine gun holes in right side and
drive back to Gioia.

January 17-20, 1944
Same. Getting dirtier. Cold. Move dispensary to another old wooden
building more centrally located.

January 21, 1944
Flight echelon arrived today at 2 p.m. from North Africa. All ships arriv-
ing. Landed on our incomplete runway without any mishaps.

January 22, 1944
Nothing —

January 22, 1944

Dearest Honey,

Here I've been too long in writing you again. As soon as I am decently set up, I will write you every day for awhile anyhow. That is until things get rougher.

At present, I am well located. I have a fair room with Zraick and Lt. Curtis (remember Curtis, he was the tall bald-headed guy). They are both good room-mates, so we are getting along fine. We have made us a stove out of an oil drum and our room is shaping up nicely now. Today we bought a small second hand table for $3.00 and I bought a comforter for $3.50, so we are even more comfortable. The toughest job still is to keep clean. There is "soot" everywhere and apparently there is no way to get around it. I am having a desk made for my dispensary which costs me $12.00. Tonight we bought some eggs to eat (fresh ones not the powdered variety) which cost 20¢ per each. Vino or wine cost 20¢ per quart — the only thing in Italy which is really reasonable. The prices are sky-high. Electric light bulbs are impossible to get almost and cost between 1 and 2 dollars when you find one — too they aren't worth a damn.

The Italian people, as a rule, are quite humble and very nice. Much cleaner than the French and the Arabs but still dirty according to our standards. The country, this part anyway, is still beautiful. The thing that strikes you most is the fact that not one square foot of land is allowed to lay idle. There is something on every foot of it. Vineyards and olive trees still predominate. It is a remarkable country and pretty in a simple, humble sort of way.

Curtis is sitting here beside me cleaning his pistol, and Zraick is just behind me with steel wool trying to get his mess gear clean. After frying eggs in mess cans it is really difficult to get them clean without soap and water. I have my helmet on the "stove" getting some warm water, so I can pull out this 3 days growth of beard.

We are going to have a very nice dispensary after all. They have allotted us a seven room brick barrack, which is going to do alright. They are plastering the inside of the outer wall so as to cover up the cracks, so we can keep it warm. Since I was here first, I picked my office first, so I got the nicest room. It's going to be nice and clean too — I hope. We are going to use 3 rooms for beds and the rest for offices. We will tap some wires somewhere and wire it, and you can rig up a nice shower out of an oil drum. So by the by we will have a nice place — I hope. We have moved the dispensary three times since we have been here. Here's hoping that this is the last time.

Honey I have received three letters from you now and one from your mother — bless her heart. Tell her I appreciate the letters from her even though I have never been too good about answering them. One of your letters was a nice 4 page letter written January 3 and one a V-mail written the 5th. In one, you sent the pictures of you and the family on Christmas, which I certainly did appreciate. You look good in the pictures — I certainly hope you feel good too. Kay looks like she has grown about a foot. And her "hams" just as juicy. Carol is much larger too and her face is nice and fat — not like she used to be. And, by the way, she looks quite a lot like Kay did about that time — don't you think so? I was slightly surprised.

Honey, I sent you $200.00 today for that $200.00 check I cashed in North Africa. So you see I'm not a total loss. I sent it to the Finance Office so you should receive a check from Washington in about 2 weeks. Let me know when you get it.

Your letters surely are a

Babe, Kay, and Carol, Christmas 1943.

help, Honey. I know that I have some more somewhere which I will get sooner or later. Letters are treasured over here. You keep them all, so you can memorize them. I can't say how many times they have been read and reread nor how many times I have looked at the pictures. I do hope you and the kids had a good Christmas.

Well, honey, I will write again real soon. Gosh how I would like to see you. You know the next time I see you, I'm not going to believe that it is true. It's like asking and expecting more than your rightful share out of life. Tell the folks hello and give them my love. Tell them I am going to write them tomorrow or the next day. I love you and the kids more than anything.

All my love,
Joe.

January 23, 1944
C-47 taking off about 9 o'clock this morning when one engine cut out and crashed at end of the runway. Burned six occupants. 1 killed. All taken into Bari (29 miles north) to hospital. Some burned rather badly and broken up.

January 24, 1944
Spitfire, one of the Limey's on the field crashed on landing this morning trying to miss the grading machine working on the runway. Curtis P-40 with another Britisher crashed this afternoon when he ran off the end of the runway. Neither hurt. Go to Italian Captain's house for dinner tonight with Evans, Young, Bernstein and Zraick. Lts. Massare and Nunes got us the invite. Had some sort of drink made out of alcohol and tangerine peel — very good. Delicious dinner. Enjoyed myself very much. Lady of the house showed her laces which were exquisite. Italian Captain was an engineer in Fascist Army. His wife was captain in the propaganda department and head of all propaganda in this region. Neither spoke English.

January 25, 1944
Lt. Ryan's B-24 named "Lamplighter" was brought in with two flats today and crashed beyond repair. No injuries.

January 26, 1944
Sick. Hellova cough and sinusitis. Stayed in bed all day.

January 26, 1944

Dearest Honey,

Honey I had the most delightful meal two nights ago. The two Italian Lieutenants in our squadron got Evans, Young, Bernstien, Zraick and myself invited out to dinner at an Italian Captain's house. The meal was simple but delicious — served immaculately in four courses. We had drinks as soon as we got there made out of alcohol, tangerine peel, and sugar which was very good — and potent. Then we ate. The first course was noodles with tomato sauce fixed like only Italians can fix it. The second was peas with some kind of meat balls with cheese in it. The third was lamb, which I didn't know was lamb until I was through with it, and the fourth was a dessert known as "cannoli." It resembled cream-puffs I guess as much as any American dish. Then all the wine you could drink of course. As far as I know Italians never touch water. They only drink wine and think something is wrong with you when you ask for a glass of "agua." Anyhow these people were very nice. He was or is, a captain in the Fascist Army and she too was a captain and propaganda minister for Mussolini in this area. Neither could speak English, but it was a very interesting night. The lady showed us all of her lace work that she had made. Some of which was priceless no doubt. She had two beautiful tablecloths, one of which took her three years to make and one 1 year. They were both beautiful or "bellissimo." (Boy my Italian is certainly improving, isn't it?) I can now say "Thank you," "You're welcome," "How are you," "good," "very beautiful," and "how much." That is the entirety of my vocabulary but it is easier than French. I'm slightly getting off the subject of what a nice meal I had the other night and how much I enjoyed the cleanliness of it, eating off of plates. Everything was lovely except the house was terribly cold — you could see your breath all the time. This too is characteristic of Italian homes. No heat whatsoever.

I am staying in today. I have had a bad cold for the past few days which was progressively getting no better fast. Last night I started having a headache, so I've been in this room all day long. I'll try to make it tomorrow if I feel like it. I have my dispensary set up now where it can run. At present, we are holding sick calls for all squadrons there — as soon as the other fellows get set up, it will be a cinch. I have fixed my dispensary up rather nice — you'd like it. It's going to be the cleanest place on the field when I get through with it.

Surely did enjoy your letters today and am glad to find out that you received the stockings, purse and things. I would get you something from over here, but everything worthwhile has long since been confiscated by the Germans, English, or somebody. Glad to know that you and the children are alright — just stay that way for my sake. Remember that I love you all more than anything.

Your hubby,
Joe.

January 27, 1944
Still sick — same as above.

January 28, 1944
To 26th General Hospital in Bari. Nice to feel sheets again.

January 28, 1944

My dearest Honey,

Here's one of these V-mails to let you know I'm still alive and kicking. Kicking more than anything else perhaps. I am now in a General Hospital in one of the larger cities trying to get over a hellova sinus infection — it's really giving me fits too. I treated myself at the camp for 3 days, but the headaches were getting too severe for me, so I came on into the hospital. Oh Boy, now I'm sleeping on sheets — so I don't mind the headaches so much. Don't know when I will get out of here but sometime soon — as soon as we get this sinusitis under control 'cause I'm really needed at my outfit.

Glad to know that Dan has the same A.P.O. number that I have. I think I know where his outfit is and will fly there as soon as I get out of here and look him up. I'm trying to get a line on John King too. Maybe he isn't so far away. Can't be very far away if he's in Italy. So far I haven't run into a soul I knew back in the States. When you write next, send me John's address just in case I run into trouble finding him.

Hit the Jack-Pot night before last when I received 20 letters — 16 from you. As soon as I get to feeling a little better, I will start answering some of them. Space is running out. Here's hoping this finds the children's colds better.

I love you,
Joe.

January 29, 1944
My birthday. Having pretty bad headaches. Have 2 Tbs sugar in urine —
will get blood sugar tomorrow.

January 29, 1944

Dearest Honey,

How are you, Kay and Carol feeling on my birthday? It would be quite a sufficient present just to know that you were O.K. Personally, I think I am better. My sinuses started draining today for the first time, so I should be back to work before too long. Still have an awful headache however. Pretty soon me and codeine are going to be steadfast friends.

Well, I found out where John was, but I also found out that he moved 2 weeks ago. I don't know whether I'll be able to get over there — rather doubt it. But will still try to contact him.

Honey, please send me a package. You can send me some air mail stationary and a pair of wings too if you have been able to find any. Last of all those razor blades too. Can't send you a picture for Kodaks are rare, films rarer and developing places even rarer. If I get the chance, I will though. Food is fair but not like my honey's cooking. Now I think I've answered most of your questions — since I just finished reading over again every letter that I've received from you overseas. Would love to see Carol — hear Kay — see, hear, and be with you. We receive no news — don't have anything but a biased opinion of the war. No entertainment as yet.

I love you,
Joe.

Mission #1: 30 January 1944
Target: Fier Radar Station, Albania.
Results: Bombs fell short and east of target.

January 30, 1944

Dearest Honey,

Boy, am I getting a rest. Nothing to do but lay here. Feel pretty good in the afternoons and nights — have pretty bad headaches in the morning. All in all not too bad however. I've written more letters in past 24 hours than in days. A couple to you, one to mother, Sis, your mother, John K. and Dan. Dan and I are only about 100 miles apart. I am trying to make a "rendezvous" with him somewhere in between us. I think I have found John's new A.P.O. number, so perhaps I will hear from him soon. I would like very much to see either of them. I'm really very well off — so don't be worrying about me.

There's one nice thing about writing. I have that picture you sent me in Lincoln in my writing pad, so everytime I write, I glance over and look at you about every three words. You are always looking right back at me and always very pretty too. It's a very good picture of you. Take care of yourself and the children, honey. I miss you!!

All my love,
Joe.

January 31, 1944
Had a hellova night last night. Didn't sleep any, even with 2 doses codeine and 1 hypo of morphine. Pain some better today but the whole

right face sore as the devil. Boys went on their first raid. Bombed radar station in Albania. Hear they missed the target however. No flack — no fighters — all returned safely.

February 1, 1944
Nothing unusual. Head some better.

February 1, 1944

Dearest,

I'm still in the hospital. Feeling fine in the evenings and nights, as a rule, but rough in the mornings. Last night was an exception however. Four of my sinuses have given me trouble since I've been here — two more and that will be all of them, so maybe I can get out then. I'm so tired of having my headache and coughing. Soon as my sinuses clear up I will be O.K.

Wagner and McFarland were just in with no particular news from the outfit. One of the Red Cross workers sent a wire over to John today that I wanted to see him, so I imagine he will be over tomorrow or so. I didn't know anything about it. I could knock her block off — John will get the wire that I'm in a hospital and want to see him and he will come a runnin' thinking something is wrong, and when he gets here, I will be sitting up probably feeling better than he is. I have already written him a letter.

Kiss the children. Regards to folks. All my love to you.

Joe.

Mission #2: 2 February 1944
Target: Durazzo Radar Station, Albania.
Results: 50-100 hits in target area. Lt. Johnson and F/O Bates, Crew 59, injured by flak. Pistol Packin' Mama crashed on landing.

February 2, 1944
Feeling pretty fair. Brought in 2 of my boys tonight Johnson and Bates — navigator and bombardier off of Lt. Formanek's ship. Hit badly by flack — while bombing a radar station in Yugoslavia. Bates' lower back shot away — lucky it didn't sever his spinal cord. He's through flying — no doubt. Johnson has many superficial wounds of thighs with one large hole clear through his left thigh but will be able to fly again. Only two wounded. However, many ships hit. Roach's ship "Little Butch" has 36 holes in it. Evan's ship had its engine shot out. One ship in 725th crashed

when landed due to a flat tire. Ruined ship but no casualties. Went up and helped operate on Johnson and Bates — both should make it, but Bates in bad shape.

February 3, 1944
Had another mission today — a railroad yard just above Rome. Flew over but had overcast so had to bring load back — no opposition. I feel better.

Mission #3: 3 February 1944
Target: Arezzo M/Y, Italy.
Results: Complete overcast prevented dropping of bombs.

February 3, 1944

My dearest Honey,

Didn't write you yesterday. Slept practically all day. Night before was a diddy. Morphine, codeine, and whatnot failed me. Last night was pretty good, however, so I should get out of here one of these days soon. Got a bed now with a "roll-up" head — regular hospital bed, which is most comfortable.

I have been of "I don't care whether I do or don't" opinion about going back into practice when this is over you know. Last night made up my mind for me. They brought in some of my boys, and I went up to surgery with them. My hands itched — I could hardly hold myself, I wanted to get into the mess so badly. It was just like meeting an old, old friend that you hadn't seen in a very long time, and you had forgotten how much you liked him. Maybe that was the reason I had a better night last night. Anyhow I am ready this morning to do about 3 years work if we could afford it. Don't know what my feelings will be once I see you and the kids. At present, I would just like to sit in our own little backyard in the middle of my "rye-patch" — drink scotch and sodas, milk, and cokes and watch my two offspring show off and do their precious little antics. Also eat steaks and potatoes for 2 meals per day.

I love you,
Joe.

February 4, 1944
Feel pretty good. Group's mission was called off due to weather.

February 4, 1944

My dearest Honey,

Last night was a better night. Right side of my face is still pretty sore, but if I have a couple of good days and nights, I'll be rarin' to get out of here. In my opinion, at least, I'm needed elsewhere (egotist).

Oh, honey! Before I forget. When you send me a package, please send me a couple of Benzedrine Inhalers and about 6 flashlight batteries, the size just smaller than those for standard flashlights. Not the ones for pencil flashlights, nor the ones for standard lights, but the ones just in between. I need some for my otoscope and can't get any of that size anywhere.

As yet, I haven't heard from John or Dan. From all I hear, they must be pretty damn busy as are we. I'll probably hear from them soon.

I think I am getting a little cleaner day by day. Possibly by now I am getting down to those layers of mud I picked up at Algiers and Constantine. Honey, take care of yourself and our "flock." I think of you all the time.

All my love,
Joe.

February 5, 1944
Rainy, damp windy. No mission today. Head doesn't feel too good. Guess it's on account of weather. Sorta feel blue for some reason — more than usual.

February 5, 1944

Dearest Honey,

Received 3 V-mails from you yesterday. Two of them were written before Christmas and one written January 11. I certainly want to commend you on your writing. You are upholding your end of this correspondence swell — as is characteristic of my wife in anything she does. I know it is quite difficult to write everyday when you aren't receiving mail. I know for I wrote for 7 weeks without receiving a line. I'm not doing as well with my end of the correspondence as you are, but really the conditions, as a rule, are quite adverse to writing. I have done better since I've been in the hospital.

I had a very good night last night. Sinuses are still draining plenty of pus and the right side of my face is still plenty sore. I have had plenty of bouts of sinusitis in my life, but none that has given me as much hell as this one. Hope it will be completely over soon and let me get back to work. I believe I can do more good there than I can here.

This bed isn't as comfortable as it once was, and I'm getting weaker. I have to stop and rest to continue writing. Just from laying in bed of course.

Later in the day — 3 p.m.

I have just returned from having my sinuses washed out again. They have been very nice to me here.

I have looked through my snapshots again — remember the little snapshot case you gave me? Looks to me like Kay's snow suit is a little large for her. Next year maybe she can wear it but, of course, by then it will probably be too small. Funny, but I have looked at those pictures that you sent me many times and wondered what was different about you, but didn't realize that it was the short hair do until today. Men are so dumb. Anyhow, it's very becoming, or appears so to me. However, honey, to me you would be beautiful if you were bald-headed. You are very pretty on the "top-side." But your real worth and beauty is a hellova lot deeper than that. That's nice, to have it both places.

It has been rainy and rather chilly here all day. The wind is blowing too. But it is not as cold as Oran or Marrakech. I damn near froze at the latter place once. Spring is just around the corner and should be beautiful here. I think the rainy season here is about the same as in the States, and I dread to think of it being any muddier than it already is back at the camp.

I think I will enclose a Lira note, so you can see what the equivalent of a penny looks like over here. Take care of yourself and the chillun'. Would give anything to see the three of you. When this is over I'm going to sit around and look at you for 6 months before I even try to make us a living. Tell all the folks hello for me.

All my love,
Joe.

February 6, 1944
Very dull uneventful day. Feel pretty good.

February 6, 1944

Dearest Honey,

Just got back from having that "internal bath of the nose," following which I feel pretty good. I'm really improving now, I think.

Today is Sunday. By the time I get home, Kay will be large enough where I can lay in bed and tell her to run out and get me the funnies. Then the cook can bring us coffee and toast, and we can lay in bed and read and doze and talk and be lazy. That sounds like it will be fun whether it works out that way or not.

I'm getting tired of this bed. Lay up here and do nothing but think (which isn't good). Listen to the B.B.C. news announcements from London, Italian gibberish, operas, and symphonies broadcast primarily from Naples by the San Carlos Opera bunch and German Military music broadcast from somewhere inland. Frequently, we hear broadcasts from Germany in English.

McFarland was just in and brought me three V-mails from you and one from your mother. Don't forget to tell your Mother how much I appreciate them.

In your letters, you seem worried about conditions here. I'm O.K., and will make out O.K., or as well as any of them. I don't write and tell you of the conditions, such as the mud and rain for you to worry about me. I just thought I would let you know as nearly as I can what it is really like. I'm gonna gripe about the mud, C-rations, etc but, after all, that's about all one can do over here. All of us have lived pretty rough in our lives, and counting my frequent hobo trips in the early thirties, I have had my share of it too. Probably will make every man overseas really realize what U.S.A. means. Knowing how well off we were is one thing, but living the other way will make us realize what it is all about.

I have been laying in bed this afternoon studying Italian. Imagine! As much as I detest foreign languages, I have succeeded in learning how to count this afternoon. It's an easy language and if I would apply myself, which I won't, I believe I could learn how to get by in a short while — but I'm too lazy.

Well, there has broken out an argument here on "socialized medicine." Why they ever start an argument like that when I'm around is beyond me. I'm trying to stay out of it, but I'm sure I can't much longer. Take care of yourself and the chillun' for us. Remember how much I love you. None of your mail is censored so write all of the love letters you want.

All my love,
Joe.

February 7, 1944
Decided it was time for me to go. Face pretty sore yet but getting better gradually and I'm just laying around. Will let me leave Wednesday so they say. Went to town today on pass to picture show, P.X. and Miramare Hotel Bar. Snowed and sleeted while I was up town. Pretty tired tonight. Little Rushing killed today out at camp. "Rush" was McAllister's co-pilot from Arkansas — a swell kid. A truck hit him. Died shortly with a Basilar Fracture of his skull. We'll miss him. Funny, I talked to his Dad a long time in Lincoln about him.

February 7, 1944

Dearest Honey,

Well I started trying to get out of this place today. If everything goes well, they are going to let me out the day after tomorrow. The right side of my face is still sore, and my upper teeth are still numb due to the infection, but that should clear up slowly. This afternoon, they allowed me out of the hospital, so I went into town. Left at 1 and returned at 4 p.m. Now I'm pretty tired even though I didn't do anything except go to a show. Saw Joel McCrea in "We Shall Have Music," which I would have at one time thought pretty rotten, but which, under the conditions, was at least entertaining. I think that's the first show I've seen since Natal and the second since I was in the States.

Since I'll be going back to field conditions day after tomorrow, my writing will probably drop off again. As I have said before, conditions out there are none too conducive for writing, so don't think I don't think of you, and Kay and Carol and our home and such just as much as I do now and always have — but, I just can't do a hellova lot about it.

In a number of your past letters, you have frequently mentioned "my sacrifices." To begin with, you are making as many as I. I have no doubt that this war, so far, has been much harder on you than it has me. You have gotten the rough end of the deal — due to no fault of ours. It just falls to your lot. Secondly, this is still a partnership marriage — remember 50-50? Even though you probably go 80% and I do the 20% — it's supposed to be 50-50. What I do over here — we are doing — just like when you correct Kay for doing something she shouldn't be doing, we are correcting her. At least that is the way I look at the situation, and the way I like to look at it. Regardless, of the living conditions of where I am, I am making but one sacrifice and that one is being away from you and the kids. Everyone over here has long since decided that the best thing in life would be to be back with their family. But, as I have said before, one only realizes what he is fighting for when he sees how the rest of the world lives.

It is pretty cold today. Snowed and sleeted some while I was out, but it felt pretty good to be in my clothes. My hospital pajamas are 49 in the waist and 45 in the length. I am 30 in the waist and 31 in length so you see I've been "lost" for quite a spell. Remember me to the folks. Kiss Kay and Carol for me.

> I love you,
> Joe.

Mission #4: 8 February 1944
Target: Piombino M/Y, Italy.
Results: Bombs strike photos show heavy concentration of hits, in M/Y, and on industrial targets. Old Tub crashed on takeoff, eight men died.

February 8, 1944
I got out of the hospital today. I still have a headache just between the eyes, but I believe I can make it O.K., and I believe I'm needed. This morning Hunt's ship crashed just after take off west of Gioia. Of course

it was loaded with 5,000 lbs of bombs. Three men got out. Seven were killed. 1 is in bad shape at the hospital now — the co-pilot. Crashed probably due to ice on his wings. Here from hospital in time to watch ships come in from the mission. Two landed with feathered props. Flak holes in many of them. Imhoff's and Brown's ships both of the 725th have failed to return. Capt. Quillen, group operations officer, was in Imhoffs plane — for some reason they had to leave the formation before they got to the target. A sergeant, the nose gunner in Winski's ship was killed — gotten by flack. The colonel was riding in the ship today. Bowen's ship shot up badly but all members O.K. Today has been a bad day for us. Let's hope there are few more like it — but I'm afraid we will see many.

February 9, 1944
Quillen and the lost 2 ships turned up in Corsica. All O.K. Had engine trouble and had to land there. No flying today due to weather.

February 9, 1944

Dearest Honey,

I've done you bad about writing since I've been out of the hospital. Things were a mess when I returned, and I'm not halfway through straightening up. My sinusitis has also returned in all its glory, but the headache has not. I can keep going O.K. unless the headache comes back.

I sent Blackie to the hospital yesterday with his back. It fell to my unpleasant duty to do so. I'm fairly convinced that he is through here and will probably be returned to the States. He certainly wants to stay here — poor guy.

Honey, I can buy beautiful lace bed-spreads for around $100 that are said to be worth $250 up in the States. Haven't bought any as yet because didn't know whether you would want that nice a bed spread. If you want one let me know. In the meantime, I will keep my eyes open. Tell everyone "hello."

I love you,
Joe.

Mission #5: 10 February 1944
Target: Allied Beachhead. Velletri Troop Support, Italy.
Results: Cisterna hit by 15 tons through town, Cori hit with 30 tons and left in flames. Three Feathers hit by flak.

February 10, 1944
Up at 4 a.m. — to briefing. Boys supposed to fly two missions. First take off at 7:40 a.m. All off. Return at about noon. Bombing troops below Rome. 5th Army apparently having difficulty holding beach head. Bomb 5 towns. Flak heavy. Most all ships shot up. Hester's crew bailed out over field. 2 broken legs as result. Brought in ship on 2 engines, belly landing. Beautiful, none of occupants hurt. Had 2 other emergency landings — no damages. Didn't have time to take off on the second mission of the day.

February 10, 1944

Dearest Honey,

Here I am back out in the old field routine. Inhaling smoke (not tobacco), being dirty all the time, and soot everywhere. There's nothing like it — filth. But I am here, and glad to be feeling like I'm doing something and not "goldbricking."

It is cold here still even though in Memphis you tell me it has been like spring. The wind is whipping across the field at quite a clip. I will be very, very glad to see warm weather. For one reason, I won't have to be so damn dirty all the time. We are now going "full steam ahead" — busy all the time. I usually rise between 4 to 6 a.m. and get to bed pretty early. Or as early as I can.

Zraick just brought in a big sack of mail to be censored. Thank goodness, I don't have to help do that. That's the only thing that my rank has ever done for me. Hey, honey, something else you can send me sometime is a pocket knife. Preferably a big heavy one with can-opener, screw-driver, etc. like a boy scout knife. Running out of paper. Take care of yourself. I love you.

Joe.

February 11, 1944
Weather bad. Driving rain and about 30 to 50 mile gale. Built desk for dispensary. Weather too bad for flying —

February 12-13, 1944
Supposed to bomb for 5th Army beachhead again. Weather too bad to get in so planes had to return —

February 14, 1944
Up at 5. Briefed. Planes off to northern Italy to bomb railroad yards. Missed rendevouz with 98th Bomb Gp and P-38's so returned to base.

Mission #6: 15 February 1944
Target: Allied beachhead at Campoleone.
Results: Scattered hits south of target, 3 hits on Anzio road and half mile south of Campoleone, 1 burst SE of traffic circle.

February 15, 1944
Up to briefing. Planes off to northern Italy to bomb marshalling yards. Target is weathered in. Planes return and bomb secondary targets (5th Army beachhead). All planes return to base. Flak only light. No casualties. Quite possible (and probable) that some of our bombs fell short of target and got our own troops.

February 15, 1944

Dearest Honey,

You know I just thought about it — yesterday was Valentine's Day. Wonder how my three Valentines fared. Hope O.K. All the ships just landed. Everybody is O.K. I have to go out on the line at all take-offs and all landings. It surely is cold out there too. I also get up for briefings every morning or so — consequently, I don't have a hellova lot of time to practice medicine (which I don't do in this man's army anyhow). Then the sanitary problem is always a problem, and you have to stay right on it continuously.

As for our finances, I think you are doing swell. I hope you got that $200.00 I sent you to replace the $200.00 check I wrote in Africa. I have some more money now, which I will send pretty soon unless I see something I want to buy for you. So far, Italy has had practically nothing of any value, or that I thought you would particularly want. I really haven't had a heck of a lot of time, however. Don't give up — I think of trying to get you something all the time and will. As for you buying a coat on time — you shouldn't have done that. I will send you the money to pay it out the next time I get into town. You are cute, honey, and I love you. I just don't understand how you intend to save money on the idea of paying for something on time.

I can get toothpaste easier than you can, so you keep the toothpaste. The same goes for shaving cream. The Army issues me a chap-stick, however, I could use another. Razor blades I can get, but I could use more. I can get 6 or 7 packages of cigarettes per week, which is plenty for me so I don't need them. I do need the Benzedrine Inhalers, and pocket knife. The fruit cake sounds swell, I would like it very much and so would Zraick and Curtis, my present roommates. Send any kind of "food" that you want to.

Certainly am glad that you are sending me the Reader's Digest. That will help a lot in finding out how the world is getting along in general. Not to hint too much but "Time" puts out a small edition for men overseas, and it comes weekly, doesn't it? As for the Commercial-Appeal, I think it best probably not to send it. It would be pretty large and only 1 days news in it. The Digest is swell though. Thanks a million.

Honey, in your letters you make me feel self-conscious sometimes. I wish I was the fellow you talk about. Unfortunately, I'm not. I'm just another guy who is over here trying to do a wee job. Sometimes whole heartedly; sometimes sorry — just human, I guess. But there's thousands of us — all of us alike, all thinking of home, talking of home, our wives, children, etc. There's lots of differences perhaps in our personalities, mine too, I guess. I, personally, know that I expect more out of my men than I did in the States. Most everyone's personality has changed slightly. For example, the major and I have come to blows about sanitary installations a couple of times. It isn't that anything is really changed or that we are being "ruined" in any manner. It's just when you are living with only part of yourself, your personality is naturally changed — perhaps "disposition" is a better word to use. Don't forget, I'm nothing but another guy who would rather be home with his wife and kids than anywhere else on earth. But due to the communal instinct, I guess, I deem it better to be doing something that demands that I be elsewhere at the present time.

I will write and tell you more about individual personalities in my next letter — sometime when Zraick is awake (he's asleep now), and he can tell me what I can say, and what I can't. You are a wonderful girl, woman, and, more important to me — wife. I miss you terribly and am only living with one thing outshining everything in the future — our future years together.

> I love you as always,
> Joe.

Mission #7: 16 February 1944
Target: Siena M/Y, Italy.
Results: Target covered by clouds, hence second Grp, including 727th did not drop bombs. Most hits of 1st Grp reported over: 10 bursts reported in M/Y and on railroad tracks. No photos of hits.

February 16, 1944
Up early. Planes off to bomb Siena again. Siena and secondary target (the beachhead) weathered in again. Many of my boys got frost-bitten. 50 degrees below zero at 20,000 ft. Only 1st flight dropped bombs which were probably ineffective.

Dearest Babe,

February 17, 1944
Snowing and sleeting all day long. Miserable weather. Supposed to bomb beachhead twice today but weather too bad. Wanted to go to Bari to opera "Foglietto" but apparently other Docs beat me to it. Weather so bad I really don't care.

February 17, 1944

My dearest Honey,

It was snowing hard this morning and it is sleeting now. That only means more slush and mud — at which I am getting quite adept in navigating.

I was intending to go into the nearest large town tonight to see the opera "Foglietto." (I'm sure that isn't the name of it but it is as near as I can come). Anyhow everyone is gone, so I guess I can't go.

Formanek gave me his wife's address today, and I will get Dorcas Haltom's address from Haltom soon. I will send them to you in my next long letter. All the boys ask about you and Kay frequently and ask to be remembered to you, especially McAllister, Zraick, Young, and Formanek. A boy just came in with a gash in his head, so I gotta get busy. Take care of yourself and remember I love you very, very, much.

Joe.

February 18, 1944
Weather still bad. Nothing happened of any importance.

February 19, 1944
Same as above. Briefed to go on mission but weather too bad.

February 19, 1944
Saturday

My dearest Honey,

Hi Honey, how are you today? I hope that you and the kids are doing O.K. Maybe the sun is still shining and you and Kay are out in the backyard — Kay playing in her "Shand-pile" and "wounning fast". Gosh, that would be swell to watch. Guess, I can watch Carol in that stage.

You seem to be worrying about my living conditions. It's true it isn't quite like home, but I'm quite comfortable. In fact, I have most of the modern conveniences. My stove is made out of a 10 gallon oil drum — has a grate in it and everything — even hooked up with gasoline. Can burn gasoline, wood or coal. Usually burn wood but start the wood with 100 octane gasoline, and can burn gasoline altogether. However, the fumes are worse then and the room is much dirtier. I have learned by experience that my nose takes the wood better. Have a little shovel to take out ashes, etc. So I'm warm. Then I have a frame built over my cot, which is partially covered by boards and I cover the rest with towels and dirty clothes, so to keep the soot and dirt off my cot. I bought me a mattress (kapok) for $5.00 — have 1 comforter and 5 blankets, so I sleep comfortably and warm. Then I have a nice wash basin made out of a 10 gallon gasoline drum that will hold plenty of water for me to wash, so I can stay reasonably clean. I always have plenty of hot water for I have a five gallon gas can that fits next to the stove, and it keeps plenty of hot water all the time. I also have "acquired" me a coffee pot and 2 cups. Can get soluble coffee and sugar cubes out of C and K rations, so I have hot coffee every morning as soon as I get up. Usually with toast because I bring over a few slices of bread from the mess tent every night. Have to get up pretty early and don't like powdered eggs anyhow, so that is usually my breakfast. Nails are in the "superstructure" of my bed all around, so I have plenty of space to hang things on. We all have a chest of drawers made out of wooden boxes that I keep towels, socks and underwear in. Then I have a large trunk made out of a box that I keep practically everything in, so I can keep my clothes reasonably clean. I have an extension cord over the head of my bed, so I have light although dim. So you see, really, though improvised, I live rather comfortably. My dispensary is the same way — you just have to improvise and you can do that very nicely.

Joe.

February 20, 1944
"Stand-off Day" — i.e. no flying. Went down to "San ———" a field below Manduria with Capt. Whipple. Did a nice Chandelle over the field before we landed coming back.

February 21, 1944
Weather too bad to fly —

February 21, 1944
Sunday, 8:30 p.m.

Dearest Honey,

I wonder what you are doing right now — probably eating lunch since it is only about 1:30 there. Hope it is a good one. I ate in town today at the transient officers mess. Had some green salad — endive. The first raw green vegetable I have had since the States. I'll probably get amoebic dysentery from it, but it just looked too fresh to pass up.

I then went over and got a shave (3 day beard) and then went to a picture show. "Air Force" was on. Sat half way through it and something must have gone wrong with the picture machine for the picture went off — we waited 30 or 40 minutes and nothing happened so we left. It seemed to be a pretty good show up to as far as we saw. I then went by the officers PX and bought me a nice Ronson cigarette lighter for $3.30 and the fountain pen I'm writing with now for $2.25. Damn good pen too. It's an "Eversharp Forever" pen — a $5.00 pen.

Had to go to town yesterday too. Those are the only two times I have been there since I got out of the hospital. Bet you couldn't guess what happened to me yesterday — while walking down a street in town a big Army truck pulled up beside me, and someone said "Hey, Captain, which way is it to ▓▓ Hotel." I started to tell the fellow, who was sitting beside him but Emil Koenig — imagine our mutual surprise. If they hadn't been lost and just happened to stop and ask me, we would have passed one another up on the same street in the same town. We met later and talked over "old times" and our wives particularly — plus our children. He has just gotten over here and was trying to find him some medical supplies. He wasn't as lucky as I — he had to ride a boat over. He is to be stationed about 90 miles from me. Boy, is he down too. Seemed to be thoroughly whipped by the situation in general and pretty downcast. Sure he will snap out of it. I saw him long enough today to just speak to him. Don't know when we will see one another again — maybe not at all while in Italy. It's hard to make "dates" for it takes so long for correspondence to get from one to another over here.

Got a letter from Dan two days ago. He is about 100 miles from here. I was going to fly up there yesterday, but the plane I was going in was temporarily grounded. I will get up there as soon as I can though. He seems to think that he has a fair chance to come back to the States in a couple of months. I haven't heard from John yet. If he is where I think he is, it's going to be difficult for me to see him anyhow. However as soon as it gets a little warmer and isn't so muddy, I think I will buy me a motorcycle — a small one-cylinder affair if I can, and then if I can get a day off, I can go wherever I please. Too, when or if we move on to some other place I can just put it in the ole B-24 and go ahead. In that way, I can have transportation wherever I am.

Too I went around town yesterday all afternoon and looked for "stuff" to buy you. There is nothing there worth having to be truthful. None of the stores have anything. I went in one supposedly 'Ladies Shoppe,' which had nothing whatsoever in the whole store but 6 brassieres and about 10 scarfs. The brassieres had elastic in them so I bought you 4 of them. I tried to get you size 34 (don't know whether right or not) but apparently they don't sell any such sizes here — so don't know what size these are. You can either alter them or throw them away and keep the elastic. I knew you could use the elastic for something. Then I got you two pair of cloth gloves (navy blue and a lighter blue, I think) cause I couldn't find anything else at all after walking around town damn near all afternoon. Also a couple of lace collars. One which I think you will like, the other may look nice on something of Kay's — I don't know. Anyhow after I had completely given up, I saw a negligee in a little "hole in the wall" store, so I bought it for you. Again, their sizes and ours don't jive at all so don't know what size this is but I'm sure it's way too short. Silk is impossible to get here — this is either white celanese or sharkskin. There are six pieces to the whole ensemble (1) bra (2) pants (3) slip (4) bed-jacket (I think) (5) gown (6) negligee. It isn't as nice as all this sounds by any means, but I think you can make a couple of the pieces fit you. It is all hand made and does have some nice lace on it. I got it so you would have something from Italy anyhow. It's really funny — the pants would fit old Mattie I'll bet. And the negligee will hit you just between the knees and ankles. Maybe with all those pants you can use the extra goods and piece on a tail to the negligee. Don't be alarmed — the Italians do everything cockeyed. I will send it to you soon. I'm interested to hear the alterations necessary. If it isn't worth the trouble, don't do it just because I sent it to you. I'm not that narrow — yet. If I can find a couple or three yards of white lace I will send it to you and if you want to add cloth to it in any places you could add the lace. The pants are really something though — or look so to me — I believe a large tent would be just as "form-fitting."

You know, I could write all night to you. I must just be in the mood, but I gotta get up about 4 a.m. and I need some sleep.

All my love,
Joe.

Mission #8: 22 February 1944
Target: Regensburg, Germany.
Results: Unknown due to cloud coverage.

February 22, 1944
Up early and briefed to Regensburg, where the ME109 factory is. Take off uneventful, went over and dropped bombs. Plenty of fighters — mostly ME210's. We were lucky only 2 men were wounded in our group. No ships lost, but this wing lost several. All our ships landed safely.

February 22, 1944
Old George's Birthday, 3:30 p.m.

Dearest Honey,

Today is a fairly pretty day. The first one in many. It's still pretty cold, however, I'll certainly be glad when warm weather gets here. I'm so tired of having colds and treating them. Guess this summer it will be the "G.I.'s" and malaria — there's no rest for the wicked.

I'm now waiting for the boys to come in — just sitting here "sweating them out." It's just about time for them now. I flew yesterday for the first time this month — went down below Taranto. Taranto is a fairly nice town. I have been there twice — just passing through. It looks about as up-to-date as any town I've seen over here — that isn't saying much.

Honey, I gotta go. I'm O.K. and will write you better tonight or tomorrow.

All my love,

Joe.

Mission #9: 23 February 1944
Target: Steyr engine works, Germany.
Results: Weather and failure to gain altitude caused formation to turn back without dropping bombs.

February 23, 1944
Briefed to go to Steyr, a town in southern Germany which makes airplane engines. Evans leads the group and he really messes things up apparently. Flew up into Germany and never dropped a bomb. All ships returned safety. I had one gunner with a frozen foot, the toes of which will have to be amputated. Also one with a frozen hand who will lose one finger. Headquarters building burnt down this morning while we were at briefing. Caught on fire from the chaplain's stove. The chaplain lost everything.

February 24, 1944
Capt. Younkin called me this morning and wanted to know if I wanted to fly to Foggia in "Tangerine." We took off about 10:30. Landed about 11:30 at a field near Manfredonia — his old outfit. Saw Johnson who is their group Surgeon and a Major now — an ex-classmate of mine. Spent the afternoon with Dan Baldwin. Left the field about 4 p.m. arrived here about 5:30.

February 24, 1944
8 p.m.

Dearest Honey,

How is my honey tonight? I hope good. Personally, I have enjoyed today more than any day in quite awhile. I took off from the field at about 10 a.m. today for the field where Dan Baldwin is. I got up there about 11:30 — walked over to group headquarters and messed around for awhile. Then saw a major walking by and thought I recognized him — asked the sergeant which I was talking to if his name was Johnson and he said yes. So I hollered at him — hadn't seen him but once since we graduated from Med School. Can't yet think of his first name. Anyhow he is a Memphis boy. He and I entered Med School on the same day and finished together. Tell Mother and Dad that I saw him — they may remember him. He is about my size, black-headed, and used to come over to our house and study a lot with Ed Leek. They probably will remember — anyhow, we ate lunch and fanned the fat for quite a while. He is Group Surgeon for that group and has been over for 16 months. About 1 o'clock, I called up Dan's squadron and got Dan to come up. Dan looks exactly like he always has — might have gained a couple of pounds. He is still the same Dan — war hasn't changed him an iota I don't believe. From all I could gather, he had worked it around to where he was a "big shot" (Of course, I don't mean any harm — I'm just trying to prove he is still the same person). He is top-sergeant over the flying enlisted men. A soft job I imagine — we don't even have such a job in our squadron. Dan seemed very well satisfied. Of course, he is quite eager to get home but so is every soldier over here, or anywhere for that matter. It's a universal disease. It isn't homesickness per se, but besides seeing their folks, there is some difference in living conditions. I was rather surprised — "when you get home" is the chief topic of conversation over here by everybody — whether they have been here 1 week or 2 years. Apparently all of them miss their wife and kids as much as I, but I doubt it. Anyhow, I had an enjoyable day. Took off from up there and flew here — landed about 5:30 — went over and ate and now I'm writing my honey.

Really shouldn't have taken the day off. But it was just one of those days that I do what I want to do regardless of what I should do. Kremers isn't here. He is on detached service for 1 week taking care of a "rest camp." Should be a nice thing for they make the environment of such a place as pleasant as possible. While he is gone I'm taking care of his squadron as well as mine. So I should have stayed right here — but I didn't. Really was nice to get away and see Dan and Johnson, though. They have a nice set up there — much nicer than ours. Ours isn't going to be too bad when the weather gets better.

Got two letters tonight. Both from Sis. I haven't received the package as yet from you, but I will soon I guess. Some of our packages got burned up yesterday I hear, when our headquarters building burned down. Doubt whether mine was in yet though — pretty sure it wasn't.

Don't know whether I have told you or not, but Blackie talked himself out of being sent back to the States and is with us again. He is doing a hellova good job, but we aren't letting him fly yet. He certainly is a swell guy. Completely unselfish, and has the outfit at heart, too. His wife is nice, but she did herself proud — he is a real fellow and a man.

Young is still around. We were practically off speaking terms for about a month, but we are coming around again. He's a nice guy, too. His only trouble is that he can't see the other fellows' point of view. I told him this at one time, and he didn't take it as I meant it — consequently, the friction. Anyhow we are doin' O.K. now.

Honey, It looks as though this letter is going to be pretty thick. I didn't realize that I had been so voluble tonight. So I'll turn over and write on this side for a while. I got Kay's letter, which I enjoyed very much. Tell Kay that her "Dadee" read every word of it (or line) and enjoyed it very much. Bless her little heart.

Don't worry about your job of raising the children. I am quite aware that you are very proficient in raising our children. I'm not worried about that part in the least. In fact, I'm not worried about anything back there that is within your powers to control. I certainly don't mean that I wouldn't like to be there and throw in my two-cents worth but you are quite capable of handling the situations. Remember — there were many times that neither of us knew what to do in raising Kay. God was exceptionally good to me on May 31, 1940. The biggest day of my life by far. I shall always be thankful. Remember I'm living to be with you again and love you very much.

All my love,
Joe.

February 25, 1944
Briefed to Regensburg again this morning. The whole wing is flying up there. The 8th Air Force from England is to bomb the same place 40 minutes after us. Take off at 8:30 a.m. Rained all day. The first half of our group returned at 4:30 led by Young. They never reached Regensburg but bombed Klagenfurt instead. Two ships were shot down. Wiersma and E.D. Johnson, both from the 724th. We took 3 wounded men off others when they landed. The other half of our group had to land at Foggia due to weather. Four ships are missing from there — one from the 726th (Coleman) the name of the ship ironically "Hard to Get." Quillen, our Group Operations officer was aboard this ship. Three ships are missing out of the 725th Kimmell, Zander, and don't know the other fellow. Our squadron came back intact. Today was a rough day — the roughest, and the blackest, yet in the history of the 451st Bomb Gp. The Germans were using everything, aerial bombs, rocket bombs, parachute bombs. We got our target however. Blasted the Regensburg Aircraft Assembly plant, where they make ME109's, off the face of the earth — so they say.

Mission #10: 25 February 1944
Target: Regensburg A/C factory and Dittmannsdorf M/Y.
Results: One of our attack Groups in 1st wave over Regensburg ruined target area. Group over Dittmannsdorf (727th) reports heavy concentration on M/Y. Lost 3 A/C at Regensburg, 2 near Dittmannsdorf.

The Regensburg A/C factory was a high priority target for the Allies. This factory produced about a third of Germany's single engine fighters and about half of the ME 109's. This factory assembled approximately 280 planes a month. The target was well defended by the enemy and success came at a high cost for the 451st. The successful bombing of the Regensburg A/C factory earned the 451st its first Distinguished Unit Citation.

NORTHERN IRELAND EDITION

THE STARS AND STRIPES

Daily Newspaper of U.S. Armed Forces   in the European Theater of Operations

Vol. 1.   No. 70.   New York, N.Y.—Belfast, Northern Ireland.   Saturday, Feb. 26, 1944.

Battle To Finish Luftwaffe Passes Sixth Day

Attacks On Stuttgart, Regensburg Climax Furious 24-Hour Drive

February 26, 1944
"Stand-off day" no mission today. Rained and rained all night. Mud and
water everywhere.

February 27, 1944
Rained harder today than ever. Mud and water everywhere. We got a
special commendation from the General of the Wing today on account
of our mission over Regensburg. We really must have blown the place
to bits, as shown from aerial photographs. Had my dispensary white-
washed today. Looks considerably better — no mission again today.
Good thing for all ships need repairing following that last one. And the
runway is damn near washed away.

February 27, 1944

Dearest Honey,

Today is one of those dark dismal days where it is gray all around you, and your disposition reflects the same atmosphere. Something tells me I'm not going to be worth a whole hellova lot today. I haven't heard from my honey in a few days. As soon as the mail comes through I will immediately feel better.

Started to go into Bari today for a picture show and concert — didn't have enough energy to budge, however. We aren't flying today, consequently the spare time. Too, I can't hang around my dispensary. I'm having an Italian white-wash the walls, and they really sling the stuff. Had all the cracks plugged up with a semi-cement mixture. I'm going to have a relatively nice dispensary yet. Soon as I get it fixed up to my liking we will probably get orders to move — or something.

I'm sitting down on my cot — semi-comfortably propped against the wall, with my feet propped up on my footstool (which is made of ½ of a 500 lb. bomb rack) writing away. Boy, is it raining outside. I mean raining. All that means is just more mud. It's already so muddy that when we are walking to mess, it's so slippery that it reminds you of a bunch of fellows going ice-skating. After this rain, it will look like a bunch of kids wading in the pond at Overton Park.

You know talking about mess — its going to sound queer when you tell Kay and Carol in future years to watch their table manners, so they will set their daddy a good example. Mine have long since been atrocious. When you are eating out of mess gear somehow your manners completely disappear.

Too, when I come in to eat and you can't see my hands for the dirt, grime, and filth, what are you going to say? Or when I don't shave for 2 or 3 days, what are you going to think of me? You know, it wasn't but 3 or 4 months ago that when I missed a day of shaving, I felt so uncomfortable and dirty that it was pitiful. Now for about 12 hours after I shave I feel like my bare face is hanging out — and it is. I won't mention bathing — I had the nicest, hottest bath last night I've had since the States, by the way — did you ever feel uncomfortably clean? Quite a sensation — don't try it for it takes time to develop. Too, I wonder if the first night I hit civilization again and try to sleep on an innerspring mattress, what it's going to be like. I'll lay you odds that I will spend a very restless night. I can really "knock it off" on this cot now. I remember the difficulty I had trying to get some rest while out at camp at Orlando. I wonder how long it will take me to get used to the "life as usual" routine? Oh well, we will have lots of fun finding out together, won't we?

Of course, you are going to have a double jolt. To begin with, you are going to have to get used to having a husband around again, poking his nose into your affairs — demanding as much of your attention as I dare — living with a somewhat converted barbarian. That will probably be a shock within itself. Boy, it's going to be fun pestering you and Kay. Guess I'll have to go easy on Carol for I'm not going to mean an awful lot to her. Maybe it won't take long to overcome that for she will be as much a stranger to me as I am to her. I suppose I can put off worrying about such for a couple of years anyhow. See how my mind rambles — who knows, maybe I'm going nuts? I love you, though.

Honey, you asked me about my income taxes. I don't know how things are back there, but it is my opinion that it is just as good to let it ride. Don't let it worry you too much. Put all the money you can away, yes. Not only for taxes and such, but also maybe I'll want to go to school again. I can't make up my mind over here. I've really got to get down to earth and think this thing out before I will know what to do. This certainly isn't the proper environment for anyone to try to think straight. It is an impossibility. But if we can see our way clear, I imagine I will get some more training. If I ever intend to get any more education, there is no doubt in my mind that the time to do it is when I get back.

I haven't as yet drawn all of my December salary, by the way. As soon as I get paid — which should be in a few days I will send you some money. Enough to maybe buy a bond or pay your coat out. I'll send your things that I have bought for you as soon as I have time to fix them up.

Now about my office furniture down in Helena. Yes, I would like very much for all of it to be moved to Memphis. I would like for it to be stored in our attic if it were possible — don't know whether there is sufficient room up there or not. You could use the reception room furniture on the front porch or out on the back lawn. I would like very much to get all the furniture that the Helena Hospital has up there especially. I believe if you got everything brought up, including your piano, it would be wise. If you go down there to do that however, you can count on it taking a couple of days or so.

Honey, don't think I have ever told you, but we are in the 15th Air Force. Whenever you see that the 15th did this or that, you can count that our bunch is in the thick of it. The boys have done exceptionally well. They fly the best formation in this wing by far, and that is what counts for the "mostest" of them coming back. They are a noble, brave bunch of boys — I am personally proud of them. Not one of them has given me a bit of trouble. They are real men.

Time out — I've had to go over to the dispensary to see a boy with a temperature of 103. He's O.K. — a strep throat. You should see the "road" from here to there. A distance of 1000 yards and at least ½ of it under 6 inches of water. Dear old sunny Italy. My heart will no doubt yearn for it once I get away. It's 9:30 now so apparently it has taken me 2 ½ hours to scribble this — with interruptions. Honey, I've written enough for one day. Take care of yourself and the rest of our family.

I love you,
Joe.

February 28, 1944
Up late. No rain today! Very nice day — seemed like spring. No flying — runway is still out. Evans and I take an ambulance and go to Bari about noon. Send flowers to Mother and Mrs. Black and go to the U.S.O. show at the "Oriente." Have lunch and supper in town and drive back to camp. Shoot the bull with Evans, Young and Zinn over two bottles of champagne (very poor) and then to bed.

February 29, 1944
Rained all day again. Runway still out.

Spring 1944

March 1, 1944
Up late and to dispensary. Call from Dan Baldwin who is in Bari. I take
ambulance and go to Bari to get there about noon. Dan, Wagner and I
go out to hospital, so we can all eat together. Also do some errands out
there. Then shop around Bari all afternoon but can find nothing of value.
Go to Dan's room, eat chocolates and gab. "Beat" Dan out of a nice
camera and 2 rolls of film. Eat dinner at transient officers' mess and then
back to camp.

March 1, 1944
10 p.m.

Dearest Honey,

This morning, I arose late and finally made it over to the dispensary. Just as I got there I was wanted on the phone, and it was Dan calling from the nearest large town. So I finished up what I had to do in short order and went up there. We had dinner together in spite of the "officer - enlisted man" stuff and walked all over town. Looked at everything — bracelets, rings, scarfs, gowns, cameos, etc but the town is absolutely stripped. There is nothing worth while left. However, I have always got my eyes open and may run into something. Anyhow, Dan and I walked and also did quite a bit of talking — mostly about Helena. Almost seemed like old times but you weren't there.

We finally went up to the room that Dan was renting in a sixth rate rooming house and sat and talked some more. Ate up a box of chocolates that Dan had that were very good. Found out that Dan had a 35 millimeter Agfa camera that he wasn't using so I beat him out of it and felt rather bad taking it, but he vows that there are two more in his tent and that he also has a "brownie" so maybe that makes it O.K. Anyway now I have a 35 millimeter camera and two rolls of film — so I can take some pictures. If you happen to see some rolls of film of this size, grab some and send them to me. Of course, I realize that is probably impossible. I wish I had known that I was going to run across one while in Puerto Rico for you could get all the film you wanted there. Film was stacked on all shelves — but no Kodaks. As you know, I bought some for you but couldn't send it home (still have it).

Well anyhow, I left Dan about 5:30 then went to the transient officers mess and ate, then home. So I've been a lazy boy again today — as I will probably continue to be until the weather clears up some. I'm writing by candle light. Since it's been raining so much, our lighting system is out about ½ the time. It will probably be repaired by tomorrow.

We are starting ground school over here for the crews like we had through training. It certainly is interesting to note the difference in their attention. Anything you try to teach them now, they lap up. They used to sleep or be restless when you lectured to them even though you knew you had prepared and given a pretty fair lecture. Yesterday I gave a poorly prepared lecture, but the boys were just on the edge of their seats eager for more. It's miraculous what a little combat will do for them. Pretty soon they will know First-Aid better than I do. They surely are a nice bunch of boys. Anyhow I wish you could see them in action — they certainly deserve a lot of respect and admiration. And to brag a little on my bunch, they are holding up under the rough stuff, physically and mentally just a wee bit better than the rest. I hope they continue to do so.

Everyone you always ask about is O.K. Wagner, Zraick, Young, the Major, the two Macs, Kremers — everyone, I think, that you remember. The Major has nothing to do so consequently has gotten lazy as the devil, but all is O.K.

Honey, one of my candles has burned out. I've got to finish this and get in bed before the other does likewise. Take care of yourself and ours and remember that I love you most dearly and shall always.

Your hubby,
Joe.

March 2, 1944
Pretty day. No rain. Runway still out so no flying. Hang around the dispensary all morning. Go into Gioia at noon and send Babe $150. Buy two chests for her which should turn out pretty. Wagner and I get in an ambulance and "just ride." Both of us are highly bored. Ride out to the small out-of-the way town of Acquaviva. Turn around and come back to camp. Meet Hargroves on the way back on his motorcycle so I get out of

the ambulance — swap places with Hargroves and ride his motor back to camp. Took some pictures today.

March 2, 1944
7:30 p.m.

Dearest Honey,

Today has been a very nice day again — no rain. It's nice just to have a day without rain over here. Whether the sun shines or not is practically immaterial. If it doesn't rain for a couple of more weeks (which is asking too much, I'm sure), maybe most of this sea of mud, mire, and muck will dry up. Only 12 more days to go — we've had two successive rainless days.

I've been terribly bored today. Doing nothing. "Piddled" around all morning. This afternoon, I went up into the little town nearby with Wagner. I went by the Finance Office and sent you $150.00. Just thought I would tell you, so you would be expecting it. It is really easy to send money home — just tell the Finance Officer who you want to send it to, more or less, and then forget it. Not any more trouble than a money order if as much.

I think I found something for you. I'm getting you two little highly polished walnut chests made with "Babe" put on the top with inlaid olive wood. One about the size of an apple box — the other about the size of a large cigar box. The first you can use for whatever it may be useful for — the smaller one would be nice for a jewelry box. I'm having many partitions and trays put in both of them. They should be real nice if they are made out of good seasoned lumber — if not, well I've wasted money before. I've made up my mind to send you ½ of the money I get over here and buy you stuff of one kind or another with the other half. We get beat on all you buy over here — but it gives me something to do, and I like to do it. It's damn difficult to find anything though. By the way, I'm getting you the chests for your birthday and anniversary even though you probably won't get them by then. I'm going to fill them and when they are full I will send them to you. I hope you don't mind too much me telling you what I'm getting — gives me something to write about and think about, and you won't know what they look like until you get them anyhow. It takes them 20 days to make these chests for me. They are real pretty. If the lumber isn't properly seasoned, it's going to crack. If it cracks, we will have them refinished. It's beautiful workmanship.

Young, Evans, Zraick, Zinn, Curtis have just come in — brought bread, coffee, peanut-butter to cook on our stove and shoot the bull. So I'll never finish this letter — consequently, I'm gonna quit. Remember that I love you very much. Take care of yourself just for me.

I love you,
Joe.

Dearest Babe,

March 3, 1944
Weather still pretty and runway still out for big stuff. The C-47's with British Paratroopers are taking off, however. Went up into Gioia early to look at some lace bedspreads for Babe. Bedspreads didn't suit me — will wait until they get more. Walked around town just looking. Bought some lace, a lace blouse for Babe and a cape for Carol. Found out that our squadron has been alerted to move south by 3:00 p.m. today. Rush out to the field and find that my squadron is moving flying personnel to Manduria, so we can fly missions until our runway is back in commission. Pack up everything including bed roll. Then find out that we are not moving until tomorrow. Go back up town in an ambulance and get a shave and look around more. Buy Kay a little velvet dress which is pretty, but too large. To camp again — eat and get in a crap game. First gambling in a long time. Win $200 so can spend more on "the harem."

March 3, 1944
10 p.m.

Dearest Honey,

This isn't going to be a long letter for I have got to go to bed and get some sleep. I am leaving here early in the morning for a week or so — going down south. Taking my old standby Sergeant Mason and a couple more of my boys with me. We will be on detached service until we get back, and I don't know exactly how long that will be. Probably will be pretty rugged down there — we will probably be sleeping in our pup-tents.

I have been out shopping again today. It didn't rain — makes three days in succession. I'm buying a lot of stuff you probably won't use, but I'm enjoying it. In fact, it's the first time I've enjoyed myself since I've been here. Today I bought you a lace blouse and Kay a brown velvet dress. Too, I bought Carol a little hat-cape affair that is cute.

I hit the jack-pot tonight. I got 3 letters from Mother, one from Sis, one long one from you and two V-mails from you. I was certainly glad to get them for I hadn't heard from you in quite a spell, and I won't hear from you until I get back here. Surely was a shock to hear that Mother and Dad had sold the shop. I feel leery about the deal. Looks like they would go crazy with nothing to do. Nothing we can do about it however. I, personally, wonder how they are going to make a living. I suppose they have found a way, and I hope they have.

Honey, I'm drawing this short on you. Tell Kay and Carol that their Daddy loves them. I love you most of all though and shall continue to do so always.

Joe.

March 4, 1944
Up at 5 am. Make usual toast and coffee. My boys bring the ambu-
lance over and leave for Manduria via Taranto. Get to Taranto and shop
around town all morning. Buy each of my honeys a dress. Then on to
Manduria for lunch. Talked to Major Gavin, assistant wing surgeon, who
offered me a majority if I will take a service group dispensary. I refuse.
No flying or flying pay. Move into tents.

March 4, 1944
9 p.m.

Dearest Honey,

   In case you get this before the letter that I wrote you last night, I moved today. Moved south from where I was. I will only be here temporarily. Tonight, I am writing to you by candle light again. I am in a tent with Evans, Bently, Young and Zraick. We are lucky for this tent has a floor and tents with a floor are O.K. I'd almost as soon live in one as a barrack if there's heat and lights.

   Bought you a dress today. The only one I could find. It's a fall or winter dress, so by the time you get it you will have plenty of time to have it altered to fit. It's two-piece. Also bought one Kay probably can wear next winter and a little blue and white checked organdy one for Carol. I'm going to get you a nice lace bedspread and then I'm going to quit — even though I enjoy looking.

All my love,
Joe.

March 5, 1944
Rain, Rain, and Rain. Boy, does it rain. Rain up on the tent floor. No
flying. Go help pick out a site for camp. All personnel to move here in
our squadron.

March 6, 1944
All our squadron and 725th to move here from Gioia. Supposed to leave
here and go to San Pancrazio as soon as 376th leaves there. No place
to operate except tents. Begin to set up our tent city on a knoll that I
picked yesterday. I go back to Gioia to pick up my belongings left there
and to get my dispensary moved down here. Garriety already has packed
the dispensary and is ready to move when I arrive. I go over and pack
personal stuff. Do not get to leave Gioia today however for while at
Gioia, one boy burns himself and a case of appendicitis turns up and
have to send them into Bari.

March 6, 1944
10 p.m.

My dearest Honey,

I'm outta ink so will have to use pencil tonight. Its lead too soft, so I hope you can read it. When Churchill said we would see much "Blood, sweat and tears," he forgot one thing — rain. Boy, has it been coming down today. We are moving from where we were by the way. My squadron and Mac's squadron are coming here temporarily — the other two squadrons will be somewhere else. So I gotta do plenty of work for awhile — starting tomorrow.

Haltom and Evans and the other two squadron C.O.s were made Majors today. I'm glad. By the way, wing headquarters offered me a majority yesterday if I would take an aviation service group. That means I would have 8 flying personnel to take care of and the rest ground men. Also that I would go off flying and flying pay — I refused. It would be nice to be a major but not making less money than I am now. Going out of paper I see. Will write a long letter my first chance. Kiss the kids for me and remember that I love you more than anything else there is.

Joe.

San Pancrazio and Manduria

By the end of February, the runway at Gioia del Colle was a complete loss because of the mud and was no longer usable. The 451st had to relocate. There was not a ready base large enough for the 451st, so they were split and operated from 2 locations, San Pancrazio and Manduria, which were about 30 miles apart. The 724th and 726th moved to San Pancrazio, and the 725th and 727th moved to Manduria, but both bases operated under one command and flew the same missions. Joe was in Manduria for the next month of his overseas duty.

The units were reorganized and shuffled throughout the war, usually based on location and administrative changes. The 451st was originally part of the 2nd Air Force, 47th Bomb Wing. By the end of the war, they were moved to the 15th Air Force, 49th Bomb Wing. The changes happened at seemingly random times but didn't have much of an impact on Joe's daily life.

A typical mission day for Joe started at briefing meetings. He would then take an ambulance to the end of the runway, ready to pick up any injured men in the event the planes crashed on take off, still loaded with fuel and bombs. At the end of the missions, he brought his ambulance back to the runway to receive the returning crews.

In between take off and landing, Joe spent his time checking on patients, updating medical records, gathering supplies, improving the camp's sanitary conditions, and "piddling" with projects. He built shelves, shopped for Babe, and visited friends at neighboring bases. Joe and his friends also built motorcycles out of scrap parts they scavenged from the base and nearby towns. The men got too involved in their tinkering and subsequent adventures that motorcycles were declared contraband and had to be sold.

Mission #11: 7 March 1944
Target: Pontassieve M/Y, Italy.
Results: Photo reconnaissance shows all tracks but one out. One not cut is blocked by debris from a train that was hit. Highways hit directly. Close coverage prevented photos by our Group.

March 7, 1944
Pack ambulance and Mason and I head back for Manduria. See our planes pass overhead headed on mission to bomb marshalling yards somewhere around Florence. Get to Manduria around noon. Set up dispensary about 5 p.m. Planes come in about 4. No casualties, no flak, no fighters. Some frostbite. Go over after supper and set-up dispensary and tent by candlelight, so we can operate tomorrow. Tired — to bed.

March 8, 1944
Spend day putting part of the floor in my tent and straightening up dispensary. Sent an ambulance to Gioia for my coal and to find lumber. Put up stoves in my tent and dispensary. Moved in my tent tonight.

March 8, 1944
Wednesday

Dearest Honey,

Three months ago today I left Lincoln. Seems like an eternity. The longest three months of my life, I guess. I have done you bad for the past couple of days. I have been moving. Had to make a couple of loads from where I was to here, so I have been terribly busy. I have my dispensary in a wall tent now — just about finished putting it in today. Tomorrow if I can "moonlight requisition" some wood I'm going to floor it, or at least partially floor it. It's about the only way you can get lumber of any variety in Italy.

You should see what I'm living in now. I have a command tent — a 6 x 10 ft. tent connected on to the back of the dispensary tent. That, at present, is my abode and will be as long as we are here. Completely private, no roommates — now if I were only an introvert, I would be happy. It is very nice however. I just moved in tonight. I only have it partially floored, but I can get by very nicely.

You know, if you could only see me now you wouldn't know me. I wouldn't hold it against you. I'm filthy. Haven't shaved for a couple of days, and my hands have been dirty for so long, I'm really wondering what they would look like if I was to wash them. I'm pretty tired, but tonight before I turn in, I'm going to try to take "one of those kind" of baths. Goodness knows I need one bad enough. Too my clothes are dirtier than I am. To make me look even worse my pants are ripped down the seam from stem to stern — about an 8 inch "gash", I guess. They're my O.D. pants and have been too tight ever since I've had them. Must have bent over for the first time while I had them on yesterday while carpentering.

I bought your birthday and anniversary present two days ago. I will send them to you the first chance I get. I'm sure you will like both presents. As I have always said before, you could give the other stuff away that I have bought if you wanted to, but if you give either of these presents away, my feelings will definitely be hurt. I'll label them, so I wont have you at a disadvantage. Now I'm flat broke. So I won't be spending anymore until next payday.

Just so you'll know, my flight surgeon's rating has bounced. The papers are not complete. So now they will have to go completely to Washington and back. All of ours bounced. Wagner got his the other day though. I wouldn't care, but I hate to be getting gypped out of $40 per month. They are supposed to be retroactive.

Well if you were here, I would get you to sew up my pants — as it is I believe I will just let them ride until I feel more in the mood. Take care of yourself honey. I miss you terribly — more than I would ever be able to put down on a mere scrap of paper, and am existing only to return to you and the children.

I love you,
Joe.

March 9, 1944
Did it pour down last night? Wind blew and it rained. Ack-Ack guns started shooting away at night. Everyone grabbed helmets. Don't know what was up yet — probably nothing. Rained all in my tent. Today I

built part of the floor for the dispensary. No flying — weather too bad.
Wheeler and Curtis came over for long visits tonight.

Mission #12: 11 March 1944
Target: Toulon submarine pens etc, France.
Results: Target hit, severe damage to installations. Some
vital spots untouched. 1 ship (724th Crew #1 and Maj.
Willhite) seen explode, killing crew. 1 726th ship missing.

March 11, 1944
Up early and to briefing. Briefed to bomb the submarine dry-docks at
Toulon, France. Planes take off without mishaps. I take off in my ambu-
lance and ride around this part of the country in an effort to find motor-
cycle parts for one I am putting together. Ride over almost to Brindisi.
McAllister goes along just for a ride. Return to base about noon. Planes
return about three. Vail and his crew, including the new operations
officer who took Quillens place, shot down just off the southern coast of
France. Their plane exploded in mid air when attacked by ME109, and
went down in flames. No apparent escapees. Vail was in the first crew in
this outfit away back at Dyersburg. Another plane of 724th missing. My
bunch escaped injury again. Tonight I boxed up and labeled three boxes
to send home to my honey.

March 11, 1944

My dearest Honey,

I cheated on you last night — I skipped writing, so consequently I didn't
get a letter off to you today. Last night I spent my time bathing part by part. You
should have seen the water when I got through. Here I am just as dirty tonight,
but I'm too all in to bathe again. I'll just have to be dirty.

For about the last three hours, I have been packing, waterproofing, and
printing addresses on three boxes to you. I will send the chest later — after it's
finished. Had to make more than one package because of duty regulations. Two
of these boxes contain your anniversary and birthday presents. I have previously told you about everything that is within them but these two articles. I hope
you like them, and I hope they reach you in time. I will try to get them off to
you tomorrow if I have time to run into town.

It hasn't rained yet today here. Apparently it isn't going to for tonight is
beautiful but cold. The moon is out in its full glory. It's so light outside that you
don't need a flashlight to make it around. These nights have been rarities since I

have been in Italy. It would be a nice night to spend with you, but so would any night, or better still, all nights.

This morning after "take-off" I took the ambulance and drove around the country-side with McAllister. Enjoyed talking to him. He is a nice guy.

Next day about 4 p.m.

Hi! Just received the letter from you containing pictures of you, Kay, the house and the snow. All that I can see about them is beautiful. For some unknown reason, the letter got wet on the way over. I could read the letter O.K., but the pictures were stuck tightly together. The one of Kay in her "shand-pile" is not distinguishable.

I have been chasing down another battery for my ambulance. I finally accomplished this feat about half an hour ago. To get something of that sort over here is like pulling some one's eye teeth. So upon sending the boxes to you and getting a battery for my ambulance I feel like I have put in a very success-ful day.

I haven't received any packages from you as yet, honey, but don't worry about it 'cause it takes quite a spell for packages to come over. I'll bet it will take two months for these boxes to get to you. I suggest if you send food that you send fruitcake because of the length of time it takes to get here. A few cans of sardines are good too. Just anything. Don't worry about sending more batter-ies. What you have sent is plenty. I think frequent pictures of you, Kay, Carol, and the house would do more to give me a lift than anything. However, I realize how hard it is to get film. In fact, for all I know you probably can't get any at all now.

You asked if we ever got the "news?" Yes there is an army publi-cation printed over here called the "Stars and Stripes." Somehow I am never able to see one though. As far as news is concerned, our "best news" comes from a propa-ganda broadcast from Germany on a program called "To the Enemy; From the Enemy." It's pretty good. I think we enjoy it more than any other program — it's screwy. Usually is the German version of what's going on — and of course what they wish us to believe. It's done by a fellow called "George" and a woman named "Sally." There are quite a bit of rather good recordings, but old (American) thrown in for good measure. Since I don't possess a radio, I only hear them occasionally.

Something interrupts me all the time. This is a lousy disconnected letter. It is now 10 p.m. so I think I'll finish this off. I will see if I can do a better job tomorrow. It is now raining down fishes for it's pouring. Nice night to stay in with my wife — so are all nights.

I love you,
Joe.

March 12, 1944
Little Lt. Prindle got killed night before last in a jeep wreck. He was an armament officer in 725th and a swell little fellow. I liked him. Briefed to northern Italy to bomb fighter field. Couldn't go due to bad weather. Sent my honey her boxes today. Spent the afternoon trying to get a battery for an ambulance. Was finally successful. Played poker tonight with Blackie, Threadgill, Sprowls, and Bernstein.

* * *

This is Joe's last journal entry. No other journals were found. While the rest of the missions are not mentioned specifically in this book, the 451st continued to fly regularly, completing 245 missions before the war ended. However, future missions that either received a Distinguished Unit Citation, that Joe mentioned in his letters to Babe, or that greatly contributed to the advancement of the Allies troops are noted.

* * *

March 13, 1944
Monday, 8:30 p.m.

My dearest Honey,

I promised in my disconnected letter last night that I would try again tonight, so try I will. Perhaps this one won't be so "unjointed." Tonight I have been making some shelves out of boxes to put some of my junk on. Now I have a place for my coffee, jam, peanut butter, sugar, salt, tomato juice, oranges, etc that I manage to pick up from the kitchen now and then. You see by that that I'm not starving. If I get a chance to sleep late, I just go ahead and sleep and I can have my breakfast just the same. Even though we are in tents, my set up now is quite convenient. In fact if I could get enough lumber to make a floor and a frame for my tent, I would like the present set up about as good as any.

Mason (my sergeant) really takes care of me. He stays in the dispensary every night. When they have to get up early to go on a mission, Mason comes in and builds me a fire. I wake up every morning by the noise he is making building me a fire. If my fire goes out during the day when it starts getting chilly in the afternoon, Mason builds me another one. They are all still good boys just as they were back in Fairmont. Mason informed me that he was going to work for me after the war — of course that was erroneous as Mason frequently is. He and Williams, another one of my boys, are in the front tent making us some coffee. Williams just brought it in so I will probably have trouble going to sleep.

After I finish writing this letter to you, I have got to get busy and shave and try to bathe to some extent. Mason and I are leaving here at 4:30 a.m. to try to get me another ambulance. It's quite aways up there, and we probably will be gone 1 ½ or 2 days. Pretty soon I will have covered our part of Italy fairly well.

By the way, did I tell you last night that Wagner was a Major now? I'm really glad. He is a very nice guy to work with. He deserves it. McFarland burnt his right hand a couple of nights ago lighting his stove. I kidded him about it for all of us doc's are always getting on the boys for getting burnt lighting these stoves. We use gasoline for fuel, you know, and if you are careless, you get a pretty fair explosion when you first light them. Otherwise they are comparatively safe. I have managed to get me about a half ton of coal, so at present, I'm burning coal. It's much cleaner. Gas gives a heck of a lot of soot. The heck of it is that I have to keep all that coal inside my dispensary, so someone won't requisition it some night and it takes up a lot of room. Room is something I don't have very much of, too.

You know I told you many letters back that I was going to get me a motorcycle. Well at present, I have acquired a Harley-Davidson engine and frame. All I need now is two Harley wheels and a few other parts. I haven't spent one penny on it either. I probably won't keep it long though for the boys are all buying one. The minute someone offers me some of the exorbitant prices they have been paying for these wrecks, I'll sell.

Well, honey, it's going to take me a good hour and a half to get off this beard and some of this filth. It's 9:30 now so by then it will be 11, and I've got to get up between 4 and 4:30, so I had better call it quits on this bit of discourse. Won't get to write tomorrow for I'm sure I will be on the road somewhere.

I love you most dearly. My thoughts are with you continuously and the rotten deal that this war has given you. I shall do my damndest to try to make it up to you beginning the day that I see you next, which I hope won't be too far away. Take care of yourself and keep your chin up. You are the best.

All my love,

Joe.

March 15, 1944
10 p.m.

My dearest Honey,

Since I didn't write yesterday, I'll take up where I left off in my last letter. You know you could put my letters together and damn near have a book of what I've done since I last saw you. I left early yesterday morning (4:30 a.m.) to go up to get my ambulance. Drove through where we were stationed and went to see about the chest they are making for you but it wasn't completed yet, and then went on to the General Hospital to see about a couple of my boys and to leave another one, and then went down into town and bought me a pair of O.D. pants. You know my old O.D. pants had ripped off of me. Also one pair of my pinks that I have been wearing (which is very, very impractical over here, but that I have needed to keep warm and cover nakedness) has shrunk up to where I just can't get them on. So therefore I had to buy another pair. Hey, would you like a good pair of $18 pinks to make a skirt out of or do you want me to get rid of them?

Then Mason and I set out again at about 10:30 to where we had to go for our ambulance. We finally arrived there about 5 p.m. It was quite a long trip. I passed the place where Dan was stationed, stopped and talked to him some on the way up and back. He has just returned from a rest camp on the Isle of Capri. Also beat him out of some more 35 mm film. Also saw Johnson and had steak with him last night.

Anyway we drove on up and got me a new ambulance — brand new, hadn't been driven but 210 miles when I got it. I'm almost as proud of it as I would be of a new automobile. The spotlight, rear-view mirror and tools are even still with it. Usually these items are missing quickly over here. After getting my new ambulance Mason and I started back — he drove the old one and me the new one, of course. Well to start our bad luck, he had a puncture. So while he was repairing the puncture, I drove on down and saw Dan a little more. Presently Mason arrived, and I got him some supper, and I went and ate with Johnson. Then we started out again. About 10 or 20 more miles and my new ambulance caught on fire — guess where — the emergency brake band, remember? I thought of you, Kay, and I out on that mountain. Well anyway, I got the fire out but had to take the whole brake band off before I could proceed further due to the fact that it couldn't be loosened any more — just wasn't done right at the factory. So Mason and I proceeded on. Few more miles and my lights went out and the motor stopped. Finally found out that the cable had slipped off the battery. So Mason and I proceeded on sleepily. Then we had to take a 25 mile detour over the damndest road you have ever seen. You have never seen a detour until you see one in Italy. Over here 6 inches down is solid stone — so these detours are rough. Anyway, we finally finished the detour and drove on

— more sleepily. Then we came to a bridge which was under repair and was either going to have to turn around and go 50 miles over another detour or wait — we waited. Soon the British had the bridge O.K. so we drove on. Got in about 6 a.m. and were really tired. You know these ambulances aren't automobiles — they are really Dodge trucks and they drive exactly like what they are.

So Mason and I went and got something to eat and I piled in bed, telling the boys to wake me for no less than the president. But at noon they had to wake me. Wing had called down — we were having a meeting of all Flight Surgeons of the Wing. So I drowsily get up and go there. To top it off, a guy (psychiatrist) gets up from the Air Force and lectures for 4 hours and fifteen minutes on how to be a Squadron Flight Surgeon. That capped the climax. I was thoroughly disgusted. Of course, that guy admitted that he had never been one. I may not be the best Flight Surgeon there is, but if I'm not one now, I'm certainly never going to be one.

After "sweating that out," I returned to my tent, and Sprowls dropped in for me to take him off the grounded list. So he stays and gabs with me until just now. He may not be able to pick out wives, but he has a pretty sound philosophy on life, by the way — interesting guy.

Well, anyway those are the occurrences in my life for the past 48 hours. So again, I'm tired and sleepy and am going to bed in a few minutes, so I can see what tomorrow brings — but, if it brings another guy who talks for 4 hours to me on how to be a Flight Surgeon either he or I aren't going to be writing our wife tomorrow night. I couldn't possibly stand it. Honey, if you will excuse me I'm going to the Land of Nod and perhaps dream of you again. Goodnight.

All my love,
Joe.

March 16, 1944
10:10 p.m.

Dearest Honey,

Did it rain today? I mean it really came down. There is almost as much mud as there was while we were in North Africa. Got your letter of February 29 today. So I'm an uncle again? Surely was glad that everything was O.K. and that you and Sis got your boy.

Today has really been boresome. Had to stay in most of the day due to rain. However during the day, I managed to get some lumber, so the boys and I floored the dispensary and my little tent this afternoon. They worked pretty hard, so tonight I rewarded them. We got a half gallon of grapefruit juice from the mess tent, and I doled each of them out two drinks of grain alcohol tonight. Mason went to town and carried in my laundry this afternoon, and while in town, he bought 20 eggs for $4.00 so we have had boiled eggs also. I have shaved and took one of those baths down to my belt, so now I am about ready for bed. Before turning in, I just wanted to tell you how much I love you.

All my love,
Joe.

March 17, 1944

My dearest Honey,

It has been a beautiful day. (Quite an unusual beginning to my letter isn't it?) No sir, it didn't rain one drop. I haven't done anything worthwhile today. Worked a little while on the motorcycle (still can't find any wheels) and chopped some wood, otherwise I've only acted as a consultant in the dispensary, i.e. when the boys get something they can't handle they call me. I've had Rosie (Rosenbaum) busy all day shortening a pair of pants that I drew from supply. He did a nice job, too. Of course, he was once a tailor. Sent two more of my boys up in an airplane for the ride and two more to malaria school. The rest of us have just been "piddling." Wrote your Mother and Dad a note tonight and started a letter to Sis but didn't have it in me so have laid it aside to finish tomorrow. Maybe I will be in a more "bull-shooting" mood by then. Got me a good light globe tonight which is worth much more than its weight in gold. Also sent a roll of film to be developed; send you the results. Goodnight and I love you.

Joe.

March 19, 1944
Sunday, 8 p.m.

Dearest Honey,

I did you bad last night — I missed writing again. Truthfully I had to take a bath and shave again. And when you are as dirty as I was and bathing out of my oil can still that takes time — you know, first wash face, ears, and neck, then rinse them off, then one arm, rinse, etc. It takes time and there is an art to it. Anyhow it was successful and I'm sure I was cleaner for a short period today. Didn't find any lice either — aren't I lucky? It ain't so bad as it sounds perhaps, I assure you.

I'm doing fine. Haven't had very much medical work to do lately — we have been exceedingly lucky — my particular squadron. I'm knocking on wood for I really want it to continue. Everyone has colds, G.I.'s etc, but not much serious stuff.

By the way, I got a roll of my film developed and a few prints made. One of the enlisted men did it for me, and we don't have any means to enlarge a picture, so I will send a wee print along in this letter. You can hardly tell it's me but it is, you may rest assured. You will be able to see that I still have a head and two arms and that slouchy old cap of mine. I'm saving the negative of a close up so if I can get it "blown-up" to the average size print, I will send you one.

Today has been a peculiar day. My dispensary and the boys have been quiet as a mouse, which is everything but usual. The boys got me irked this morning, and I had to "let them have it." I was such a surprise to them that they have just about been talking in a whisper all day long. They will be through pouting by tomorrow, and everything will be normal again.

Take off was early this morning, so after take off I got Tony, my ambulance driver, to drive me around the countryside to look around and also to try to find me a couple of motorcycle wheels — no luck. I don't know, but I'll bet you are glad. Anyway, it gripes me to have a perfectly good motor which runs like a top but no wheels to put on it. Maybe I'll find some yet but I doubt it. If so, it will solve my transportation difficulties.

Hey, it's been clear again today! Three days in succession that it hasn't rained in Italy! I'm sure this is a record that hasn't been met since the time of Caesar. Don't worry it will be raining before morning. I pulled off my rainboots for the first time since I have been here today. My feet felt so light that they almost flew up and hit me in the face. Last night, I washed the mud off my shoes and looked at them for the first time since Africa. I can still see them tonight. This probably sounds foolish to you, but it means quite a bit to us.

Blackie is still the same Blackie. Still with the same bad back. He hasn't flown yet even though he would give his right arm to be able to. He told me today that he was about ready to give up. Of course, you know from a friend's point of view, I would hate to see that happen — professionally it's the only thing to do which I have been trying to tell him ever since Fairmont.

Well, honey, Lt. Love just came in to talk over his troubles with me I'm sure. I do more work from 5 to midnight right here than any other time.

Goodnight — I love you with all my heart.

Joe.

March 20, 1944

My dearest Honey,

The day has passed much quicker than usual. This morning I wandered around with Evans inspecting the area, later held inspection of all the men, and this afternoon I flew for a couple of hours with Lt. Stenning. When you are busy all day, it really does help in passing the time.

Got a V-mail from you today written March 2. Sent you a half length "portrait" of me in an air-mail this morning. They are so huge that I hope you don't have to dig out my microscope to tell that it's me. Space is short — so I'll tell you quite frankly that I love you with all my heart.

All my love,
Joe.

March 21, 1944
7 p.m.

Dearest Honey,

Today I received 2 real long letters from you. Surely was nice to get them too. One was written February 22 and the other the 24th. Sorry to hear that you tore yourself all up pruning the rose bushes and got yourself all sore. I'm sore too — my right arm and shoulder and back. We have a baseball mitt and glove and the last couple of days I have gotten out and caught some with the boys. I had no idea I was getting so old. When I used to get sore from taking exercise, it would wear off in two days at the most. This soreness in me is just hanging on and on. Ain't it hell? When I get home, we will grow old together and then I won't give a kick.

Well, I knew it couldn't last long. We had about 5 days without rain, but today it had to start all over again — just as it was getting where you could walk around without getting mud up to your knees. I wonder how long it will last.

I was supposed to get up this morning at 2:45 to go to briefing. When I awoke however it was around 6:30. Naturally, I thought I had slept through it, so I got busy wondering what time I was supposed to do my little bit for this war effort. Called up operations to find out the times for everything and found out that we were "standing down" (not flying). Had I known that I could have slept late — doggone it.

March 23, 1944
Thursday, 8 p.m.

Apparently, I started something and didn't finish it. I really thought that this letter was finished and on its way to the good ole' U.S.A. Guess I'm getting weak in mind as well as in body. I am strong in love though. Gotta be strong in something.

Yesterday I took off and drove about 100 miles for some motorcycle parts. Got them from a British Salvage yard — didn't cost one cent. Got back about 9:30 last night. The boys had me a good fire going and a cup of hot coffee waiting. I drank the coffee realizing that I shouldn't and you know the results. Well I finally got to sleep about 2 a.m. but one hour later I was awakened to attend briefing. Consequently, not a whole lot of sleep last nite.

Today has passed like lightning, fastest day yet. Got up at 3, went to briefing and take off — then came back to my tent and began working trying to make one motorcycle out of 2 halves. Worked all day long. Part of the time, I had the motor in the dispensary because of the fact it was pouring down and cold. Well about 5 p.m., I got it all together and started it up. It runs too — needs some adjusting to the carburetor, tightening of the clutch, a few more minor things and it will be A-1. Had a lot of fun working on it though. Tonight I'm dead tired. I bent over so many times today that my back feels as if it's going to pop any minute. My hands are sore due to barked knuckles, burnt fingers, etc. I enjoyed the day, but am I sore? I told the boys that they could flip coins to decide which one of them would have to help me dress myself in the morning.

Tonight, I am writing to you, shaving and bathing. Also I'm boiling some eggs on my stove for me and two of my boys. Got 4 on now (80¢ worth), but no coffee for me tonight — you can bet on that. Blackie was in for a short period. I gave him a drink of American Rye Whiskey that I had acquired. Real "Bottled in Bond" Rye Whiskey. I never liked Rye, but this tastes better than the best Scotch at the time being. Blackie was pretty busy and couldn't stay but for a very short time. We spent it talking about our respective wives, as usual. You see, Blackie has a lot of respect for his wife too. He sure is a nice guy. I wish his back didn't bother him so much. I do love you. Take care of yourself and ours.

All my love,
Joe.

March 28, 1944
10:30 p.m.

My dearest Honey,

In the past week, I haven't done so well about writing as I had since being out of the hospital. No excuses. I will try to get back on the ball. I have been writing a V-mail about every other night, which isn't too bad — a couple of those were Air Mails too I think. You of course, are still upholding your end of the correspondence in the usual "Babe fashion" of doing everything — right on the ball without variation. I went to our old field yesterday trying to get your chest and jewel box that I was having made for you. They had been completed but not to my satisfaction, so it will be another week before I can get them.

I have completed my motorcycle now. I now have the best one around, and that is saying something for we all have one almost. Young is getting one tomorrow and so is Prewitt. Wheeler, Hargrove, Threadgill, Cook, Moore, Jergens, and many others already have one. Blackie and I are going south tomorrow on mine and Hargroves if we have time. Don't worry, I know how to ride one — I'm not going to get hurt. Will write a long letter tomorrow.

All my love,
Joe.

March 31, 1944
9 p.m.

My dearest Honey,

I had the very best intentions to sit down and write a real long letter last night, when I found out that at 1 a.m. Evans, Curtis and I were going to pull out of here in my ambulance for distant points to pick out a campsite. Maybe we are going to move again — who knows? Anyway we got back at 7 tonight after riding all day on Italian secondary roads or worse. When I say Italian secondary roads, you have no idea what they are like for I don't suppose you have ever been on any to compare. The closest comparison I could make is the road we were on in California on that mountain the day the car caught on fire. If anything these are rougher, though a wee bit wider. You can readily understand why I am not only tired tonight, but I am also sore where an auto ride always hurts me the most. After this long dissertation about something that only

vaguely is about the original point, I didn't write last night for I had to get a couple of hours sleep before pulling out, and tonight my letter will probably be pretty sorry cause I'm too tired to think in one channel for any period of time.

I had a real loss today. Some Italian kid stole my camera out of the ambulance — damn it. It's my fault for being so careless. I should know the Italians better by now and do — it was plain forgetfulness and carelessness.

Major Bentley isn't with this squadron any more. We are all glad from Evans down to the soldiers — particularly the flying personnel. Bentley was transferred to the 724th as their executive officer to take Major Reynolds place. Major Reynolds has been sent back to the States — he had pneumonia which became unresolved (didn't clear up as it should in his lungs even though the fever, sickness, etc go away) and was sent back. Major Elliot and Captain Reynolds both from the 725th have been sent back also — I don't think you knew either of them. All of them were in their 40's and had something ailing. This goes further to prove my point that this is especially a young man's war. In fact I am so old, I think that I shall request to be sent back as over-age — I'm sure they would grant my request, aren't you? Jack Bernstein took Bentley's place as executive officer, and Curtis took Jack's place as adjutant. Both of them are doing excellent jobs in their new capacities. Now only one man in the squadron out ranks me, and that's Evans, of course. I'm the oldest captain here in the 727th — you know how big that makes my head, rank means so much to any "doc" in the Army. The one fellow that I was glad for in the shake-up was Curtis — he is now eligible for a Captaincy, which he decidedly deserves. Evans is very well liked over here by the squadron, so is Bernstein. Bentley very much disliked. In the States the whole picture was just reversed. The Colonel is much better thought of too. Blackie is still the best liked man among the flying personnel as he has always been.

Haltom has done well over here. He received the Distinguished Flying Cross a couple of days ago for leading our most successful raid since we have been over. He has been doing a splendid job as have all the boys.

I'm getting gypped by the way. I was going to get to go down and be the M.D. at a rest hotel for a week beginning April 7. Now that we are moving I won't be able to do it. That would have been a nice rest with innerspring mattresses, eating out of plates and all the trimmings. Oh well, I wouldn't have known how to act and a tie would feel uncomfortable anyway — much less a blouse. I'm pretty comfortable as it is, and I don't work much anyhow. Rosenbaum has become such a good first-aid man that I am practically excess baggage. Mason is a good "taker-carer" of me and my "trouble-shooter" and good generally. Williams, with Mason's assistance, makes out all my reports. Wilson is my carpenter and mechanic and is good at anything. The rest of them just do what they are told and anything that comes up. They all really are doing a good job and are taking care of me.

A couple of days ago my tent almost blew away. I had to make a trip, which took all day. That night when I returned, a frame had been built just to cover my tent, making it much more stable and much roomier. Since I'm such a sorry housekeeper, every 3 or 4 days I walk into my tent and it is all cleaned up. I don't even have to go after a canteen of water. Since Mason, my sergeant, looks after me like I was his "only baby chick" I kid him a lot. I'll say "Mason, how did you ever live before you had me to tell you what to do, how to drive, etc." Mason just grins and says "Well, Captain, someone has always told me what to do so there is no need for me to think for myself." He is in right now boiling some eggs that he buys for 20¢ each for us to eat. This has gotten to be a nightly occurrence. Williams makes us coffee every night. They are a good bunch, and I don't believe they mind working for me a bit — army or no army. I know and they know that they don't have to do the things they do for me personally. In fact, they do too much for me. Enough of my 8 boys.

Honey, did you get your flowers Easter? I sent you some roses and some gardenias, Mother an azalea plant, and Mrs. Black some carnations. Did they come at all and if so did you get what I ordered? Let me know. I just wanted to tell you, so if anything happened where someone didn't get theirs, you could tell them I tried and maybe their feelings wouldn't be hurt.

By the way, I can tell an effect on Mother already since they have sold the shop. Perhaps it is my imagination, but I don't think so. All my letters consist of now are her visits to one doctor or the other and how badly she feels. Seems as though she found a "gold-mine" in Doctor McInistion who apparently sympathizes with her. True, I'm sure she doesn't feel too good with her blood pressure up so terribly high. But if he continues to tell her how bad it is, it is going to make her think about it more and consequently probably make it higher. You just can't sympathize too much with a neurotic — it's no good. This may be a hellova way to talk about one's mother. As you know it's strictly between you and I, and I don't mean to be disrespectful. If I were there I would handle the situation, but I can't do it through mail without hurting Mother's feelings, which I don't intend to do under the circumstances. Apparently, Dad has a mild heart failure and should see a doctor and get on Digitalis. I'll bet he has shortness of his breath, tires easily, productive cough, frequent urination, and swelling of the ankles at night which go down by morning. Tell Sis to get him to see a doctor, but don't tell Sis anything I said about Mother. Always remember Sis is a swell Sis, but she cannot keep a secret at all — neither can Dad, and as you say perhaps I can't either. Mother can, however, so that's something in her favor.

April 1, 10 a.m.

Honey, looks like I got wound up last night and couldn't stop. Finally the lights went out, and I had to quit. I feel better this morning. Had a good night's sleep which always helps. I want to get this off. I love you more than I could ever tell you.

Joe.

April 4, 1944

Dearest Honey,

I didn't write to you again today, so tonight you deserve a real nice long letter. In fact, you deserve one of those kind all the time, but I fall short. Bless your heart, you never do — you are still the best wife on earth and the only woman as far as I have been concerned for just about four years. You know about four years ago, I was having my first date with you — quite by chance I'll admit, but gosh, weren't we lucky. It was next to the luckiest day of my life — the luckiest coming about 6 weeks later May 31. The past four years with our various trials and tribulations have been my happiest by far and I believe our happiest — even in the midst of all the movings, upsets, departures, etc. I have been so very fortunate, and I can assure you that I am thankful. I guess I am really and truly thankful for the first time in my life, even though I have always had every reason to be so. You have been a good teacher directly and indirectly and have set me a wonderful example of everything that is good. I am a fortunate man.

Today has been beautiful. One of those days that you feel that spring is really possible in Italy and not a myth after all. It hasn't rained for a couple of days — you can walk around camp without your boots. In the early mornings, it is still quite chilly but up in the day you can get by with a light sweater. At night and in the mornings, you certainly still need a fire though. Italy is loaded with fruit trees, and there have been blossoms on the trees for quite awhile. Maybe the rainy season is drawing to an end. I suppose yours is just getting started.

Today Hargroves and I got on our motorcycles and rode to Taranto and looked the city over again. It is nice to have a mode of transportation all your own, and then it's nice to have the cool breeze hitting you in the face. You have to work on it all the time, and doggone if the time hasn't passed a whole lot faster since I've had it. Do you realize it has only been four months since I left the States? Seems like an eternity within itself, and they tell me that the first 6 months pass the fastest.

Blackie and I went to our open air theater tonight but left in the middle of the picture. It was strictly "no bona" as the Italians would put it — one of Humphrey Bogart's pictures — something about the "North Atlantic" was the name of it. We have shows rather frequently now. So far I have been to two and almost froze to death both times. Tonight it was quite nice out though, but the picture was no good.

You always write about spring cleaning — it must be catching. I shined my shoes yesterday. It was the first time they had been shined since Natal, Brazil. That was the extent of my spring cleaning. Yes, I did put on clean underwear — on a dirty body of course.

Honey, I'm going to stop early on you tonight and go to bed. I have to get up early tomorrow and I'm very sleepy. Lately I just haven't been getting enough sleep. Good night, honey. Take care of yourselves. I'm O.K.

All my love,
Joe.

Mission #24: 5 April 1944
Target: Ploesti oil refinery, Romania.
Results: Highly satisfactory. One column of smoke up to 15,000 feet. Six ships were lost – 3 727th ships to enemy aircraft, 1 to flak (with 724th C.O.), 1 725th ship on take off, and 1 727th ship declared missing.

The Ploesti Oil Fields were some of the main oil refineries that supplied Germany's war effort, a very important target for the Allies. There were multiple raids on Ploesti before the campaign was over.

During mission #24, the 451st was the last group in the formation and encountered a strong force of German fighters and flak to and from the target. The successful mission earned the 451st their second Distinguished Unit Citation, but it came at a great loss.

The 727th sent nine planes. Only five returned. The Craven Raven was one of the planes that was lost. It was piloted by Lt. Wil McAllister, one of Joe's good friends. Lt. McAllister's last words were reported to be, "We are fighting for our lives, men."

179

Castelluccia di Sauri

The third and final move for the 451st was to Castelluccia di Sauri, located between Foggia and Bari. The base was arranged in an oval with the four squadrons spread around the runway. The headquarters were located at one end of the base in an old Italian villa. The runway went down the middle of the base and was made of steel planking over limestone, allowing it to be used in any weather.

Each squadron was equipped with medical dispensaries and mess halls, surrounded by living quarters. Some of the men built "homes" from scavenged concrete blocks and wood. Joe supervised the construction of the squadron's water and sanitary conditions, and he built a hot water system with "acquired" parts. When the men first arrived at Castelluccia, they ate outside under a tent, but Joe prioritized the construction of the mess hall— along with the Officers Club.

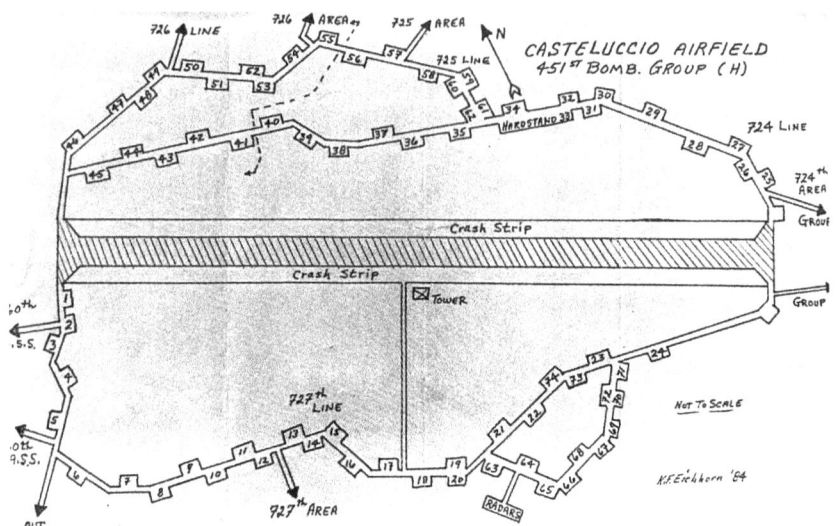

Castelluccia Airfield. The 727th was located in the southwest corner of the base.

Spring 1944

Joe spent a great part of the spring setting up his dispensary and continuing to care for men both at the base and in the hospital. Often men would come to the dispensary to talk to Joe about their fears of flying. It was his job to evaluate their mental capability and talk them back into the air. Within the dispensary's medical supplies was a post-mission 2-ounce shot of whiskey for each man. The drink helped calm the flight crews, so they could go to the post-mission briefing.

Joe also had the unpleasant task of determining the men's psychological health and flying status. At times, he accompanied men to Medical Disposition Boards. The Medical Disposition Board was established to directly handle the psychiatric problems associated with aerial combat. Prior to the establishment of the Board, a soldier was "treated" or sent to a rest camp before returning to combat. With the Flight Surgeons' testimony, the Board was able to assess each soldier's situation and had the authority to immediately reassign or discharge men.

* * *

April 7, 1944
Good Friday, 9 p.m.

My dearest Honey,

Every day or so I think I will send you a V-mail, but somehow I write an airmail instead. Hope it doesn't leave you without hearing from me too long. I haven't been too consistent in writing again this week. I'm sorry, but we have just moved again and when you are on the road or in the air or all torn up all the time, you just can't write. That's the way it has been lately. I have put 2200 miles on the ambulance in the last month and 700 on my motorcycle since I fixed it up — figure it up, that's almost 100 miles per day.

Today has been a busy one. I arrived here at 3 a.m. after traveling practically all day yesterday. Put my bed roll down and slept out. Sleeping was swell except there was really a heavy dew this morning. Got up early and began to set up a dispensary again — and I have gotten it all set up and ready to go. I have three tents now. One large pyramidal tent for the dispensary, another large one for my enlisted men to stay in, and then my small one that I use as an office and place to stay combined. Evans wants me to move down with him, Young and Zraick, so I suppose I will move in with them in a couple of days. That will be OK in one way for I'm not so available at night time, but I don't like to room with Young at all — but I will move down I suppose.

The weather looks like it is going to be much nicer up here. Today has been beautiful. There is always a breeze coming off the snow capped mountains. It is cool at night time — in fact, right now my feet are nearly frozen. But all day long it has seemed like the "sunny south." The sun shone down brightly. We all worked today with our shirts off. We all got a little sunburned, but it felt so good we didn't mind. In fact right now, I feel a little itchy across my shoulders. I'm going to do the same thing tomorrow if the sun shines as brightly. Boy the sun surely feels good when you haven't felt it in a long, long time.

I rode my motorcycle up yesterday. Wheeler and I rode up together and 3 officers from the 726th. Started up about noon and although we didn't ride over a couple of hundred miles, I arrived here at 3 this morning. Wheeler had trouble with his all the way. Mine just "putt-putted" along as smooth as velvet. I am much closer to Dan and Johnson now. They are about 40 miles due east of me. I will run over and see them pretty soon if they are still located there and I suppose they are.

By the way, I passed through the place where I was having your chest made yesterday. Picked it up and also the jewel boxes I had made for all the ladies of the family — you, your Mother, my Mother, Sis and Katy. Didn't leave anyone out, did I? I'm afraid they are made out of green lumber, but if they aren't they will be nice. If they are, then they will be all warped by the time you get them. I will try to fix a way to send them as soon as I can get to it for if they stay here, they will get all scratched and scarred, or wet which would be worse.

I'm having to write by candle again tonight. I have been trying to get some benzene for my lantern all day, but there's none within a radius of ten miles. I'll go hunting again tomorrow. I'm tired, so I think I shall retire to my lovely boudoir — of course, after I get my butler to draw me a nice hot bath. I love you with all my heart.

Goodnight,
Joe.

Although warped by time, the boxes Joe had made sat on Babe's dressing table for the rest of her life.

April 8, 1944
Saturday before Easter, 8 p.m.

My dearest Honey,

My theme song tonight and every night is "Oh How I miss You Tonight." Somehow tonight just seems to be a little worse than usual. I'm lonesome — I feel all by myself even though all my enlisted men are sitting within spitting distance of me playing blackjack. What I would give to be with you.

Honey, in almost every letter I get from you, you mention that you will make "it" up to me for sacrificing this or that. You don't owe me anything — you, no doubt, are going through more actual hardships than I am. We are separated, yes that is the only real hardship for me. I, like you, can't get used to it — but, wouldn't our wedlock have been a weak thing had either of us found it easy to be separated. In that way, I'm certainly happy that neither of us has found it too easy. You not only have to be separated, but have all the responsibility dumped on your shoulders — raising Kay and Carol, taking care of them 24 hours per day. Taking care of the house, paying the bills, etc. You have a million things that makes the going tough. It is I who owes the other, not you. The outdoor life, the tents, the rain, the dust, the absence of baths, the dirt, etc, aren't really bad. Sure we gripe about them. A soldier is going to gripe. I don't care what happens, you can rest assured of that. Since I'm primarily the chaplain for this outfit (secondarily the flight surgeon) all the guys gripe to me — since there is no one for me to gripe to, I suppose I have been griping in my letters to you. But what I'm trying to put across is that this is not bad. You can get used to anything like this. As you know, I never did like to bathe anyhow, I never cared much about eating, and so I get along rather nicely. It isn't half so bad as it sounds. If I weren't an old settled down married man at heart, I don't believe I'd mind at all, however the single men say otherwise. Anyhow don't worry about me; physically, I'm sound; mentally, I'm getting by; morally, I'm above reproach other than playing cards occasionally and taking a drink whenever I can find some, which is too rare. So at least I'm a good boy if you already didn't know that. You know this war might be good in another way for me, I shall always remember and know and realize how sorry I get along without you. I think I have known all along that I wouldn't do very well, but now we both know. Perhaps if everyone had similar experiences many marriages would be happier and many people would be more tolerant of one another — however, I don't believe such a lesson had to be taught to you or I. It's a cinch the war can't last forever — sooner or later someone is going to give out of resources.

Tomorrow is Easter. Surely hope you get your flowers. You will have to consider them for Easter and Mother's Day. There are only certain given periods that you can send flowers, and the time passed for sending flowers for Mother's Day before I knew it was time. It is now too late. I'll be thinking of you, however, even though there will be no way I can show it.

Well, honey, I've rambled a bit. But I feel like I've talked to you so I feel better. I got a V-mail from you today saying that you were having the garage doors fixed for which I'm glad. If there is anything else that is needed to be done, go ahead. It doesn't seem like it costs as much to do it a little at a time as it does all at once. It's about time for you to go to town and buy you some clothes. Hope you have already done it, so you will have a new Easter outfit that is so necessary for a woman to feel like she is dressed. When you get something, get something nice regardless of cost — I want my honey to look her best.

Remember how much I love you so you will take special good care of yourself if for no other reason than for me. Kiss Kay and Carol for me.

All my love forever,
Joe.

April 9, 1944
Easter, 8:30 p.m.

Dearest Honey,

Today was a red letter day. I went into the nearest city on my motorcycle and finally found a place to take a bath. I had to steal my way into it more or less, but it was a tub bath with hot water. The first tub bath I have had since in our home in September and one of the very few I've had since December. It was divine!

I'm writing this V-mail because the last two or more letters have been long ones. Personally, I don't like to write these things at all — I feel defeated before I start and I have just got to hit the high spots of the day's occurrences. Haven't gotten any mail from you in a number of days because of the move primarily, but that only means I get more later. By the way, I sent you $125.00 today through the Finance Office. When you get it let me know. Take it and buy you some nice clothes (not maternity dresses this spring) .

Will write more tomorrow. All my love to you, Kay and Carol.

Happy Easter,
Joe.

April 11, 1944

My dearest Honey,

Tonight I am low. I feel O.K. physically but am just down psychologically. I had two boys to turn up today with that old "hush-hush" disease, and I've felt bad ever since. Just sorta gets you down — you preach, teach, show, talk, and everything else trying to pound into these guys a few facts, but either they think they are smarter than you are about such or they are too dumb to comprehend. I just hate to see them ruin themselves and their lives — that's all.

Been pretty busy today. Besides giving a lecture to the boys, I've increased the floor space of my tent and built me a clothes rack. Made both out of the steel matting we use to make runways. It works fine too. Now I've got to get my clothes out and hang them up, since I have a hanger for the first time.

Well they aren't going to let us ride our motorcycles anymore due to some mix up about them. I'm going to try to sell mine. Don't know whether to sell it or make a four-wheel vehicle of some sort out of it. No, I'm not completely crazy — just need transportation.

I haven't heard from you since I have been here. Suppose the mail will start in a couple of days — hope so — that will make me feel some better within itself. Honey, I'm in no mood to write tonight. Will do better tomorrow. Tell Kay and Carol hello for me.

All my love,
Joe.

April 14, 1944
Friday, 8 p.m.

My dearest Honey,

Got a long letter from you today. It was an especially good one that I enjoyed very much. You were speculating about the future once we get back together. I'm glad you are fixing up the yard and the house. Not only does it make it sound more homey, but it also enhances its value. I only wish that I were there to do the work. I would really enjoy it. I would really like to plant the flowers and the fruit trees.

I've been very busy today. I have been putting in a water system to my tents — sewage and all. We now have running water in our tents — not hot and cold yet, but give me time. I have a 55 gallon oil drum up on a 10 ft stand for my tank, and my pipe is all aluminum, which are hydraulic lines from wrecked airplanes. My sink is made out of half of a 5 gallon oil drum. It all works pretty nice but needs some finishing touches. My "sewage" system goes out to a deep pit out from my tents that we have dug and blasted with dynamite. Of course, the hole is filled with stones, so there won't be any standing water around. My own little municipal water system. It's a lot of fun working these things out and trying to get (one way or another) the things you need. It sometimes taxes your ingenuity. Soon as I get time, I'm gonna start on a shower for us. Of course by the time I get it working, we will move as we did last time.

I've gotta go tomorrow to see the 15th Air Force surgeon — probably will leave about 7 a.m. and be gone all day. Sorta dread it — would rather be around here especially when my boys are flying.

We got some replacement crews yesterday. They surely are green. I think they are much worse than we were. So far they have asked such fool questions as "where is the officer's club and service man's club," "we want to go into town to buy a gasoline lantern," "does the flight surgeon know how many men should be in a tent," " where are the showers?" When the latter question was asked and he was told to find a can and go to the water tank we have in the area, he thought we were joking. War is really going to be hell on them until they change their attitude, which they will pronto, I'm sure. Living conditions really aren't so bad as they sound. It's sorta fun to see what you can make to make things more comfortable.

By the way, we all got a coke today. First since the States. You couldn't have bought one for $10.00 even though they, of course, were all hot. They really were good — funny how values change so much over here. A coke really made you think of the States.

I have your chests all boxed up now. All I've got to do now is get some paint and address them and take them into town to the Post Office. Don't know how many days it will take me to accomplish that. I have them wrapped up in blankets (all wool) to try to keep them from getting scratched up on the way over. I'm sending the big chest over by itself and all five of the small ones in the same wooden box. Speaking of boxes, I still haven't ever gotten any packages from you. You can send me some candy if you can get some and a fruitcake. Both would taste good to me. Maybe you can't buy candy now, and fruit cakes are probably out of season. Don't go to too much trouble. If you send them, be sure you pack them quite well. It's a long way, and some of the packages that the boys get are all torn up. No doubt many don't get here because they aren't well packed.

I'm sending Prewitt, Wheeler, and their crew to the rest camp tomorrow. There are about 3 men on that crew that need it. This is the first bunch that I have sent. The boys are doing a wonderful job. They are a swell bunch.

Well, Honey, I'm going to draw this to an end. I love you most dearly and miss you just as much.

> Goodnight,
> Joe.

> April 16, 1944
> 11 p.m.

My Honey,

Just a shorty tonight, so you will know I'm 'alive and kickin' as usual. I didn't get to write you last night, and I'm so tired I just haven't got enough energy left to write anything else.

Night before last I was up all night. Then early yesterday morning, I had to go to Bari which is quite a little ways. Got down there and attended to all my business and started back at 3 p.m. Had trouble with the car I had borrowed from Group Headquarters and didn't get back here until 3 a.m. Mason and I have been no good all day long — he was with me. This morning I had to get up early for briefing and take-off, so I haven't had any sleep. So as soon as I finish this I'm going to bed. Should write to Mother but she must wait.

By the way, Garrity and I addressed your boxes this afternoon. Hope to be able to get in to send them soon. Well I'm going to bed. Even though this is a sorry attempt at writing, I still love you as always. Take care of yourself and the kids for me, please.

> All my love,
> Joe.

April 17, 1944
Monday, 2 a.m.

My dearest Honey,

"Life ain't never been so good." That's a favorite saying of all of us over here when we get something for our own comfort. It is a phrase that is worth quite a bit within itself for upkeeping the morale of the American Army. That being the case — "Life ain't never been so good." — I am sitting on top tonight (If I just had my wife and chillun') I now have running hot and cold water in my tent and dispensary. I just finished putting in my hot water unit at 1 a.m. I have since taken a partial 'one of those kinds' of bath, shaved, combed my hair, and brushed my teeth. Hot water running out of a faucet (really an airplane hydraulic valve) just like back in the States. You have no idea how much running hot water uplifts one's morale — it's really sumpin. I bet I keep 100% cleaner now (which won't be very clean). I believe you would get a kick out of it.

I'm enclosing a picture of our hot water heater and our tank, and I'll draw you a diagram of the "King Water Works Co, Inc." How do you like that setup? I realize that it looks like something that Rube Goldberg thought up, but it works considerably better than most commercial types. Thought you might like an idea of what the gadget and the plexus of pipes looks like.

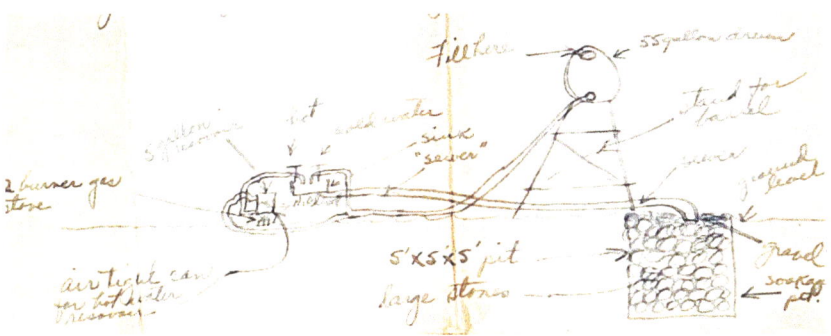

The above probably doesn't mean a thing to you — but it worked out almost like the water system in any house and the same principle as any hot water tank. If I could just make my hot water heater thermostatically controlled, it would be swell, but I have to light a burner to have hot water, which of course is no job at all. I do wish you could see my setup — believe it or not, with my hot and cold water running in my tent and dispensary, I just about have the conveniences of modern civilization. The peculiar thing about it is that it works. Don't know what I will start to work on now in my spare time — probably a "front porch" for my tent and a "beach chair." It's really fun playing around and seeing what you can think up to make the surroundings more comfortable. I dare say that

King Water Works Co, Inc.

mine is the first running hot water in this outfit at least.

I also have changed "sheets" on my bed tonight. First change in about 3 weeks. Of course my "sheets" is a mattress cover slitted down one side. My pillow case is a towel pinned on a pillow that I "requisitioned" while in Dakar. By the way, Honey, when you are sending some of those packages over sometime, please send me a couple of pillow slips for a rather small pillow — unbleached domestic is perfectly good. I certainly don't want anything nice — it wouldn't go with the surroundings.

Tomorrow I am going to take out my stove in my tent and rearrange my bed, trunks, etc. I'll have lots more room. It's still cold here in the mornings and at night, but I can get by without the stove.

I've been out today looking over the area for water holes and oiling them. Trying to get the jump on Mr. Mosquito before he gets the jump on me. Also gave a lecture on malaria today to the squadron. This afternoon, Evans and I have been "making the rounds" trying to find some pipe somewhere to make some showers for the men. We got some but have got to find some more. He has certainly learned a lot since he has been C.O. of this outfit and is now a damned good one. He certainly has the men's good at heart.

Honey, I sent your chests today. If they are scratched up it is partially my fault because I was supposed to sign my name to the certificate tacked on the outside and I forgot to, consequently, I know customs will unpack them. If I had signed my name they might have gotten through without ever having been opened up.

All my love,
Joe.

April 21, 1944
9:20 p.m.

My dearest Honey,

Well I haven't written you in three nights so I guess I'd better make this a good one. I have been on a two day pass. Wednesday morning Major Bentley came in my tent bright and early — woke me up and asked me to go on a two day pass with him. Since he had transportation (a jeep), I thought I would see if Evans cared. He didn't mind, so on a 48-hour pass I went. It was the first pass I have been on since Fairmont, and it certainly felt good to leave it all behind. Passes over here aren't so good, however, for it is impossible to get 'away' from the army over here.

We left here Wednesday morning about 11 and headed for Naples. When we arrived there, we had to find a place to stay and for a while it looked as though we weren't. Finally we were allowed to stay at the Transient Officers Hotel even though we didn't have orders. That night we went to a picture show and saw Roy Rogers in something — the only one that we could get into. It passed that much time, and believe it or not, wasn't too boresome. From there we went to the hotel and to bed. The following morning we awoke about 9:30 (banker's hours) and went out to the ruins of Pompeii. Of course since Vesuvius has acted up again, it is covered with plenty of lava dust, but it is about what one would expect — quite interesting nevertheless. I borrowed a Kodak and took 4 rolls of film, so you could see everything of interest. Really did wish for you there for I realize how much more kick you would get out of visiting Pompeii and seeing Vesuvius than I would. Vesuvius is just barely smoking now, but you can see the fresh lava down the side of her and of course the countryside is covered with lava dust. Yes, you would enjoy seeing Pompeii, Vesuvius, and Naples. But I'm afraid it's one pleasure you will never have for I never intend to let you out of my sight for a 24-hour period again once I come home, and I'm positive that I never will want to see Italy again once I get away — selfish critter, aren't I. Maybe I'll change my mind later — they say that time does things like that, but we will have to wait and see.

Well we left Pompeii about 2 p.m. and went back to town — back to the Hotel and lounged around on our cots a while. I layed there and thought of you — how many times we had been in hotels together, how much you would have enjoyed the morning and 5,000 other things concerning you. We then drove out to a lovely place where you could buy drinks, bought one and then back to town to the Red Cross. There we had a snack and last night sat around the piano and sang until about 10 and then to the hotel again. This morning up and back to camp. There is the happenings of my 48-hour leave.

By the way, there is plenty of merchandise over there, or it looked like there was. But as you know I sent all my paycheck home this month but $12. With my next paycheck, I may buy this Kodak "Bantam." With the following paycheck, I'll go over and buy you and the children anything you want if you just let me know what you would like. I bought you a necklace — a bracelet set at Pompeii yesterday — just costume jewelry of course but something that you may wear. It is made out of seashells, but you would never know it by looking at it. As I have told you before there is no decent jewelry over here for there is no gold or silver left in the country. I can buy you some cameos though if you want them. I'll be buying you something from time to time so you might as well let me know what you would like if I can find it. There is much more there than anywhere else I have been, and I've been all over our part of this country at one time or another.

Pompeii, 1944.

Got a short note from Burr Craycroft yesterday giving me his A.P.O. number. He is over here too apparently even though I haven't found out where he is. He said that he knew where I was and would be over when he had a chance. The letter was mailed before I moved here, however, and he had the wrong A.P.O. number in my address, so I really don't know whether he knows where I am or not. When I get caught up a bit I will try to find out. In the meantime, I will drop him a note.

Got a letter from you and your mother today. You were worried about buying a new hot water heater. Of course, you did the right thing. I don't believe you were as worried about doing the right thing as you were paying cash for it. I'm glad you bought a new good one, and I'm glad you had the money to pay for it in cash. Now see, I told you that you always would do the right thing and so far you haven't let me down a bit nor will you. The letter from your Mother was very nice and very much appreciated as usual. She wrote me right after Easter and thanked me for her carnations. She mentioned that you received the gardenias but did not mention whether you received the roses. Neither did she say whether Mother's flowers got through. Certainly hope they all did. She told me how cute you dressed Kay and Carol for Easter — certainly wish I could have seen the three of you.

Well honey, I hope this letter has made up in part for skipping writing a couple of days. Take care of yourself and tell yourself that "your man" loves you just as much as he ever did (I would say more if I believed it were possible), that he admires you even more, respects you likewise, and misses you more than words can tell.

Goodnight,
Joe.

April 24, 1944
7:30 p.m.

Dearest Honey,

I received two letters from you today — one written March 21 and March 27. They were good letters as usual — long and interesting. They came in handy today 'cause I've been down. I'm hypersensitive to Atabrine, so I suppose that's why I don't feel too good.

In one of your letters today, you asked me how much I weighed. Honey, truthfully, I don't know. I haven't seen a pair of scales since I left the States, but I dare say that I don't weigh as much as I did. However, I am probably in better shape physically, so it doesn't matter. In fact, it probably is a pretty good thing for I would be without clothes had I not fallen off some. There is no such thing as "dry-cleaning" in Italy, I don't guess. I take my clothes in with a quart of gasoline for them to clean. But they are always washed when I get them back and shrunk just a little more than the time before — it never fails. Every rag I have is ruined, but it doesn't matter. Over here, one doesn't give a damn whether his clothes fit nicely or not. I would have to buy more probably when I get home. Guess we will start wearing khakis pretty soon.

However, it's none too warm even yet. I have had my fire going all day long, and it's felt good. Thought for a while that it was going to turn warm, but it's still cool enough to have a fire and wear your sweater and leather jacket. I once started to take out my stove but I'm glad I changed my mind and didn't. Guess it's pretty warm there by now — probably raining quite a bit though. I have had enough rain this year to do me quite a while. We have had very little since we have been up here though — thank goodness.

Guess what I have been doing most of the day? I have been saving all my mail. Well, it has just about moved me out of my large, spacious boudoir, so something had to be done. Today (in fact for the past two days) I have taken all the letters out, glanced over them and putting them in my firewood box to start fires with. Paper over here is scarce, and one usually starts his fires with wood shavings — now I'm a member of the elite. Beginning today your letters serve a dual purpose — keep up my morale and give me a way to start fires. That doesn't sound nice, does it? But I can assure you that I mean no harm, and if it wasn't for your letters I would go nuttier than I am. Don't tell anyone but the others are nice but are just correspondence — yours are the ones which do the good, and when the mail gets all messed up and I miss hearing from you for a few days, I can really tell a difference.

Besides sorting letters I have cleaned my tent up somewhat today. Also remade my bed from bottom to top. Took off a few of my blankets. I have been feeling "things" crawling on me every night lately, so I examined my bedding — found some fleas, also a flea on me today, but no bites. They must not like my flavor, for they crawl on me but don't bite. Since I found them in my bed today, however, I have "felt" them all over me. Imagination, of course (I hope).

I have rearranged my water works. Yesterday I made me a built-in sink and hot water heater all in one unit. Painted it today, and it's really nice. Now if we were to have to move again, I could just box it up and carry it along — I wish you could see it. It's very nice, and I'm proud of it as you would be of a brand-new bathroom. I now have a real washbasin in the dispensary just like home. It's very nice too.

Tomorrow night I have to go to a meeting of all the Flight Surgeons of this wing. I'll probably see Koenig for he is in this wing too. Haven't seen him since that day in Bari. I haven't answered Burr Craycroft's note yet but will in the next few days and will also try to find out where he is. Don't know where I'll get some clean pants for that meeting. I'm wearing a borrowed pair now.

Went into the nearest town a few days ago. Carried in my cleaning, laundry, my pair of boots to be repaired, my watch to be repaired. My boots had ripped out in the back and my watch (the old one — not the one you gave me) had a bent balance staff. Will go back in about a week and gather up all the stuff. I also went to the Red Cross to have some cake with chocolate syrup all over it, which was very good.

When I get back I want you and me to take some time off together. In fact, I'm looking forward to that already, and we will certainly do it if we can even if it is five years from now. Boy, you don't know how much I am looking forward to that. Well, honey, this is enough for tonight. Young just dropped in and sends his love which doesn't half tell the tale. Take care of yourself and Kay and Carol.

I love you,
Joe.

April 25, 1944
11:30 p.m.

My dearest Honey,

I've just returned from a meeting of all the surgeon's of this wing and I'm down in the mouth, disgusted, and maybe even homesick, so I shouldn't be writing you tonight. I'm just in one of my weaker moments tonight and will snap out of it tomorrow. At this moment if I could see and be with you and Kay I would do it no matter what.

I've been listening to a 'brass hat' telling us how things should be done and run over here tonight even though he just got here and doesn't know what the score is — slightly disgusting. I would like for him to come out and let me show him what war really is — I'll bet I could convince him damn quick. Saw Koenig tonight. He's fine and says his family was as well the last time he heard — nice guy. I like him. As I said before, I'm down tonight so don't pay this letter too much mind. Even so, I still and yet and will always love you with all my heart.

Joe.

April 26, 1944
Wednesday night, 10:15 p.m.

My dearest Honey,

I received three V-mails from you today — all nice — made me feel better. I could get a postage stamp from you with writing on the back, and it would make me feel better. You will always be the "light of my life."

Went into town today and looked around. Went to the Red Cross Theater but the projector broke in the middle of it, so I left before they fixed it — Wallace Berry was playing in "Rationing." Formanek went with me. He has been 'down in the dumps' lately due to a variety of things, so I made a point of going into town with him alone. We then walked around town to see if we could find anything. I bought you a cameo bracelet with seven cameos in it (of course the "gold" is brass, but we can have the cameos reset into something nice back in the States). I also bought you a cameo ring and some small cameo earrings. We went by and had some coffee and doughnuts and then returned to camp.

Tonight I went up to Group Headquarters to see Wagner and went to a picture show while up there. It was George Raft in something — all about International spies in Turkey. I think I had seen it before with you — however, having nothing better to do, I sat through it again. Somehow or another no picture show can hold my interest anymore. Guess I'm out of the habit.

Sorry to hear about all the storms. Knowing how you dislike storms, I wish even more I was there with you during them. Of course, I'd probably be scared to death but we could be scared together, couldn't we?

Hey honey, take yourself and Kay down and weigh yourselves and write to tell me what you weigh. Also tell me whether Kay is as chubby as she used to be, also whether Carol is chubby like Kay was, and whether she has blue eyes or not. I'd like to see Carol and "see" her personality. She must be her Mama over, and I'll bet she is in personality, if not also in looks.

It rained all day, but not hard. Can't gripe though for this is the first rain we have really had here. It's certainly not like the other two places in that respect. Well, honey, my fire has gone out and my feet are pretty cold, so I'm going to bed — wish it was with you instead of by myself. Don't know when that event will be but it's one day nearer than it was yesterday. Goodnight and I love you.

Joe.

April 27, 1944
10:15 p.m.

My dearest Honey,

It's been raining all day long, so I have stayed in more than usual. And I've gotten a lot done too. I've gone completely through my "trunk" and my bags and rearranged my clothes. Some of my clothes that were in my trunk were wet, so I got them out. Got my underwear out that are old and worn badly and put them on a shelf so they will be easily accessible, so I will wear them instead of my reasonably good stuff. I want to wear them out. All in all I've had a busy day "spring cleaning" my tent. My tent is leaking like a sieve. Got a bad leak right over my cot. The bad thing is I can't move my cot until I get my stove out — and it's too chilly to move my stove — so I guess I will put up with the leak.

Wrote you quite a long letter last night and a V-mail the night previously. I was pretty depressed in both, but I am now in better spirits. Forgive me 'cause I was just "down." Everything is O.K. I'm just looking forward with the greatest amount of anticipation possible to the future years with you — knowing that we will be as happy then as we have been in the past.

All my love,
Joe.

April 28, 1944
7:30 p.m.

My dearest Honey,

Gotta make this a good letter for many reasons. Firstly, I received a very long letter from you today written exactly one month ago. Secondly, I received the package with the wings and the foot warmers today that you mailed three months ago. The letter was swell — 9 pages long and "just what the doctor ordered" to make this a brighter day.

The wings were exactly what I wanted. And arrived here in plenty of time for orders from Washington making us Flight Surgeons (with $40 more pay per month) have still not come through. However, some of the boys are already wearing their wings. Personally, I think I'll just wait until I get paid for carrying them around. And the foot warmers — believe it or not may still be of practical value even though it is quite near May 1. I have had both my stove and the stove in the dispensary going all day long. The foot warmers are going to be especially nice to wear around my tent at night after I have already pulled off my shoes and socks and am wearing those G.I. slides that you sent me.

I have a couple of big projects on now. I am making an Officer's mess, Enlisted men's mess, kitchen, mail room, barber-shop and storage room out of the old walls of a building which was in our squadron area when we got here. It is really going to be sumpin. I have the kitchen almost finished now. With the help of about 20 Italian laborers, and the Engineering Department, we are going to get it built. Gonna have steam heaters for the food as it is being served, tile floors (really), and everything. It's a big project, but it will be the nicest thing around when it gets finished — and it keeps the time passing. The wing surgeon came around today and was impressed. However, that within itself is nothing — he's a tenderfoot, and, as yet, doesn't know what it's all about. Wish I had him in the mud of Telergma, North Africa or the first place I was in Italy — would he go nuts! There I go talking like a veteran, well, I am quickly becoming one. You say Hugh has mentioned that he hopes I turn up in his part of the world. Don't believe that will ever happen — believe I would go the other way first, but who knows? Hope Hugh is getting along fine and am glad that he has been situated fairly comfortably. I'm sure after the North African and Sicilian campaign that he deserved better.

Next morning —

Didn't finish last night apparently. Got in a bull session on venereal diseases which lasted until 1 a.m. instead. Sorry. You should see our barber shop. Got two Italian barbers in half a tent who will shave you or give you a haircut. No hot towels and cold water but just for someone else to shave you is a treat. We will make them a good place once I get the building for the mess hall finished.

Honey, I want this to get in the morning's mail, so I had better get busy and send it. Tell Kay that her Daddy loves her and remember that I love you "mostest" of all.

Joe.

April 29, 1944
Midnight

My dearest Honey,

I'm sorta late getting started tonight so I will make it a "shortie." I have been over playing poker with a bunch of the boys, and I lost — doggone it. It's the first time that I have played in a long, long time too. That's bad.

The wing medical inspector came around today. He was quite "taken aback" with all my projects. I think my water works really struck his eye — said he was going to come back and take pictures of all my stuff. Been living in high cotton today. Got my rations and got 2 bars of candy and a coke. So at present I'm drinking the coke and have eaten 1½ bars of candy — gonna save the other half.

Thought we were going into khakis the day after tomorrow but it has been postponed. I'm glad for it's much too cold for khakis unless you wore your "long Johns." Goodnight, my honey.

As always,
Joe.

May 2, 1944
8 p.m.

My dearest Honey,

I haven't written you in a couple of days. It hasn't been because I haven't been thinking of you — far from it — you may rest assured that I think of you often every day, regardless of whether I write or not. The reason I have not written is because I have been on a 2 day trip.

Yesterday about noon I took off with Captains Younkin, Bowen, and Byers and we flew to Bastia, Corsica to pick up a crew of mine who had been forced down there. We went to Corsica via Sicily and Sardinia, however, we never landed at either place. Corsica is very picturesque. You can't say it's beautiful for it isn't, war has left its taint. But it was a beautiful place at one time, and the mountains are still very pretty. There is less over there than here, however. I hunted all morning to try to buy something for you — you know, just anything. Bastia is a large town, much cleaner looking than Italian towns are, and it was nice to hear French spoken instead of Italian for the change. French without the Arabs give a much better impression. These French looked like they bathed occasionally and tried to live decently which is more than I can say about the Italians or the French that I had seen previously. I took a few pictures that I hope turn out. Also some pictures of a couple of Islands in the Tyrrhenian Sea, also a couple of Vesuvius as I passed over it on the way back. The four rolls that I took while in Pompeii are being developed, and as soon as I can get them, I will send you a small print or so. It is impossible for me to get enlargements, and all I'm taking now are of the smaller variety. I'm going to try to get a small collection for you to see things that might be of interest to you. Of course, I wouldn't hint to you, but I am now using a Kodak "Bantam" and it uses film size FX 828. It's a sweet little camera — has a swell lens on it.

Honey, just after I had completely given up about getting the packages from you, they both came. I received the package with the fruit cake in it yesterday. All of the things are appreciated, the batteries, Arid, book, newspaper and all. The best however was the fruit cake. It was good. Note I'm using the past tense. My boys enjoyed the little that they got. Thanks a lot. Any reading material, like the Pocket Book Editions will be appreciated anytime. Not only will I read them, but everyone around here will also.

I'm sending Young and Threadgill's crews to rest camp this week. Young has had a mild sinusitis ever since he has been over here. I'm sending him more to give me a rest than to give him one for I'm tired of listening to him. Threadgill and his bunch are in serious need of a rest. They have been doing a lot of flying. Zraick is still his own egotistical self. I hear he is now in for his captaincy, which he certainly deserves 'cause he is just about the only Intelligence Officer in the Group worth his salt. Even so I still don't like him as a tent mate. His hands are no good to him at all — he can't even cut kindling. It's a good thing he's a lawyer where he works with his head and mouth for if he had to make a living with his hands, he would starve to death. Young is a nice boy, but selfish to the nth degree — not about going his half on something, but Young is always right regardless of whether he knows what it's all about or not. In the U.S., he was very popular with the boys, but over here it's quite the reverse. He is still a nice guy nevertheless, but I'll be glad when it's time for him to go home. Perhaps by the change in my opinions you probably think that I'm the guy who has changed. I may have changed some, no doubt. I certainly would have no "bedside manner" anymore. Guess I can develop it again, or at least hope so.

I sent the kids a present apiece today. Sent Kay a bear, and Carol a couple of plastic toys that will float. There was nothing more appropriate to send Carol — perhaps she can play with them while taking a bath. Had to send Kay the bear for it reminded me of "Panda." I shall never forget how much she cared for Panda about a year ago. However, it may be too young for her now. Of course, it is difficult for me to remember her except as I last saw her and impossible for me to remember Carol as anything except a poor, scrawny, little thing that was apparently half starved. Gosh, am I going to be surprised one of these days.

Honey, you had better have all of that female company that you are having now, for you aren't going to have time for them when I get back. I want to be with you for a long time without friends, foes, or kinfolk. Mentioning kinfolk — I received a letter from Mother today and she is quite down. Of course, I know Mother pretty well, but I do believe she is getting worse than she used to be. Maybe she is rid of Aunt Lis by now and if so (for Dad's sake primarily) maybe she will do better. My Dad is a real man — I admire him very much.

Well, honey, this has been a very rambling letter. If the censor happens to open this one, he will go nuts before he is finished with it — may not be fair to him but that's his tough luck. I have enjoyed very much talking to you tonight however — probably even a little but more than usual. I believe I can now go to bed and sleep just a little bit better than I could have, had I not have gotten all of this out of my system. Again thanks for the two packages. Kiss the children for me and remember that I love you more than anything in this world.

Goodnight,
Joe.

Mission #38: 5 May 1944
Target: Ploesti Oil Storage Tanks, Romania.
Results: Severely hit several large explosions caused by good coverage. 727th ship piloted by Lt. Mike Boyle and 1 from the 725th lost to flak. The Shilay-Lee caught fire, 8 crewmen, including combat photographer, became POW's in Romania.

May 5, 1944
Midnight

My dearest Honey,

There is very little to write about today, so I really shouldn't start an Air-Mail at this hour of the night but those V-mails just don't give me time to get started. So consequently I'll write an Air-Mail anyhow.

Honey, I finished my shower today. Took a nice hot shower too. Now if I so choose, I can bathe daily. That's really something, too. I've got it so it heats 55 gallons of hot water. Can fix it so it will heat 110 gallons just as easily and can do it in 15 minutes. However, 12 of us took showers today with only 55 gallons of hot water, and there was still some left over. It's really the berries. I'm quite proud of it (like a kid with a new toy). Now I won't have to take one of "those kinds" of baths all the time.

Kremers and Wagner dropped over today for me to take a peek at Kremers. He has Sciatic (inflammation of the nerve in your leg) and is having quite a bit of trouble. We are going to send him into the hospital in the morning. I hope it turns out that it isn't one of those tough ones. I believe the war is tough on us "docs" in this outfit. We are naturally the "serious" type and realize our obligations. I know the other four of them get down in the dumps too. Perhaps

McFarland does better than any of us, I should gripe.

I started enlarging my tent today, and honey, I'm going to have a dandy. Fixing it where it will almost be twice as large as it is now. And I'm going to "screen" in one side — then I will have a screened in porch, more or less. Of course I can't get any screen, but I'm going to the nearest British Camp tomorrow and get some stuff that is near to screen, if I can beat them out of it. They have a material that is a mesh cloth painted with airplane dope which lets light in, will knock wind out but of course let in no air. Nevertheless it should make me a nice sun-porch. No kidding, it is gonna be nice and roomy too. Then I will have a roomy tent, hot and cold running water, a hot shower 10 feet from my front door and a good cot to sleep on. I'm gonna confiscate some bricks somewhere and put in a brick floor in the sun porch part. All I'll have to do then is build me a beach chair.

One of my boys walked up to me tonight and said "Whatsa matter, Captain, don't you like to practice medicine?" Of course I don't really practice medicine out here to begin with; and secondly, I enjoy piddling with something for it makes the time pass so much faster. Today went by like greased lightning — although it's been a hellova day as far as the squadron is concerned.

Tonight I received two more air mail letters from you with pictures of yourself and the babies at Easter. Gosh, honey, you have no idea how much both of them have grown. Well, honey, it's "nigh on to" midnight. I'm gonna stop for today. I love you so doggone much, so please take care of yourself and those two babies for me.

Your Joe.

May 6, 1944
Midnight

My dearest Honey,

Just a note tonight. I have been a chaplain for the past 5 hours letting one of my boys get his troubles off his chest, consequently I haven't gotten to you earlier. However, I think in that 5 hours the guy feels better now, so I've accomplished my purpose, I guess.

Done quite a bit of real medicine today. Did an autopsy on a young Italian child about 15 miles from here this morning, which gave me an empty feeling.

This afternoon I built on my new house some more, however, it isn't finished yet by any means. I sorta hate to finish it for then I will have nothing to do in my spare time to help pass the time away. Guess this will keep me busy for a week at least if I try not to completely finish it too quickly. Got two V-mails from you today. Take care of yourself, and I love you.

Goodnight,
Joe.

May 7, 1944
10:30 p.m.

My dearest Honey,

Doing you bad again tonight. V-mails two nights in a row. I won't make a practice of it. Last night, it was Sprowls that wanted to get something off his chest, so he stayed till past midnight. Tonight it was Kavanaugh and he just left.

I am dead tired. I have finished (almost) my new "house." I laid a brick floor for it. Took me almost all day to make a door and hang it. It's very nice, and do I have room? I suppose my living quarters are now about 11 x 13 ft. — plenty of room. I've worked hard from 7 this morning until about 9 tonight, and my back is killing me, and my hands are sore. Guess it's from handling bricks. Got a lot more little things to do, and then I hope I stay right here until things are over in this theater. Got a V-mail and an Air mail from you today — out of paper again. Will try hard to write you a long one tomorrow.

All my love,
Joe.

May 10, 1944
10:30 p.m.

My dearest Honey,

About the time that you get this it will probably be your birthday or our anniversary. Whether close to one or another, I take this opportunity to wish you the happiest of birthdays and a pleasant anniversary. May I be by your side forever more on either occasion, so I may give you the same wishes personally. As far as the anniversary goes, on May 31, I will have completed four years of more happiness, pleasantness, and satisfaction than I ever expected to, or knew existed on earth. I am a blessed man. I hope your boxes that I sent you while down south have reached you and, if so, I hope you liked your birthday and anniversary presents. You are a lovely woman, and I am fortunate to have you.

My mess hall is almost through, we have done all we can to it. The Engineers started putting a top on it today. The officers club is coming along fair too. The enlisted men held a meeting a couple of days ago and elected me to sponsor them in building a club 40 x 75 ft.

I have just finished building me a 10 by 12 ft wooden storage room next to my tent. All in all you see I've turned into an engineer instead of a doctor. In importance, my jobs in this organization are: First, I am the squadron chaplain without a doubt. Second, I am a construction engineer. And third, I also dab in medicine occasionally. I'm busy as heck though, and the days have been passing very fast lately. Let's hope I stay busy and the days pass even faster. I am egotistical enough to know you miss me, but, honey, I'll bet you will never realize how much I miss you. That's the only thing that makes this war tough

727th Mess Hall

for me, and I should be ashamed to gripe about that. There are so many guys that are doing so much.

I finally got paid for last month today. There was an error on my vouchers, so I wouldn't accept any money until it was straightened out. You know the extra paperwork that entails. Anyhow I will send you most of it as soon as I can. I am going to try to go into town tomorrow, but doubt if I can make it. Use the money as you see fit.

You should see my left thumb and forefinger. I thought I was a pretty fair carpenter. I've knocked the nail off my thumb and got a large blood blister under the nail of my forefinger — hit my fingers more than the nails apparently. Some carpenter — I used to be able to do better than that. Italian nails have round heads on them, and it really makes nailing more difficult. (Gotta have some alibi).

Kiss Kay and Carol for me. Tell Kay to be a good girl and I will buy her a "trishickle" once I get home. I love her with all my heart.

Joe.

May 11, 1944
6 p.m.

My dearest Honey,

Today has been almost a typical summer day in the South. Sun shining brightly, hot, but not sultry. Too I feel about like I would down there — lazy, and with spring fever. We didn't fly today, so everyone felt rather lazy, I think. I piddled around here (my tent) some, took a shower, put a roof on my storage room and in general have just been taking life easy. Wanted to go into town, but I had no transportation, so I will have to put that off for a while.

Got to go over in a few minutes to have a meeting of the enlisted men and decide a few things about their club. Didn't hear from you today, but that means I will probably get two letters tomorrow which is O.K. by me. Take care of yourself. Tell the kids hello.

All my love,
Joe.

May 12, 1944
1 a.m.

My dearest Honey,

Perhaps you are wondering just what the heck I'm doing writing to you at such an ungodly hour in the morning. No, I don't think I'm nuts yet. It just so happened that I haven't written earlier tonight because I have been over in Lt. Roach's tent playing poker with him, Curtis and Bernstein. Since I didn't write last night due to sheer "sorryness," I thought I would write you a long letter regardless of the time of morning I started it.

Besides writing you a letter I'm sitting here drinking tomato juice that we got in our rations this week. Lately our rations have been much better. For about the last month, we have been getting one can of fruit juice per week. We have had cokes twice (one a piece of course), and now we can get some candy. When we first got here those things were unheard of. Yes, rations are much better and since I have worked on my tent and made it larger, etc, I'm living alright.

I went into town today primarily to get my rations and to send you some money, also get my watch that I was having fixed. Anyhow I got my watch and my rations but completely forgot to go by the Finance Office. I sat up there and knew there was something else I wanted to do but couldn't think what it was.

I went into the nearest small town today to see about a chair that I am having made. The Italian that is making it asked me to go to his house to see his "bambino" who was "malatta" — or baby who was sick. I went by but couldn't tell anything about it for I had no instruments, nor could I ask questions for my Italian is still as weak as ever. Anyhow I told them that I would "returne." So tonight I picked up one of the lieutenants who slings this lingo and went up to see the baby. It so turned out that the baby had bronchopneumonia, and I left them some medicine. Before I could get out of the house, I saw two women with two different diseases of the breast (all women over here have many small children — Mussolini's influence), and a man with a bum knee. They must have thought that I came forth to hold a clinic. For a few minutes I felt completely lost.

Got a V-mail from you today, which was written May 1. It was Monday and you were sorta blue and down in the dumps. You hadn't received any mail from me, and it was a bad dreary day anyhow. I'm sorry, honey.

Well, honey, I suppose it's quite bedtime by now and besides I'm pretty cold. The nights here are still pretty chilly, and I have no stove now since I have "remodeled." I feel less guilty now that I have talked to you tonight. I love you so terribly much, and it seems as though the time over here just drags by minute by minute regardless of how busy you try to keep yourself. Kiss the kids for me.

All my love,
Joe.

May 14, 1944
Sunday, Mother's Day

My dearest Honey,

Today was your day, and I was not there to do anything in particular to let you know what a swell mother you are to our children. Soon Kay and Carol will be large enough, so they can show you themselves. I just wanted you to know that Kay, Carol, and I appreciate and respect you as a mother.

The past couple of days have been somewhat less monotonous. Yesterday I got the framework made for my "beach chairs," and early last night I put the canvas on them. They are very comfortable. I almost feel guilty sitting in them and propping my feet on the window sill of my screen porch. My house is the envy of all my neighbors — it really is nice.

Last night Prewitt and his navigator, Ramsey, dropped in early in the evening to talk and stayed until 1 a.m. We made some coffee, cooked some eggs and toast to pass away the time. Also shot a lot of bull of course. Prewitt with his mild sarcasm is quite interesting. Ramsey is having a fairly hard time of it but never has said a word to me or anyone else. His nervous system just doesn't have the stability of some of the other boys. Consequently, to do the same job is much tougher on him, but he's in there pitching nevertheless.

This morning after take-off Young, Blackmon and I took some new bombardiers down to Bari. Blackmon piloted, and Young co-piloted. I think it's the second time Blackie has tried to take-off and land a ship since he has been over — poor guy. "Never say die Blackie" — carried to the nth degree. Young got back from rest camp the day before yesterday. I think his sinuses have cleared up some too. Enjoyed the trip down and back, however the air was sorta rough.

By the way we flew down in "Hop Scotch" which was McAllister's old ship — remember. Upon the directional compass — taped up there — is the good luck bird that you gave Young while we were in Lincoln that day. I suppose he must have put it there while flying over here. (I still have mine also although it shows wear.) Anyhow we flew to Bari and dropped this fellow off, had a very nice lunch and came right back. We managed to land just before the flights landed. In fact, they let me off at the end of the runway.

Tonight I went with Young, Zraick, and Evans to the outdoor picture show up at Group Headquarters. It wasn't very good — John Carradine and Margo in something. Had a bunch of old film shorts that were at least 3 to 10 years old. Every time I go up there I promise myself that I won't go back. Then within a week or so, someone asks me to go to the show with them and I weaken.

It's getting mighty late and I want to drop Mother a line also. I love you so very much and hope that we shall never spend another Mother's Day apart.

> All my love,
> Joe.

> May 15, 1944
> 11:15 p.m.

Dearest Babe,

Got two letters from you today — one air mail and one V-mail. I haven't done much today. My ambulance is off the line for the first time in two weeks, so this morning I got in it and went to town. Went to the Red Cross for coffee and cake, and to the Finance Office to send you $150.00, which I didn't forget this time. Let me know when it arrives. Got me a shoe shine, bought some matches and a couple of handkerchiefs and then came home. Also went by the hospital for a few minutes to see Kremers, who isn't getting much better.

Came back from town about mid-afternoon and found out that I was supposed to go to a Wing meeting of Flight Surgeons tonight, which I did. Saw Koenig while down there, and he is getting along fine apparently. However, he too is getting that same "don't give a damn" attitude that now I have had for quite some time. Honestly I've gotten where I don't care what any of the "higher ups" think or do. I do just about what I want to when I want to and let nature take its course. When I leave camp I don't tell or ask anyone I just go — when I take off I tell them I'm going and don't ask. Sooner or later I'm gonna catch the devil for it probably, but I don't care very much — I have practically but one thought and only one ambition and that is to get back to you.

You asked today about Young and me. No, we haven't completely called it "quits." However, I don't have the admiration for him that I once had, and he has been quite a problem for me directly and indirectly. The boys come to me with their gripes naturally, and Young is responsible for many of them. Consequently, I have made it a policy to steer clear. As far as tenting with him, I'd rather tent with the devil himself. He is impossible for me to live with as I found out in Lincoln. So I have tactfully evaded that ever since I have been over here and intend to continue doing so. For the first time in my life, I have found that I enjoy being left alone and living alone. Perhaps I have more time to think of you that way.

I put up my mosquito net today. I was supposed to put it up exactly one month ago but had never found time to go to the trouble. For the last couple of mornings, however, I have had a chance to sleep late and couldn't for the flies, so I decided to eliminate them. Therefore, I probably won't get a chance to sleep late for a long time now.

Well, honey, I'm gonna cut short tonight and quit and go to bed. I'm so damn blue and down in the dumps, and tired of it all. Wish we would move again, so I would have something urgent to keep me real busy for a while. Will do better tomorrow. So goodnight for now. I love you with all my heart, body and soul.

Joe.

May 16, 1944
Wednesday, 11:20 p.m.

My dearest Honey,

Sorta hit the jackpot again today. Got 2 letters from Sis, one from Mother, one from your mother, one from Lt. Love who is on his way home (I sent him — he had a ruptured intervertebral disc) and last but far from least, one from my Honey.

The day has passed away comparatively fast, which is far too slow at that. Got up late, about 8:30, went over and had chow and shot the breeze with a few of the boys, then found out I had to attend a meeting of C.O.'s, executive officers and flight surgeons at the Colonel's place. Went up there and as soon as that was over Young, Massare, McFarland and I took Sands by air down to Bari. We had lunch at the airport (they serve delicious meals there in comparison to ours) and took off for here. I had a good time for I co-piloted for Young. Did all the flying both ways and co-piloted for him while landing, which is only the second time I've ever done such. Got back here about 3 I guess and have been "piddling" just about ever since.

In your letter today you told me that we owe only $3,000 on our home. You know to me that sounds pretty good, particularly considering the fact that it cost around $7,000 at the beginning, didn't it? That being the case, we have saved quite a bit since we bought it. You know honey, if you ever get down to

Dearest Babe,

Helena, I wish you would open the safety deposit box and see how many bonds we have. Personally, I have forgotten. I don't know anything about our business anymore. I haven't the slightest idea whether you have $500 in the bank or $1000, nor do I care — just so you have enough cash an emergency.

In your letter today you mentioned about Carol sitting on the "little brown seat." It is impossible for me to realize that she is that old and large. I can't even imagine that she can sit alone. I'll bet that's a sight — particularly with Kay's help and comments. What I'd give to see them!

Well, honey, it is time for me to say "nite-nite" — "my eyes are sheepy." Let's hope tomorrow and the remainder of tomorrow's pass even faster than today. Tell the kids hello from Daddy. I love you with all my heart.

Joe.

May 18, 1944
About 10 p.m.

My dearest Honey,

Perhaps you won't be able to read this letter. I'm propped up in one of my beach chairs on my "screened in porch," and I thought I would write from this position rather than clean off my table. The table has needed cleaning off for over a week, but I'm too lazy. Yes, I'm quite comfortable.

Haven't done much today. I finally got me some screen (real screen wire about 4 or 5 yards) and have screened in my porch — it's very nice now. I can have light, air, and keep the flies out at the same time. You know, screen wire costs $5.00 per yard over here and nails cost $1.00 per pound. I am sitting here in my foot warmers and the army slides you sent me and am very cozy. After I finish writing to you, I am going to go out and take a bath in our private shower and then head for that well-known Land of Nod.

Had a chance to go back to Corsica for an overnight stay again tonight but thought I would give the ride to Wagner instead. He hasn't been on a long hop since we have been in Italy, I don't believe. Since Wagner went I thought I'd better stay here — that makes me hold the group, Kremers squadron and mine. Wagner will be back sometime tomorrow, and then I will have no job at all.

You asked about Prewitt in your last letter. Prewitt is doing fine. I'm hoping that I will be able to send him home in 4 to 6 weeks. He is just about first on my list to try to get back. His whole crew is O.K. for that matter — Wheeler, Ramsey and Haldane. I'll be glad to be able to get these old crews back — I know them too well. Most of your other acquaintances are O.K. too — Young, Zraick, Evans, Threadgill, Kavanaugh, etc. they're a good bunch.

I'm sending Coulter and Formanek to rest camp early in the morning. Sprowls and his crew will be returning. Think I will send myself to a rest camp in the next month or so. Won't that be something — "Flight Surgeon Sends Self to Rest Camp." Just the same, I believe I will. Too I'm gonna take even more time off.

Most of the fellows are still in pretty good spirits and morale is good. Speaking of morale — I received a letter from Mother today, and she seemed to be in better spirits. Her whole letter was about her chickens, which is much better than her usual letter about her ill health.

I received the pictures of the folks, Sis and Royce and the children today. Thanks a million. They all look quite good to me. By the way, I have acquired some larger snapshot negatives of me. Large enough so you can tell who it is. As soon as I can have some prints made of them, I will send you a couple. Well, honey, I believe I will be getting up pretty early in the morning.

Goodnight.

Next day, Friday, 11:05 p.m.
My dearest Honey,

It looks as if I goofed off on you last night doesn't it? Well, I will at least finish this letter to you tonight, if I don't do anything else. I received a V-mail from you today and also one from Sis.

Today, finally my authorization for Flight Surgeon's Rating came through, which means I should be able to send home $40 more per month. The thing that gripes me is that it is only retroactive to March 13 instead of December 9 as it should be. That means the government gyps me out of about $150, and I don't like that even a little bit, but I guess there's nothing I can do about it. At least it is a raise from now on — which means I make about $480 per month — not counting the $21 deduction for my meals and the $7 deduction for insurance. Really I can do very well on $5 per month I guess — not that I have ever gone to the trouble of keeping up with it. I did mean to have a sizable check to send you of back flight pay, but apparently I won't get most of it. Whatever I do get, I will send you to use as you see fit. I suggest if you have no use for the extra money, put the surplus in bonds. They will come in handy when they begin to mature for it's my opinion that there will be some relatively lean years ahead. But always keep yourself a surplus in the bank of $500 to $1,000 if possible. Just telling you how to run your own business, honey — that's unnecessary, and I know it, and I'm sorry.

I had a nice thing happen to me today. One of the boy's mothers sent me a package in care of her son. It contained 4 sticks of chewing gum and two bars of candy. I really appreciated it although he felt like the devil giving it to me. By her writing, I gather she is about middle aged. It was nice of her. The boy (her son) just got a fractured skull a couple of weeks ago. He is now back on the job with me. It was Williams, my clerk.

Tonight Young, Zraick, and myself got in the ambulance and drove to town to see a picture show that is supposed to be good (Danny Kaye in something) anyhow we did leave in the middle of it for the sound effect was so bad you couldn't understand a word that was being said. Before the picture they had an Army Orchestra on the stage, which was very good. We all enjoyed it. After leaving the show we went by the Red Cross for coffee and doughnuts and then came on home. We are 15 to 20 miles from town and about 8 miles of it is the roughest road (gravel) that you have ever experienced — just like driving a Model T over cotton rows.

To see you, Kay and Carol is almost too much to think about. Tell Kay that Daddy wants to play with her when he gets back, too. In fact, my hands are itching for it. All my love to you — the best woman, mother, and wife there is.

Joe.

May 20, 1944

My dearest Honey,

Today I received 2 Air-Mails from you. Both nice as always and more than welcome. Haven't done very much today. There was no flying and that usually cuts down on my work considerably. I have been "spring cleaning," dusting, etc around my tent again. Finally managed to get all the junk off my table — now I can find room to write. Also cleaned the box off that I use for a clothes closet, built a door for my clothes closet in an effort to keep out some of the dust.

Honey, in one of your letters today you gave me a report of our financial status. I think you are doing swell. I have only one suggestion to make. Don't buy anymore bonds until you have built up the balance in the bank to around $500.00. That will just make me feel like you are more secure.

It was raining this morning when I awoke. It looked pretty most of the day however, but now it is raining again. This is nothing like it was this winter though. Our camp is in the middle of a large wheat field. Instead of mud and muck everywhere there is nice green wheat from knee to waist high.

I've been here sitting in my relative mansion watching Evans, Young, and Zraick trying to build them a front porch on their tent. As usual, Evans was doing all the work. I'm still glad that I live alone — not very choosy am I?

Take care of yourself and remember that I love you with all my heart.

Joe.

May 22, 1944

My dearest Honey,

I believe six months ago was that fateful morning that I stood there on the drive of F.A.A.F. and watched the grey Chrysler pull away from me in the semi-darkness with my life in it. Six months is over, and I wish there would never be another six months, but I'm afraid that is asking too much.

I received a lovely letter from you today. A descriptive one which hit me just right — though I know you were in a melancholy mood when you wrote it the night of Mother's Day. I appreciate the nice thoughts expressed about me and us — my feelings exactly. So glad that you got some flowers for Mother's Day. It's the least I could do. Kay and Carol should be getting their toys soon. Hope they come through.

Enclosed I am putting a picture of me and the motorcycle — taken when I first got here and was allowed one. I still have it, by the way, but it's hid. So far I haven't been able to sell it 'cause they are outlawed in this outfit. I'll tear it down and play with the engine before I turn it in after all the work I put on it.

I love you,
Joe.

May 23, 1944
11:20 p.m.

My dearest Honey,

I've had company all afternoon, evening and night. Consequently, I'm just now getting around to writing you. Young, Sprowls, Whitmarsh, Zraick, Bernstein and Roach have, one or another, been in my tent since noon. I had intended writing you early and either going to the show or reading tonight but such has not been possible. Roach finally just left, so now I'll just write a few words and then go to bed.

I am enclosing another snapshot of me standing in the front door of my dispensary. The posters on the door are not very nice, but they are somewhat effective and that's what counts. (Note I said somewhat). All three of them are some lesson in venereal disease control. That is quite a problem over here for the percentage of infectivity seems to be much higher than in the States. G.C. does not react favorably to the "sulfa" series of drugs at all over here. In the States, we got quite a high percentage of cures. This picture was taken by a very small camera. I have another enlargement that one of the boys did for me that

Joe in the door of his dispensary, featuring VD posters.

is about 9 x 11" but the posters are much too plain. This picture was taken a couple of months ago while I was in South Italy.

By the way, got me another wrist watch today. It is a BOUET. Never heard of one before but it is a very good watch. Bought it through the Army. They will only sell them to Medical personnel. Cost me only $17.50 but probably cost about $60 to $75 in civilian life. It is a Swiss, water-proof, shock-resistant, watch with a sweep second hand. Looks very nice too. I have put my two Hamilton's away now.

It's been much colder today than it has been for quite a while. I couldn't stand it any longer, so I went and dug my stove out again and put it up this morning. I have had a good hot fire going ever since, and it really feels good. Well, honey, I'm gonna stop short tonight. This isn't much of a letter but it is better than a V-mail. Even so I love you just as much.

All my love,
Joe.

May 25, 1944
Thursday, 11:30 p.m.

My dearest Honey,

I skipped writing to you last night. First time in quite a while though. This isn't going to be very long either — just a note for I just found out that I gotta arise about 3 a.m. tomorrow. Sounds like private practice, doesn't it? Only now I know when I'm going to have to get up and then I didn't. Didn't get any mail from you today but yesterday I received 2 V-mails from you. I think lately the air mails have been coming as quickly as the V-mails. The mail service up here has been very good. Much better than before. I'm so glad to know that 2 of the boxes have reached you. I sent three at that time, however, so you should have gotten another one by now. If not let me know. There is another bed spread in it which I think you will like also, plus some more junk.

I've been piddling some more today. I haven't really accomplished anything in such a long time that I'll never be worth a damn anymore. Have been looking over the area some today and have found some discrepancies that I think we can improve upon. Trouble is they all entail quite a bit of work. Guess I will arrange for some Italian labor.

I've got to go down to Bari for a couple of days at the beginning of next week. I'm going to try to get Blackie to run down with me if he can. I gotta meet some Medical Disposition Board with a couple of my enlisted men.

Have been with Wagner the whole afternoon and evening. He helped me look around. I wanted to show off my new kitchen — we have moved into it now, and it looks like some swank hotel kitchen. Away out here away from civilization with nothing but tents around it really looks funny, and I'm real proud of it. I'll guarantee that there's nothing in the Air Force that will compare with it in field conditions. I have redone my plumbing too. I have a 450 gallon aluminum reservoir (an airplane wing gas tank) for all my water systems now. The other gasoline barrels were rusting keeping my water dirty all the time, so we changed yesterday.

Well if I got in bed at this minute I would have only three hours of sleep, which isn't enough for me anymore by any means, but I had better try to get some. There is nothing that I actually own that I wouldn't give to be with you. Until then I will love you with all my heart as I do now.

Goodnight,
Joe.

May 26, 1944
Friday

My dearest Honey,

Went to the picture show with Evans and Roach tonight. It again was so terribly rotten that we left at the end of the first reel. "Moon over Vermont" it was called and did it stink. It seems that we have run into a series of terrible pictures. Every time I go it seems to be a little worse than the one before. I know that tonight it sounds like I'm a terrible old grouch. Oh well, griping to you isn't going to help a lot — just make you more uncomfortable — which is very selfish on my part.

I received a V-Mail from you today. You had received the other box. Glad it came through. Which bedspread do you like the best, the heavier one or the light one? That's the only heavy one that I have ever seen over here. Glad that the lace blouse fit and hope the bras did too. If I remember they buttoned and if I remember again yours never did. Funny thing, when I bought those four I completely bought the store out. That was the only thing that was left in the whole store. Sorta hard for you to realize a store staying open just for the possibility that it might sell 4 brassieres, isn't it? Anything may be seen in Italy.

Tonight is the first real clear night I think I have seen on this side of the Atlantic. There isn't a cloud in the sky and the 'milky way' is filled with billions of stars. Not very chilly either. I have my stove going, but I could get along without it tonight and be almost comfortable. In four minutes it will be midnight again. Then it will be your birthday won't it. I wish you a Happy Birthday, but I'm not going to take this opportunity to do it. Tomorrow when I write I will try to be in a better frame of mind so I can do a better job of it. Tonight I will just say goodnight and pleasant dreams, and that I love you.

Joe.

May 27, 1944
10:50 p.m.

My dearest Honey,

To me, today should be a holiday. Even I have celebrated it partly as such. Tonight we opened up our new squadron Officers Club. I went up early and had a few drinks with the boys. When I left a few minutes ago, I left over 98% of them completely saturated in alcohol of one description or another — vino, cognac, Italian whiskey, grain alcohol, etc. I even saw one quart of Scotch and managed to get one drink out of it. Due to the fact that I guess I'm getting old, the party got too loud and noisy for me so I came home. I dare say it will last until the wee hours of the morning, which is O.K. by me. We aren't flying tomorrow, so it's alright if the boys want to get drunk. They probably won't be able to tomorrow night if I can help it.

As I started this letter, today is a holiday to me. I am thankful for May 27, also for May 31. I began to really live — to really enjoy life, to get more out of life than I thought it held four years ago. I owe that happiness to you for I had never had it before. It is worth much to me, however, to know that you are there taking care of things, looking forward to that day as much as I. I truly have much to live for and much of a reason to be where I am. The happiness that I have had in the past four years must be preserved. Thanking you from the very depths of my feeling for this happiness is much too formal. I only wish there was some way I could express it.

In conclusion I want to wish you the happiest of birthdays. I hope from the bottom of my heart that I am with you on your next one. Good luck in the coming year and God bless you. I love you so very much.

Joe.

May 28, 1944
Midnight

My dearest Honey,

It has been a long time since I've written one of these short ones. It's so late this is the best I can do tonight. Went to town today with Evans to get my rations and to go to the picture show. Show was terrible as usual.

After returning I began getting my clothes and myself in shape to go to Air Force Headquarters for a couple of days. Blackie, Evans, and I are leaving for there early in the morning. Blackie is going down for a brace for his back. Evans is going down just for the trip and to take a couple of days off. I'm going down on business for the group to present some men to the medical board. We will be there two days. Had much to do to get myself into shape. Shine shoes which hadn't seen polish for months. Wash a belt which hadn't been washed since the States. Wash a cap which has never been washed before. And find a complete uniform that is wearable. That's getting harder and harder. I haven't replenished at all, and I'm not going to as long as I can help it.

Probably won't get to write you tomorrow night but I will the next. Take care of yourself. I love you very very much.

Joe.

May 30, 1944

My dearest Honey,

Got two letters from you when I returned today — one was written 1 month ago and the other was written May 19. As I have told you previously, I had to go to Bari for two days and a night to meet a Disposition board about some boys in the group. Blackmon had to go down for a corset (to try to help his back), and Evans just went along. We had a nice trip and enjoyed ourselves.

Went to three picture shows while down there — all were good. That's damn unusual. Don't understand why all of ours are so bad up here. Saw Paulette Goddard and _____ Mc_____ in "Standing Room Only" (good); Robert Young and Marsha Hunt (I like her) in "Joe Smith, American" (good), and another one with Robert Young about orphans in England which was very good too, but not as good as the other two. I'm getting pretty good remembering about picture shows, aren't I? By the time I get back to you maybe I'll be able to talk to you about a show after we return from it that night. There's no doubt I'm improving.

Last night we went to the Air Force Officers Club and it was really nice. They had a four piece orchestra and an Italian girl that sang as good as any American I ever heard. She sang in both Italian and English. She had the most perfect control over her voice I have ever heard and she hit her notes right on the button. I really did enjoy listening to her. The funny thing about the orchestra was that they would swap instruments with one another every few pieces, and they were just as good.

I also got over to the Officers Red Cross, which I never pass up there, and got some ice cream. Two helpings yesterday and only one today. So maybe I will be able to curb my manners somewhat when I see you after a few months. We will have to take it gradually though or it may throw me into shock.

One of my old crews, Larson's, finished up his missions today. I'm grounding him as of tomorrow, and will send him home as soon as I can make the arrangements — lucky stiff. Of course, he deserves it and then some. I have some more who should finish up soon. Larson doesn't know I have grounded him and his crew — they think they are going to have to fly a few more. Better to do it this way if possible, so they won't sweat out the last few so much. I'll bet he will get the surprise of his life when I tell him tomorrow.

Being away from you gets more and more difficult. Some days I can do nothing but think of you.

I love you,
Joe.

May 31, 1944

My dearest Babe,

Today is that day. The red letter day of my life. In fact, I know that it is the red letter day of both of our lives. I've thought of today four years ago many many times today. I would look at my watch and try to think what was happening to us exactly four years ago.

This morning about 9:30, I had the prewedding fright. About 10:30 we were getting ready to go there, and it really dawned on me that I had a part to hold up for the rest of my life, and how important to both of us especially that I hold up my end of the bargain. About 11 or 11:15, the fright was still there as you remember. We are lucky that I hit the right finger. I still believe the preacher said "Love, Honor and Obey" in spite of your protests. Since it never has been a real problem, I suppose we can let it go either way. Then within 15 minutes I heard my father-in-law threaten me for the first time (and last, I think) — "You'd better be good to her."

Then we took off. Beginning to live, weren't we? Wondering whether it would be a successful venture — now knowing how damn successful it has been, and how successful it will continue to be. Having faith, one in another, has made it so. Then we had to turn around and go get your coat. Then stopping at Three Corners Drug Store at Waldron and Jefferson for Dentyne gum and on out highway 51. Remember almost running over the horse and wagon? Yes, I know you do. Then stopping at Jackson to fill with gas. We ate a bite there too, didn't we? I don't quite remember. Possibly we didn't, but you will have to admit it didn't take me too long to learn my first lesson — when it's eating time stop and eat whether hungry or not. Then on to Brookhaven and John's and his remark — "You have just done the best day's work of your life." At the time he said that, John didn't have the slightest idea that he had just made the most correct statement of his life. I will always be grateful to John for that remark alone. Then on to New Orleans and how uncomfortable you were while I was spending all that time looking for Frank. I didn't use a lot of sense, did I? But I learned a little bit more — remember the Jung? — It was a happy day, and I am grateful for it. To think about it in detail makes me happier than I have been in quite a while. You have been the essence of my happiness ever since.

Before I married you I thought that medicine would be my real life. How foolish I was? Here I've been away from medicine completely for months, and I don't even miss it. I've been away from you for six months, and I'm nuts. My real life is my life with my wife and children. I don't know whether I can tell you how thankful I am for our lives together, how grateful I am that our married life has been so harmonious and successful. We have been terribly lucky — we haven't had a real set back since we started. I hope our blessings continue.

You have been largely responsible for all of the above. You have been the most excellent wife and mother I have ever seen. The man that couldn't make a go of it with you would be a fool and a ham. I realize all of this, and realize how fortunate I am. So I'm taking this opportunity to again tell you how grateful to you I am, how proud of you I am. I know of no better time to tell you than on our anniversary. Nor of no better time to tell you that with all my heart.

All of my love,
Joe.

Joe and Babe's honeymoon in Mexico City, 1940.

Dearest Babe,

Summer 1944

Rome was liberated in the beginning of June, but the 451st continued to bomb Romania, Hungary, northern Italy, and France. Each crewman was required to fly 50 missions in order to return to the United States. During the summer of 1944, Joe had to say goodbye to the original crews that had completed their missions. These were Joe's friends, whom he had known since his training days in the United States.

* * *

June 1, 1944

My dearest Honey,

Today has been one of the laziest and longest days since I've been in the Army. Lazy because I haven't turned my hands to a thing, long for the same reason I guess. We didn't fly today, and therefore I had little to do.

Got a letter from Burr Craycroft yesterday. He is as you know in England and living like a king apparently. Says he hasn't been "under canvas" yet and for the time being lives in a mansion with a private swimming pool, tennis courts, etc. Sounds like it must be a rugged war up there. Says a W.A.A.F. even serves him coffee or tea in bed. Tough, ain't it?

Today was pretty hot. Reminded me a lot of Dixie, however it is somewhat drier. Didn't hear from you today but am sure I will tomorrow. Remember today four years ago? The coffee shop, the curb market, the gardenias, the big turtles, the smell of the market. I love you with all my heart.

Joe

June 3, 1944
10:30 p.m.

My dearest Honey,

Today hasn't been quite so monotonous. Young and I flew down to Bari and carried three other officers with us. I had someone I wanted to see at headquarters and they did also. Enjoyed the trip and the lunch that we had at the transient officers mess down there. We only stayed down there for a couple hours.

Did I tell you that Young received the Distinguished Flying Cross a couple of days ago? Evans is also in for one and should get it any time now. Jones, who was more or less an "eight-ball" back in the States, got the Silver Star which is a very high decoration. Those are the three highest decorations yet received in this squadron. However, more will be coming. Of course, all the boys flying on missions have the Air Medal plus a few Oak Leaf Clusters. They are still doing a wonderful job, but it is now telling on them. I'll be glad when the old crews finish up, and I am able to send them home. Shouldn't be so terribly long.

Starting another building project. Zinn, Reese, and I are building a stone house about 50 feet from the dispensary. Not that I'm not quite comfortable as it is, but there I will probably be even more comfortable, and it will be something to occupy my time and mind.

Sent you and the babies a little something from Bari today. The babies some more dolls. I'll let you find out what I sent you when you receive it. It's a little different this time. Something tells me that my honey is working maybe too hard out in the yard. I know that cleaning out those shrubs is real work. Remember, I did it once.

The colonel from the air force (Air Force Surgeon) came around today to inspect my area. Seems he was very pleased with it and somewhat in awe of all of our contraptions. Honey I haven't got anymore writing within me tonight, so I'm gonna quit. Tell the kids that their Daddy still loves them, and tell Kay that she need not worry that I will buy her a "trishickle" when I get home. If I can't buy her one, I'll make her one. I love you with all my heart and can hardly stand being away from you. Goodnight my honey.

Joe.

THE STARS AND STRIPES
CASABLANCA DAILY

Vol. 2 - No. 25, Tuesday, June 6, 1944   One Franc   Daily Newspaper For U. S. Armed Forces

ALLIED FORCES OCCUPY ROME

June 6, 1944
10:45 p.m.

My dearest Honey,

I did you bad yesterday and didn't write but this time I had a good excuse. Been sicker than a dog for the past 48 hours — can't make anything stay up or down. Haven't eaten anything yet, but my gut feels a little bit better — it's still pretty sore though. I'm not the only one — we must have gotten hold of some bad food or my mess personnel might have gotten careless in handling it. I guess we will all live though.

Young had a wreck in his jeep a couple of mornings ago. I took 7 stitches in his forehead just at the hairline. Today I removed all the stitches but two. It looks pretty good and should leave practically no scar at all. Kremers has gone back into the hospital with his sciatica. He has since been transferred to another hospital. I wouldn't be at all surprised if his days of combat are over, and he would be coming home pretty soon. This is just a guess of course. For his sake, I hope he doesn't come back for I see no reason for his leg to improve any here.

Tomorrow I am taking Blackie into the hospital again. The third time since he has been over. He says he is ready to go home this time, but I'll have to see it before I believe it. He has been replaced in the group, however, by a "yes man" of the lowest calibre. So he has no particular job over here with this group now. I certainly feel sorry for him. He has, in my opinion, gotten a raw deal. He should get an E for effort anyway. Also gotta take a couple of my boys down for consultation along with Blackie. One of them is having trouble with his eyes, and the other one has suddenly begun to get deaf due to no particular reason at all. I'm getting more screwy things lately.

By the way, I heard today in a roundabout way that I was being sent in for the "Soldiers Medal." If I get it, I will send you the citation so you can read it. Had I known it was going in, I would have put a stop to it, but I'm afraid it is out of my hands now. At least it will look good on my Army 201 file.

I haven't gotten a letter from you of any sort in quite a while. Guess all the hullabaloo in this theater is holding everything up. Enclosed I am putting a Yugoslavian bill to aid our collection. This 500 dinar note is worth exactly 50¢ at the present rate of exchange. This bill has quite a history behind it, so don't lose it. One of my good friends gave it to me.

Well I hear that I am getting up at 2 or 3 in the morning, and it is now 11:15. That being the case I don't have a heck of a lot of time left for sleep. You know four years ago tonight, you and I were either at Monterrey or Valles — remember? We will have another honeymoon one of these days. One just as happy as the first one if we can afford it. Well, if we could afford the first one we certainly should be able to afford our next one. Goodnight darling, I love you very much.

Joe.

THE STARS AND STRIPES
MEDITERRANEAN

Vol. 1, No. 175, Wednesday, June 7, 1944 · DAILY EDITION · TWO LIRE

FRANCE INVADED
Beach Defenses Pierced 10 Miles

June 8, 1944
3:30 p.m.

My dearest Honey,

Yesterday I took Blackie down to the hospital. He took all of his stuff with him this time. He is whipped, beaten, defeated and finally ready to call it quits and return to the States. He is really down. I certainly felt sorry for him and hated to bid him goodbye. I wasn't the only one that hated it either — everyone did. I suppose he will be on his way home pretty soon. He told me if he got near to Memphis, he would call you up.

Upon returning here this morning I had two packages, a V-mail, and an Air-Mail from my love — also a V-Mail from your mother and a letter from your Dad. Was glad to find out everyone was doing O.K. for lately mail has been very irregular. Thanks a lot for all the nice things you sent. I'm really eager to open up a can of those sardines. The knife sure will come in handy too. I had one but lost it about a week ago. The Reader's Digest will be read and digested by everyone around. They are prizes. Everything was nice and arrived in good order.

Thanks too for the pictures of Kay, Carol and Janie. That Carol looks like she is a giant. I wish I knew her — I just can't feel like she is ours, which isn't right at all but which will

Carol with Janie, 1944.

221

right itself when I'm around again in 10 minutes. The picture of Kay laughing is priceless and precious. That's one of our better pictures of her. Gosh, but she is growing by leaps and bounds too. If there had only been a picture of my Honey enclosed, the collection would have been complete.

I am glad to hear that the box arrived with all the jewel boxes. I hope they were in good shape, not scratched and not warped too much. Was the blanket still wrapped around them? Has your chest arrived yet?

Bought me a new summer cap while I was in Bari. My other one was just so filthy that I've got to throw it away. Don't intend to buy anything else until I have to though. I haven't sent you any money this month. Don't think I am cause I'm gonna get me up a little reserve. Since all the new fronts have opened I may need some cash any time — one can never tell. After I get a little reserve, I will again send you my paychecks for I have no need for them.

I saw Kremers while down at the hospital. He is still having quite a lot of pain in his leg, and so far they haven't found sufficient cause for the reason. Don't know whether he will get back to us to stay or not. I hope he doesn't for I believe his leg is going to bother him for quite a while, and this is no place for a man that is ailing. Too Kremers has a pretty nervous disposition, and I don't think this has done him much good. He is a nice guy, and 6 months of this is enough for him. I hope he is sent home. He is too old for this racket anyhow.

Wagner asked me a couple of days ago if I wanted to go on 10 days detached service to one of the Station Hospitals. I refused, so I guess McFarland or Quinn will go. I don't want to get that near medicine again until I'm ready to stay in it. That would sorta be like passing through Memphis on a train and being able to see you but not being able to get off. Consequently, I think I will stay out here in the field, and look after my and Kremers' squadrons.

Honey, I know since the invasion of France started you and your Mother and Dad are worrying like everything about Hugh. I wish there was some way that I could reassure you. Words concerning such are always in vain. However, apparently, that operation is coming along as well or better than was to be expected. The men up there will probably get rest quickly, so maybe the going won't be too tough. Hugh certainly should have had enough training to know how to protect himself. If none of these thoughts are of any help, then just have plain faith. That will probably be of more help than anything. Again there is no need to worry about me 'cause I'm O.K.

Since we have taken Rome now, I've got to figure out some way to get up there sometime. Don't know how I'm gonna work it but give me time. They surely did get around that hurdle in a hurry. Well, honey, thanks again for the packages and the pictures. Tell Kay to be a good girl — take care of yourself.

All my love,
Joe.

June 9, 1944
12:15 a.m.

My Honey,

Today has been the usual. Nothing very exciting for which I'm thankful. Excitement is something that I don't crave. I just like to see things go along slow and easy. This morning I went out with Zinn and Reese and helped to tear down an old German barracks. The three of us are having a stone building built, and we needed some tile for the roofing and some lumber. It gave me something different to do, so I enjoyed it. I enjoy doing stuff like that because it helps to pass the time. Too it amazes "the boys" so. They can't imagine a doctor being able to do plain old manual labor — sorta tickles me. I don't think they would be too surprised at what I would do now. I've done so many different things.

After the planes got back from their mission and we had supper, I got hold of Evans and took him into the hospital to see some of our boys. Then after making the rounds he and I went on into town to see the movie. It was Judy Garland and Van Heflin in "Lily Mars." I had seen it once but didn't remember too much about it so enjoyed it again. Then we came back to my tent and had a couple of drinks, and he has now gone home. Combat is sorta getting him down, so I try to spend as much time with him as I can. He's a good boy. He knows his responsibility and doing an excellent job of being C.O. The conscientious guys are the ones this stuff hurts. The happy-go-lucky ones and the egotists take it very well. When Evans finishes his missions, I will hate to see him go 'cause he has learned a lot and has learned to put himself in the other fellows boots. That's worth plenty. I've learned to like him quite a bit. Too, he is very easy to work with. Through him I could almost run this outfit if I wanted to. He has been very considerate of my suggestions — he doesn't interfere with anything I do or start, and I am able to be my own boss completely. So as far as Army life goes, I suppose I'm pretty fortunate.

The nice thing in my life right now is the knowledge that you will be there waiting when I return. I've rambled enough. Goodnight honey.

Joe.

P.S. Glad to know that you received your chest also.

June 12, 1944
7:30 p.m.

My dearest Honey,

I haven't written you in two days. There's not a lot to write about. Things are about the same. The old boys are "sweating-out" their missions now like nothing you have ever heard or seen. Most of them will finish up within a few days, and I will be so glad to see them finish and go. They have certainly earned it. So far no one has finished but Larson and his crew. Before the week is gone a number of others should be done.

I haven't heard from you in quite a spell. Got a letter from Mother and a note from Burr Craycroft today. Every time there is a new offensive somewhere, it really does mess up the mail situation. I'll probably hear from you tomorrow.

We didn't fly today so this afternoon I went into town and saw a stage show and a picture show. Both were good. The stage show was a performance by Arabians. The Arabian dancer that they featured was quite a dancer if somewhat on the vulgar side. She was unique to say the least. The picture was also good. "Jack London" was the name of it. I don't know who the leads were, but they were someone I have never seen before and both were good. The guy that portrayed Jack London was especially good I thought and somehow reminded me of Spencer Tracy even though he didn't look like him at all.

I hear that General Twining who is C.O. of the 15th Air Force will be around tomorrow to inspect the Group. Colonel Eaton has already told me that he was gonna guide him to my area for the sanitary inspection, so I guess I had better get out in the morning and get things done. It's a left-handed compliment in a way by the Colonel, but I had just as soon see the General go elsewhere.

I've got some good news. I'm going to put my dispensary in part of the new stone building. It is 25 x 30 feet which is much more room. Our stone house should be finished pretty soon. Boy, it's gonna be a better winter this year if I'm still in this theater. I have a couple of 64 exams hanging over my head right now. I wonder if I will ever really be able to get back into the harness again.

I still love you, want you and need you as much as ever before.

All my love,
Joe.

June 14, 1944
Wednesday, 2 a.m.

My dearest Honey,

Well I have some news for you for a change. This evening I found out that there was a ship leaving here early in the morning for Egypt. So I immediately started pulling strings. I didn't have to pull but very few (tell Evans that I wanted to get on that plane if possible and ask Wagner if it was O.K. with him) to get myself on board that plane. I'm more or less thrilled for I've always wanted to go there. I had to get quite a few guys up at group headquarters to get some orders cut for me. Anyhow in the morning at about 5, I'm supposed to start out with a crew from the 726th. I'm pretty thrilled over the prospects. The heck of it is that I haven't got a single clean thing to wear — I just sent a big laundry out and I have one pair of pants left and no shirts, no socks, but plenty of underwear. I surely am glad I didn't send my money home to you this month.

Started to do it too. I've got one roll of film to fit the size "828" Kodak and I'm going to try to borrow a 35 mm camera to use the film that you sent me. So I should be able to get some pictures for us.

I'm sorta leaving Wagner and the boys in a lurch with Kremers gone but I've been taking care of two squadrons, so the others can for a change. Due to various and sundry reasons, I don't feel too bad concerning that score. My orders read that I will be down there for four days, but I wouldn't be too surprised if a week didn't come closer to the time that we will spend down there. I will try to get you something of some sort. They say it's almost like the States at Cairo and Alexandria — hotels and everything. I only hate to go away for it means that I won't hear from you until I get back.

I am going to enclose three snapshots that I think you will be interested in. One is of Young squatting down working on that sorry motorcycle that he obtained. He never did get it running before we were ordered to get rid of them. The other is of Lt. Roach and myself taken in front of the Officers Mess tent while we were in south Italy. The other is of me and the dispensary. Note the "mudders" that I have on and the mud. I wore those mudders every day from the time I got here until we moved up here. They are worth their weight in gold.

Haven't done a heck of a lot today. Took care of my sick call and the 726th this morning — went to early take off like I always do and to landing this afternoon. This afternoon the colonel called a briefing of all C.O.'s, operations officers, and flight surgeons, so Young, Evans, and myself had to go up there to listen to him. Sprowls and most of his crew finished up their missions today and will be coming home soon, I hope. Tonight we threw a party at our club, and I've been busy ever since. As yet I haven't packed nor borrowed any clothes to wear and it's getting pretty late.

I'll write you tomorrow night from Cairo. I just want to tell you that I love you more than anything else on earth. Take care of yourself, Kay and Carol.

All my love,

Joe.

Dearest Babe,

June 18, 1944
Sunday, 10:15 p.m.

My dearest Honey,

Well here I am in Egypt, and it is everything that you would expect and then some. It's just like any large cosmopolitan city in the States almost, with a few glaring exceptions.

We arrived here two days ago. Landed at a very modern airport outside the city and rode a truck in at dusk. The first thing we noticed of import was automobiles and plenty of them, and most all of them American. There is any kind of automobile you ever saw here. They all drive from the left side of the seat too, just like in the States. Most all the way into the city we passed the time away calling out the make of autos going by and the year it was made.

We came into town and went to the nicest private hotel in town and asked for a room. They were filled up, but we did manage to get a room at one of the three better hotels. The hotels are still a far cry from an American Hotel. I'd say this hotel reminded me somewhat of the Colonial Hotel in Monterrey — that is, in calibre. It is quite a trip down here — about 1500 miles. The first night we were so tired that we just hit the bed. The next night we went to a cabaret which was very nice and drank and ate and watched a floor show. Last night we did the same thing.

Yesterday we hired a guide and went around to visit the various Mosques in the city — or, at least, some of them. We also went out on the edge of the desert to the pyramids and the sphinx. I took plenty of pictures of them both. They both are a little bit disappointing in that they look exactly as you would expect them too. By the way, I rode a camel out to the pyramids — that's an experience.

Here you can buy steaks, highballs over the bar, cameras, film, shoes, or anything your heart may desire. Of course, I didn't say what you have to pay for these things for prices are exorbitant. I came down here with $170 and I'm almost broke now, and I haven't bought anything except some trinkets for you and 48 rolls of old out-dated Kodak film. Here too they play the American for all he is worth and then some. Boy there is one thing we are famous for and that is that we are rich and suckers. Personally, I don't know about the former, but I'm sure they are right about the latter. The unit of currency here is the Egyptian pound which is worth $4.13. That pound will buy you just about what $1.00 will in the States.

I wish I could elaborate on these places enough to give you a clear picture of what they are like, but I just can't. I will notice 100 things per day that are different that I intend to tell you about, but they are so trivial that it is difficult

to put them down — such as the difference in the way the French, Italians and Egyptians shave you. All of them give you better shaves than the Americans, but you don't enjoy them as much. I'd rather take a beating than let one of these guys shave me again.

Another thing you should know about is the Nile. The Nile isn't such a large river. I'd say it's about the size of the Arkansas at Little Rock. But it's very clear. There are many irrigation ditches going off from it. From here to where it empties into the Mediterranean, it is literally filled with many large islands, making it look like a plexus of smaller rivers rather than one large one. The Nile delta which you have heard about all your life is only about 10 miles wide. That was a surprise to me. And on either side of that delta there is nothing but desert. When I say desert that is what I mean too — sand, sand, and sand. The boys that fought this war from El Alamein to Tunis fought a rugged war. It might not have been muddy and cold, but I know it was sandy. You can still see deserted camp sights, airports, tank tracks, abandoned tanks, wrecked airplanes, from here on up to Tunis I guess. I know they are from here to Benghazi. I flew right over that one road along the northern coast of Africa from Benghazi to here. It is really a lonesome affair — not a house or inhabitant in sight. I wish you could see it from the air, but I would hate to see anyone have to drive it. That is a real desert.

This is just about the first place I have been that I would like for you to see. I haven't been to Alexandria as yet and don't think I will go there, for I haven't the time nor the money, but they say that it is nice too. I think that we will start back to Italy about the day after tomorrow, but instead of flying directly back we are going to take our time. We will probably stop in Benghazi and perhaps Tunis before hopping the Mediterranean.

On every little vacation that I take, I miss you the 'mostest' of all. I get more homesick on these trips than any other time. Maybe after I'm home for a while and get used to the luxuries of civilization and the blessing of you again, and if we are able, I can retrace my steps and let you enjoy this part of the world.

All my love,
Joe.

June 20, 1944
4:45 p.m.

My dearest Honey,

Here I am in Cairo — the largest city I have been in, I guess, since the States — 1,300,000 people and I am so lonesome. When I'm in the field helping to fight a war, I can manage to get along. Just barely, at best — but I can manage. I have sense enough to realize that it is no place for you — but when I am in a place like this — where it seems there is no war going on, loneliness grips me.

You know, I've been to many places and seen many things since I've seen you. In fact, I was thinking only yesterday that I have flown the Atlantic, flown the Mediterranean twice, flown the Tyrrhenian Sea four times, the Ionian Sea twice, over the Adriatic some and some of the smaller gulfs, bays, and seas. I have been perturbed to some extent that this war might change my opinion about things in general. Mind you, not about our feelings toward one another but about my outlook on life itself. All of this is something that is bound to leave an imprint upon one's mind. I probably have developed different points of view about people, politics, and everything other than moral ideals, since I left you. I want you to help straighten me out when we are together again.

We were supposed to go back today — so we packed and caught a bus out to the field. When we arrived there we found out that they were doing some rather major repairs to our plane and that we are forced to remain here until they are through. In fact it looks as if it is going to be about 10 days before she is going to be in flying condition. That being the case, I am going out to the airport tomorrow afternoon and try to hook a ride with the Air Transport to somewhere near my home base. I can't afford to wait down here for the duration, I'm sure. The outfit didn't expect me to be gone but for four days, and I've already been away five. It will probably take me a couple of days to "hitch-hike" (by air, of course) back. I have but one thing in mind, and that is to return to you.

All my love,
Joe.

June 20, 1944

10 p.m.

My Honey,

Just wrote you a letter this afternoon. Don't forget that I too get down and have a rather meager outlet for my excess steam except in your direction.

Tonight I feel some better after going out and filling my stomach with boiled shrimp. If I had just had some of that sauce at Arnand's, they would have been even more delicious. Last night I had squab and the two previous nights I had steak (one of them double), so you see my appetite is none too bad with decent food. Anything looks good after C rations. Too I can get a couple of Scotch and sodas prior to my meal which makes my appetite even more vicious.

I have wished a thousand times today alone that you were here with me.

I love you,

Joe.

June 23, 1944

10 a.m.

My dearest Honey,

Well, I'm still on my way to my home base. Yesterday I flew from Cairo to Tunisia where I am now. I am leaving here about noon for Italy. I'll bet the other doc's and my squadron are really going to be griped at me. I was supposed to be gone for four days and it has already been 8 or 9 days. I couldn't help it 'cause the plane broke down though. Yesterday I flew over here in a C-47 — don't know how or what I will fly the rest of the way in.

Cairo was certainly interesting but too expensive. Boy it really set me back, and I didn't do much of anything. Went to a couple of shows and to a couple of cabarets to eat and watch the floor shows and that was the extent of my enter-tainment (except drinking scotch and sodas). It certainly was a diversion though 'cause it was so different. Watching people dance made it even worse being away from you. I wonder how long it will be before I see civilization again.

I love you and miss you,

Joe.

Dearest Babe,

June 23, 1944
11:30 p.m.

My dearest Honey,

Well I finally got back "home." Arrived at the airport about 5 p.m. and got a truck to come pick me up. Got back out to the base about 6:30. Looks as though they got along O.K. without my help. Luckily the boys haven't flown but a couple of times since I've been gone — they have been having bad weather up in these parts. Sorta glad to be back too. Wonder how many days it will be before it begins to get on my nerves again.

You should see my stack of mail. I'll bet I had twenty letters waiting here. Apparently mail broke though while I was gone for I had hardly been hearing from you at all just before I left. I certainly did enjoy reading them — some twice or three times — funny how some letters strike you — they just make you feel so good when you read them that you can read them over and over and still get a boost from them.

Apparently, I fooled you. The flowers you received June 10 weren't for Mother's Day especially. They were just flowers that I wanted to send. They weren't for anything special except for you. You like flowers and I can send them occasionally — don't you know how much I love you whether it's Mother's Day, Birthday, Easter or what? I like to send you stuff 'cause I can sit here and muse on what your expression will be when you get it — particularly if you don't expect it.

I received a letter from Henry Jones and Frank Posey today also. First time I'd heard from Frank in a long, long time. He was finishing up and was in the Navy. Somehow his letter didn't produce the same feeling of friendliness and comradeship that they once did. He also told me that they have a big baby boy named Frank III born Dec. 26 (the day I left the America's, by the way).

Seems as though my wife has turned very domestic. (1) Putting up jelly, and so much too (2) Always cooking up some new recipes (3) Always working in the yard, or sewing, or something. It's all very good. But if I were you I wouldn't overdo the yard work. As I have said before, I know that is work. I believe I would do it very, very slowly or hire me a man to do it, or let it go. I'll fix the yard when I get home even if the stalks of Johnson grass have to be sawed instead of cut. I don't want my wife to work herself to death you know — I need her too badly.

I wrote you a "shortie" from Tunis this morning. I'll try to catch up for the few days I missed while traveling. Have you ever noticed how much more difficult it is to write when you aren't receiving mail? Honey, your letters did me so much good. I feel so much better after reading one from you. If this war would

230

end as soon as you and I are praying for it to, it would have ended long ago. Maybe something will happen, and we will be back together before too many moons. Until then I want you to know that I love you more than I can tell or show you.

Joe.

June 24, 1944
10:05 p.m.

Dearest Honey,

Today I have been walking around and seeing the boys. I have told at least 50 guys what kind of a place Cairo is. In fact, I've told the same thing over and over again until I'm tired of Cairo. Saw Wagner today — he flew over to Corsica and back today. Marion is doing fine — taking pregnancy in her usual stride, and is expected to deliver about the middle of July.

More news — Blackie has already left the hospital for home. He is either at the Port of Embarkation or on his way now. I heard today that Kremers is not expected to return to his squadron — that they think he has arthritis of the Lumbar Vertebra. I, personally, doubt that but for his sake I hope he is sent home nevertheless. There's no sense of a guy being over here half sick.

To prove to you that physically I am as good as ever, I will give you my weight. I weighed while in Cairo for the first time since leaving the States. I now tip the scales at 158. One pound more than ever before. Truthfully, I thought I had lost some. In fact, I'm sure that during the winter I was down quite a bit, but you can tell by my weight that I'm O.K. now.

Today it really rained. I don't think I have ever seen it rain any harder. Got all in my tent. Tonight I have a fire going to try to help to dry things out a bit — too it feels good. Hey, did I tell you how I had to fight the bed bugs in Cairo? I would turn the lights off and in a few minutes turn them on again and start swatting. This procedure I did many times — bet I killed 50 a night. Another surprise — I got another ambulance while I was gone — a pretty good one and I really needed it.

Well, it's goodnight again I guess. Tell the kids that "Daddy still yoves them very much" but I love their mother best of all.

Goodnight,
Joe.

June 27, 1944
Tuesday, 10 a.m.

Dearest Honey,

Yesterday Wagner and I went into town to get our rations. I mailed you a package of knick-knacks and do-dads. If you keep the jewelry, I will tell you where each piece came from when I get home. Some came from Egypt, some from Pompeii, etc. The plate came from Egypt, of course. You and Kay can divide the gum — sorry I couldn't get Dentyne. Enjoy that gum — don't forget that it has done a lot of traveling — made in Chicago — sent to Italy, then back to Memphis. Quite a trip for a piece of gum.

Last night Evans came over and sat down for quite a spell just to talk. That's the reason I didn't get to write you. He is flying quite a bit now — he is even flying when Young is supposed to fly (much to Young's chagrin). Seems the Colonel is trying to finish Evans up. All the old crews are just about through. Sprowls and Larson have finished, so has Monniger. Threadgill, Prewitt, Roach and Kavannaugh each have a couple more to go. All of them should be on their way home soon. I will certainly be glad.

Bernstein is leaving tomorrow on a three-day pass to Rome. Young is going up there as soon as Bernstein returns. I'm trying to make Zraick and Evans take some time off but so far haven't succeeded. Evans doesn't want to leave until he has completed his flying.

Saw Haltom yesterday. He has become much less popular since he has become a Major for some reason or another. His boys seem to think it "went to his head." I haven't been around him enough to tell. Davis is coming home for good on the next boat. You remember Davis. He was the C.O. of the 725th — married that little brunette from Jackson, Mississippi while we were at Fairmont. He didn't last but a few missions before it got him down — that's no disgrace however — some people can take it and some can't. I'm quite aware if it were me that I couldn't. That's just about all the news about the boys that I can think of at the present time — 6 months later I'll probably bring you up to date again.

Well, I'll be moving my dispensary again soon. The squadron has built me a nice stone building about 50 yards from here for my dispensary. I've got to share part of it with the orderly room, but it's gonna be pretty nice and cool too. It's got a nice tile floor and I should be able to keep it clean easier. My boys are putting in the plumbing now and building some tables and shelves. I suppose it will be about a week before we really get in it. The boys are doing all the work — I'm too lazy.

I think I will enclose a couple of pictures I have that I got while in Egypt. They are good pictures. I'll have gobs more when I get them developed. Honey, I want to get this in this morning's mail, so I gotta hurry. I'll drop you another line tonight. Tell the kids hello for Daddy.

I love you,
Joe.

Kodochrome photos of Joe in Cairo, 1944.

June 27, 1944
3 p.m.

My Honey,

I wrote you a letter this morning, but I just received two Air-Mails (June 16 and 17) from you so I'll at least start another to you now. You mentioned in one of your letters about all the company. Honestly, I believe more people drop by to see you than anyone I've ever heard of. You know, that within itself is quite a compliment to you. When I get home we aren't going to have any company for a while — just us. I'm going to be too selfish.

No, I haven't received any packages from Sis as yet. It takes about 2 months for a package to get over here usually. Don't worry about trying to find another pocket-knife for me. The one you sent is a very good knife. I don't need all the "gadgets" on a knife anymore. In fact, I have just about everything I need (except my family of course). Even while I was in Cairo, I could hardly sleep due to the springs. When one gets used to a cot without springs, he rests just as well there as he does on a Beautyrest. Don't worry about the way I live. I don't mind it in the least. A tent is a very nice place to live — ask Mother and Dad — if you have water, heat, lights. I have all of these things, so I live very comfortably. As I have told you many times — the toughest part of this war for me is the absence of you.

By the way, Honey, sometimes I think that perhaps you are under the impression that letters coming to this theater are censored. None of your letters are. You can ask anything you like but I'm not so sure I can answer because of military reasons. What percentage of my mail is censored anyhow — just curious? I imagine about 1 out of 10.

11 p.m.

Didn't finish this afternoon but at least I made a good start. This evening Evans and I went into town to that picture show that I promise myself each time I go there that I won't go back. I have never been there yet that something wasn't wrong with the loud-speaker and you can't understand what's going on. We saw something about Roger Toney's Gangsters. It wasn't even fair. Then we went by the Red Cross. Saw a Red Cross worker there that was in the bed across from me when I was in the hospital. He is leaving for home in the morning after being overseas for 27 months. He has a 30 day furlough and is then slated to come back. Of course he could stay if he wanted to — you gotta admire guys like that.

Funny how much more I miss you in the afternoons and evenings than any other time. The radio was playing "I Love you Truly" while ago. It sure took the words right out of my mouth. For I do love you truly and always shall. Goodnight.

Joe.

June 28, 1944
Wednesday, midnight

My dearest Honey,

Do you know that is the way I practically always begin my letters to you? Sometimes you probably think that that is just the way I'm starting my letter, but really, every word of it means exactly what it says. You know it's an awfully good feeling to know you are back there waiting for me to return.

Today has been a rather boresome day — nothing interesting happening at all. Late this afternoon I got in my ambulance and drove up to Group Headquarters just to sit and talk with Wagner. The day passed so slowly. Wagner, Rhors, and I had a couple of bottles of Italian Champagne which helped out some. Then I returned to here and went up to our Officers Club to help Threadgill, Kavannaugh, and Roach celebrate — they finished up their missions today. Tomorrow we aren't flying, so I'm letting them "take their hair down" more or less tonight. Surely will be glad when all the old boys finish. I surely want to get them out of here and on their way home, and they surely deserve it.

Next night, 10:30 p.m.

I received four V-mails and an Air-mail today. One V-mail from your mother, one from Kay, one from Carol and one from you — the Air-Mail was from you too. The ones from Kay and Carol were for Father's Day and your Air-mail was also. Very nice things you say about me, honey, and I appreciate you saying them and feeling what you say. It's a wonder that my chest doesn't swell up to such an enormous size that I bust the buttons off my shirt. It's always pretty big because of my wife and family. The children's letters were precious — I liked Kays very much — I could just see her scrubbing the "Toidy" seat with her "toofbrush." I couldn't get along without your letters — they are the only thing that helps to pass by these extremely boresome days.

This morning I "piddled" around doing nothing. About noon, Evans and I got in the ambulance and went up to Group Headquarters and picked Wagner up and drove into town to get our rations — also for me to try to get some white paint, which I was unable to do. We then drove out to another one of the nearby bases to see a friend of Evans. Then we came on back here and all had supper at our mess. Tonight we went up to the Group Headquarters and had a couple of drinks with Wagner. One almost has to keep on the move to try to keep from being bored stiff. There is simply nothing to do sometimes but sit and the days are much too hot to do that. The evenings are still very cool though — thank goodness. The past two days have been our hottest, and they have really been sweltering. It isn't too humid here though, and we are able to take the heat better than you can there in Memphis.

I just had to go in to see an Italian who hurt his leg in an automobile accident. They are the biggest babies you have ever seen. Most of them moan, and groan, and carry-on like a five year old child — what a people! How Mussolini ever expected to fight modern warfare with them is beyond me. I wonder how Caesar did everything he did.

Well honey it's time for bed. I know that I've got to be up with the sun in the morning and that doesn't leave too much time for sleeping.

I love you,
Joe.

June 30, 1944
11 p.m.

My dearest Honey,

I have just returned from the best concert I have ever heard. Jascha Heifetz playing his violin. If you ever get a chance go listen to him even if it costs five or 10 bucks. He is the best — never hits a sour note — I mean he is real good. Makes people enjoy classical music that never thought they would. He opened up by saying "I don't know whether you like classical music or not but it's like spinach — even if you don't like it, it's good for you."

I'm sitting up here with my shirt off, with calamine lotion gluing the hair on my chest down. Yes you guessed it — I got too much sun today while I had my shirt off. It didn't bother my back for it has already been exposed enough to take it — but my chest was not able to take care of itself. I'm certainly glad that I have finally learned to sleep on my back, sides, stomach or whatever way that strikes me as being the best for the night. Quite an accomplishment considering the many years I slept on my tummy and no way but on my tummy. My wife taught me to sleep on my side, and I think I've picked up sleeping on my back due to canvas cots.

The night is beautiful tonight. It's just as light on the outside. There is a bright half moon and all the stars are out — a beautiful night. Today has been extremely hot even though we have had a very brisk wind most all day long.

This afternoon Evans and I got in the ambulance and went up to Group Headquarters for him to have his picture "struck." He just won the Silver Star, so they wanted his picture again — makes him have the S.S., Distinguished Flying Cross and Air Medal — he's doing alright for himself, isn't he? When I dropped by to see Wagner, the Wing Surgeon was there. He has come down to take care of Kremer's squadron for the next few days. We are still not sure what Kremers outcome will be. It seems that they change their mind almost daily. Quinn is leaving tomorrow for 10 days detached service at a hospital — we won't miss him — he is still not too energetic. However, due to our isolation, he has to take care of his own squadron now which is something he had never done previously. There I go gossiping again.

Honey there's nothing more to say tonight except the same thing — that is I love you very very much. Take care of yourself and the children for me.

All my love,
Joe.

July 2, 1944
2 p.m.

My dearest Honey,

I'm writing on my knee, so if you aren't able to read it, that is the reason. Back where my desk is it's burning hot — like that heat that can only collect under a tent. Here on "the screened porch" it is much more comfortable, especially with one' shirt off. You should see my torso. From too much sun the other day, I'm cherry red. It's not sore or anything, but it surely is red — looks like I'm on fire. I sat out in the sun this morning trying to get my back as burnt as the front — don't know whether I succeeded or not but no doubt will by night.

Yesterday I did you bad again — I didn't write last night. We didn't fly, so Evans, Wagner, Peterson (the Dentist) and myself got in the ambulance and drove over to the seashore. There is a beautiful white sand beach there. We didn't stay in very long though. Came back and we all had showers under my shower and then went up to the club and had a can of American Beer. I'm not so fond of beer but it did taste good. Yesterday I moved my dispensary across the road into the nice little stone building that we built. Tile floors and everything now — it's very nice and much easier to keep clean — also it's much cooler.

This morning I got up early and went to take-off, then came back and worked in the dispensary for a while. All morning I have had a new visitor — a little puppy that is a wee thing and is as cute as he is small. He's a little devil though for he has already torn up a sock of mine completely. He is just a dog but he has such sad, mournful eyes that he's cute. I don't know where he came from, but I do know he was hungry. I went over to the mess hall a few minutes ago and got him a couple of cooked cakes of C-ration meat that I turned down for lunch, and you should have seen him go after it. He is laying down now on the floor sound asleep.

Later — 10:30 p.m.

I've got a lot of fresh hot news — just out. Evans will leave for the States with these old crews who have finished up. Young will stay here as Commanding Officer, i.e. take over Evans job. If Young makes as good a man as Evnas has, we will be lucky. Found out today also that Young is in for his majority — which will be nice for him. I'm sure glad for Evans. That guy has really done a good job, and all the boys know it and respect him for it. Haltom will be the only "old" C.O. left after Davis and Evans go home.

Going to bed now, honey — wish I were going to bed with you. I need you for thousands of things — just now you could put some calamine lotion on my back if you were here. Don't know how much longer we can continue to exist but hope it won't be too long.

Goodnight my love,
Joe.

July 3, 1944
10:30 p.m.

My Honey,

You know what I've been doing all afternoon? I have been looking at the best snapshot that was ever taken of one of the most beautiful women I have ever laid my eyes on. The pictures of the cherubs are good too. As you say the one where Carol has her lips puckered up like she is going to whistle is particularly good.

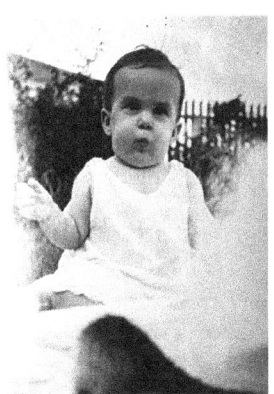 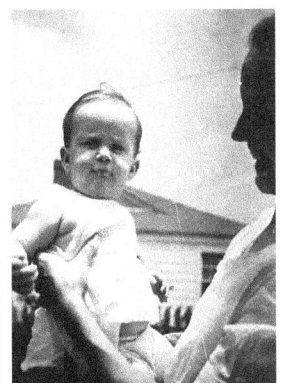

Carol *Kay* *Carol and Babe*

Since I have started this letter I have had quite a few distractions. Bernstein has been in talking over his troubles (he just returned from Rome, by the way, and he says it is both beautiful and interesting); I've had to go over and help calm down a bombardier that went "haywire" today; I've talked to a boy who is having an extremely hard time adjusting himself to this environment; and I've talked to Scarlata (my little extravert) who finished his missions today. I started this letter at 10:30. It's now 2 a.m. Think I will go to bed.

Again I love you with everything I have and appreciate you. I am thankful that I can say that I am —

Your hubby,
Joe.

July 5, 1944
Midnight

My dearest Honey,

Today has been a banner day, I have been very busy all day. I have written 95 letters — seven carbon copies to each; answered 95 questionnaires where it was necessary to write page answers — with four carbon copies to each and signed my name to all these things. If you figure it up, that's singing your name quite a few times. So this is only my 96th letter I have written today. I got a call from the Air Force yesterday telling me I had to write a complete letter on each man I was trying to return to the States. It was something new. I asked them when these letters were due and he said "Today." It was 5 p.m. when he called, so I got them out today. Had every typist in the squadron typing though.

After finishing up late this afternoon, I caught the ball with Young for a while for a little relaxation. Caught one ball on the end of my thumb, and I either sprained it awfully bad or broke it. I do know it hurts awfully bad, and I can hardly write with it.

Prewitt and another fellow just strolled in. They have arranged to fly up to Rome tomorrow and stay 24 or 48 hours. I hear that they have some silk hose up there, so I will try to get him to get you some. They say there is everything in Rome. The Jerries didn't have time to take everything with them. Boy, if there is anything there it's the first place they have left anything. They are extremely adept at cleaning a place out.

The wind was so strong last night that it blew off the top of the enlisted men's club. We had only had it opened for about 10 days, and it was a beauty. Was doing a thriving business too. I'm almost afraid to start building it again for I don't know whether we would have time to finish it and make enough money to pay for it before we move again. You know one can never tell in this man's Army.

Well honey, I've written enough letters for today. I think I'll go to bed now.

This day has passed a little faster than usual for which I'm thankful. Hope the rest of them pass fast regardless of how many there are, so I will be with you again.

All my love,
Joe.

July 6, 1944
6:10 p.m.

My dearest Honey,

Can you read this? I'm writing without the use of my thumb. Try it sometime, it's quite an accomplishment. I broke the thing yesterday, sure enough but not bad. It'll be O.K. in a day or so. When you get so old that you can't even catch a ball except on the end of your thumb, it's pretty bad.

Today Young got his majority. Evans should be leaving anytime now, and Young will take over. A fellow named Mahon is going to take Young's place and he's a real swell guy. I like him as well as anyone I have ever known in the Army, I guess, even though I haven't been around him very much. He is quite a bit older than the rest — about 32 or 33. He taught school for 5 or 6 years and is a very nice guy. He will be able to handle the fellows nicely.

I have been in town twice today. Saw "Bombers Moon." I guess I saw it with you once before, but I had forgotten it until I had been in the show for a while. I need you so you can tell me what shows I've seen. Tell the kids hello!

All my love,
Joe.

July 7, 1944
Midnight

My dearest Honey,

Tonight we had a grand opening of our Officers Club. We have been using it for a couple of months but tonight we had the "opening" and invited everybody. It was quite an affair, and I think most everybody got tight on cognac or grain alcohol. (I don't write like this 'cause I'm tight now — I just can't wiggle my thumb yet.) I wouldn't have let them have the party 'cause we are going to fly tomorrow, but it was probably the last chance to have one with the old boys present. They should be going in a couple of days.

I received three letters from you today. First I had gotten in three or four days. Your letters surely do help me. I wonder if you really know how much. Glad you got your perfume — hope Kay and Carols presents came in too. I love you with all my heart.

Joe.

July 8, 1944
Midnight

My dearest Honey,

Tonight I will see if I can write something other than one of those "quickies." My last two letters to you have been of that variety. My thumb is much better, and I can at least hold a fountain pen now. Honey, 'fore I forget — I sent you $225.00 today — tried to make up in part for not sending you any money last month. Still haven't saved any money though for possible "emergencies" 'cause I spent last months in Cairo. Do whatever you wish with the $225 — you know it's yours. Just let me know when you get it.

Tomorrow we aren't flying — I have got a chance to fly up to Rome for a few hours but don't know whether I will take it or not. In fact, I don't believe I will. The boys haven't been on many trips, but I have and my conscience would hurt me. Pretty soon I am going to go up there however. They say that prices are going up and up. That's typical. I've told you before what suckers we are, and I suppose I'm one of the biggest. In another month, you won't be able to go up there unless you own the mint.

Went over tonight to the 726th area and saw Brian Aherna and Rosalind Russell in "What a Woman." It was pretty fair. I think it could even have been considered good if I had just been in a more receptive mood. After the show I came back by our club and had a couple of glasses of tomato juice (imagine me drinking tomato juice as an evening drink).

Hey honey, haven't you heard from Hugh yet? I'm interested in how he's coming out and whether he is still in England or in France. The next time you write him give him my regards and tell him I hope the going for him isn't too rugged. It must have been rugged enough while in Africa.

Well, honey, I gotta go to bed. One of these days, maybe I won't have to go to bed by myself. I'd much rather have some cold feet near me.

All my love,
Joe.

July 10, 1944
1 p.m.

Dearest Honey,

Seven months ago today, I took off from the continental limits of the U.S.A. Remembering something that happened back in the States almost is like remembering something back in childhood.

Yesterday started out by being one of the dullest mornings of my life. There simply was nothing to do nor did it hold prospects. It turned out to be a very enjoyable afternoon and evening. I had 12 bottles of beer that I brought back from Cairo with me. Evans got us some ice. He had some crackers that someone sent him. So we packed the beer in ice and had ice cold beer and sardine sand-wiches (using 2 cans of the sardines you sent me). Roach, Evans, and I enjoyed it very much. Then last night we had tickets to Irving Berlin's Soldier Show "This is the Army," which is excellent. It is a scream from beginning to end — one of the best bits of entertainment I have ever seen. I wish you could have seen it.

This afternoon, Mahon (my newly acquired friend), Evans, and I drove over to the beach for a swim. You know, I don't care much for swimming, but you gotta do something to pass the time away. Then late this afternoon when I returned to camp, I went over to the mail room and got a letter from you, a package from you, and a letter from your Mother. It is the third package I have received from you — contained magazines, fruit juice, Kodak Film and two very nice ribbed T-shirts. Thank you, my honey, you are very thoughtful.

Tonight, Young and I went to see the Group movie. It was sorry as usual. The name of it was "What's Buzzin Cousin." That by itself should have told me that it was going to be sorry. At present, Young is sitting on one side of the table writing his fiance. He's just about finished now.

Well, the first of the boys left today. Eleven departed about noon for U.S.A. What a happy thought that must be. Certainly was glad to see them get on the way and will be glad when the rest get going. I want to send all I can back. I hope never to see any of them again until the war is over. One of the boys, Lt. "Judge" Bryan, says when he gets to Washington (his home) he is going to call you up. He is a hellova swell guy, and I'm pretty sure he will call you just to tell you that I am fine and in the best of health. He is also going to send you a roll of Kodachrome film I took. They are of bright colored flowers in Egypt and Italy.

Honey, I started this letter at 1 p.m. — I'm finishing up at 1 a.m. — three different sittings, but I've enjoyed talking to you. I'll say now that I love you with all my heart. Tell the babies hello for Daddy.

Goodnight,
Joe.

July 11, 1944
6:30 p.m.

Dearest Honey,

Hey, I didn't find the pillow slip in that package that I got from you a couple of days ago until today. It's already on my pillow, and I can hardly wait to go to bed tonight. It looks so clean and nice. I've been using a bath towel wrapped around my pillow and that does pretty good, but you always have both ends of the ticking sticking out that way. Thanks a lot! — I notice it's made out of good durable stuff too. Did you make it?

I have managed to keep myself partially busy today. First we flew — that always takes up some time. Then I had to put in another door to my porch. The wind has been blowing my doors off faster than I have been able to keep them up. I really have it fixed now though. I've got the door frame made of solid "plexi-glass," which is really tough stuff. You know it's the stuff that airplane windows are made of. I almost dare a good wind now. Then this afternoon, I started to work making me an icebox. We are getting ice pretty regularly at the mess now, and I should be able to get a few pounds every day or so. Ice water tastes so much better than hot water. I haven't gotten it completed yet but I already have ice in it. Built me a tin box with a hinged lid — then a thick wooden box around it that fits tight. Now I'm going to suspend that wooden box in another box and put sawdust between the two boxes. That should insulate it very well and keep ice pretty good. I know it works for I have three large ones in operation now at the mess hall and the two clubs. Also have your juices that you sent me on ice and some for Evans — plus some candy bars.

I took out my stove today for the fourth time. As you would expect, tonight has turned out to be the first chilly night in weeks. I'm sitting up here now with my leather jacket on. The days are nice and hot though. I have gone without a shirt all day long. Don't even have to heat the water for my shower any more — the sun takes care of all that for me. Well honey, I haven't got a lot more to say except that I still love you as much as ever.

I shall always,
Joe.

July 12, 1944
6:30 p.m.

My dearest Honey,

Today has been a rather gloomy one in that it has rained continuously since about 10 a.m. It has really rained too — reminds us of Italy in the winter time. We have a tin roof on the dispensary, and I haven't been able to hear myself think in here today.

When the boys landed this afternoon, they had quite a rat-race on account of the weather, but when it was all over no harm was done. Everything was O.K. Late this afternoon I have been very busy writing letters about a couple of my men and doing a 64 examination on one of them. I think I will take them down to the Air Force in the morning and try to get them transferred to some other kind of work and some other outfit. They just aren't psychologically suited for this sort of stuff. If so, I will be gone for two days. If I go, Evans is going with me, if he can. (Notice there are two if's in that sentence so it probably won't all come to pass.)

One of the sergeants over at group headquarters just called up the dispensary for Mason or Garrity. Neither of them were here at the time, so I took the call.

What he wanted was my address, so he could send it to some of his kinfolks in Memphis so they could look you up — he didn't want me to know about it for some unknown reason. His name is Irwin so if someone looks you up for him, that's it. I don't know why the secrecy. Maybe he wants to tell you things about me. Oh well, just something else for me to "sweat-out."

Well, I had better finish and go over to my tent and take a bath and dry out. I have been soaked through and through twice already today. Hope my clothes didn't get wet in the tent, 'cause it's been raining inside there almost as hard as on the outside.

I'm busy worrying about getting home so I can live with my wife and children so we can be happy. Isn't it a swell thought?

I love you so very much,
Joe.

July 14, 1944
10:15 p.m.

My dearest Honey,

Last night I was in Bari and couldn't write you a letter, however, I got punished for it for today. When I returned to my squadron, I found that I neither received a letter yesterday nor today. Bet I get some tomorrow though.

Got rid of one of my men while I was away. The other they sent back with me but it's O.K. for he is alright. Surely was glad to get the other one off my hands though. Evans and Wagner both went with me. We were so busy tending to business (or at least I was) that we didn't even have time to go to a show or go by the Red Cross for some ice cream.

While I was down there I saw Bassett who was an acquaintance of mine at March Field. He and I reported there about the same time. I'm sure you wouldn't remember him for he wasn't around long enough. He is a nice guy though, and I liked him a lot for the short time that I knew him. He has been stationed within ten miles of me ever since I have been located here, and I didn't even know he was in this theatre. Only last week, he was transferred to the Air Force to a better job. He has been overseas for 18 month.

Had a swell bed to sleep on last night — nice soft springs which didn't roll too much to the middle of the bed to be uncomfortable. However I didn't get to enjoy it much. I had the "G.I.'s" and had to run to and fro down the hall all night. Today I'm doin' pretty good however. Evans was driving back, and before we started, I gave him instructions that if I said "Stop" I meant for him to stop instantly without questioning. Luckily, no such occasion arose.

After I wrote you day before yesterday, it turned cold. Remember I told you the day before that I had taken my stove out? Well anyway, I had to put it up again. My tent was so cold and everything in it was wet. So I again (for the third time) had to reinstate it, and boy, it really felt good. I got out some of the sardines you sent me and some crackers and some tomato juice and sat down all by myself and really enjoyed the evening. Also wrote Katy, Otha and the kids while sitting there. Tonight it's quite comfortable. I have my leather jacket on though, but it will be good sleeping. Truthfully, I don't know whether it will or not — my blankets are half on and half off. I haven't made up my cot in two weeks. I would rather do almost anything than have to make up that cot. Glad that isn't considered my job at home, 'cause I'm just no good at it. My tent hasn't been swept out in a month I know. It's filthy. Every night I say that I'm going to clean up the next day, but somehow or another I always find some half-suitable excuse to get out of it. There is nothing like procrastination. It's a damnable thing.

I did receive something in the mail today. Your Dad sent me a shoe-shining rag. I'm gonna write him a private letter and thank him and also kid him. I say 'private' letter for it isn't going to be fit for mixed company to read.

Well, honey, I gotta "hit the sack" for I'm tired and I know I'm gonna have to arise in the morning about daylight. Wish I could see you and be with you. Take care of yourself and the kiddies, and don't let the eldest one get you down.

I love you,
Joe.

July 16, 1944
10:50 p.m.

Dearest Honey,

Today has been a long day — a happy one and a sad one. I arose early for take-off. After take-off, I came back and had breakfast. I then started shaking hands with the boys and telling them goodbye, which I did until just after noon. Finally it happened. All the orders on the old boys came through yesterday. Everyone that I was trying to send home was on the orders. I certainly am glad. That really cuts the "originals" down to almost nothing. Some of the boys dropped by last night for a chat before they pulled out. Among them Formanek (who got your address), Threadgill, Prewitt, Adler.

Yes, I got rid of all the old boys with exception of the ones on the staff. They all took off from Evans on down — five truck loads of them plus one of my ambulances full and one jeep. The 727th has changed its whole personality today, and outfits do definitely have personalities. Some of the boys were so surprised that they were on orders and pulling out. Few of them really got a good sleep last night. I don't know how many of them got your address and said they were going to call you. Of course, most of them will be over that "gracious" feeling by the time they get to the States. I was really glad to see them go, but it surely felt funny when I sat down at the dinner table tonight, looked around and didn't see anything but the faces of neophytes. I hardly knew any of them.

This morning they brought everything — world almanac, clothes, clothes hangers, kodak film, tables, water coolers — everything. The tent is so crowded I can hardly move around in it. They even left $265 dollars worth of debts for me to collect for them. Sprowls walked up just before he got on the truck and handed me $40. Said "Here, doc, I want to pay you 'fore I leave" — and honestly, I don't remember ever lending him $40. It was a happy occasion and still sorta sad — we had been through a whole lot together and each of us knew it — but I'm sure glad they are on their way home.

Three of them elected to remain overseas and are going to transfer into fighters. They were afraid if they went back to the States they would get slammed back into the "Flying Boxcars." They are Hargrove, Wheeler, and Malone. Guess the orders transferring them should come through soon. That's what you call game, brave, and also devil-may-care. You really have to hand it to these guys.

This afternoon, I sold my motorcycle to Hargrove for $150.00. It was worth more than that, and I could have sold it down in Bari for $225, but Hargrove likes it so much, and I'd really rather have him have it than someone I didn't particularly know. Anyway, it didn't cost me one penny, and all it has been doing for the past few months was sitting hid in the back of my tent. Tomorrow or the next day, I will go into town and send you the proceeds from it — or part of it. That's clear profit, so buy yourself something with it. I don't believe you ever got that spring outfit — if you did, I never did hear anything about it. Hated to see the motorcycle go, for while I could ride it I surely did enjoy having it and working on it. A motorcycle can really take up a lot of your time.

As of this afternoon, Young is now C.O. and Mahon is Operations. Mahon will really be alright, but I'm still skeptical about Young. Hope he learns quickly, but I remember how long it took Evans. Evans was a damn good one since we have been overseas. He always took up for the boys no matter what. Evans turned from just a mediocre pilot to a very good C.O. in 1 year. When he left here this morning, I dare say he was 10 years older than he was 1 year ago and a whole lot wiser.

This afternoon after the planes returned, most of the old boys (staff personnel) came up and sat in my tent. We shot the bull for a while and then some of us went in and took a shower. Then all of us got in my ambulance and went up to the picture show at Group. It was a very entertaining show for a change too. "Fired Wife" was the name of it. The only thing wrong with it was the husband and wife couldn't keep from having misunderstandings which, to me, finally got slightly boresome (knowing all the time that was no job at all if you just had the right wife). In the end, it turned out O.K. though, as usual. Hope I haven't bored you with all the above stuff. Sorta sounds like Eleanor in her "My Day" column. That's about all I can write about though — what happens to me during the day and the effects it has on me.

Got two V-mails from you today. Certainly was glad to get them too for I hadn't heard from you in three days. They boosted my morale considerably. It's pathetic how low that morale can drop when I don't hear from my honey for a couple of days. Gosh, how I love you

Joe.

July 18, 1944
11:05 p.m.

My dearest Honey,

I finally got a letter from you today. It was written July 3. Just one finally filtered through. One of these days soon I'm going to get a whole stack from you — Oh Happy Day! I also received one from Mother. There's much difference between her letters now and while Aunt Lis was there. I used to dread reading them. The last couple have been very interesting though. Again today, she said some mighty nice things about my Honey.

Today some of the boys finished up, and I should finish up some more tomorrow. These boys have really finished up fast. They are some of the first replacement crews that we got in. Some of them I don't even know their name yet. Can you imagine? I'm the guy that is supposed to even know whether they like cabbage or spinach, how they think, how many kids they have, etc. and I don't even know their names — some flight surgeon. I am doing something for them. The ones who finished up today I am sending to Capri for a week. They are leaving early in the morning.

Talking about leaving — tomorrow night Zraick and I may start out for Rome. Zraick, Young, Mahon, and I are the only Squadron Staff officers who haven't been up there on a three day pass. They say it's very nice and also educational but is now high as the very devil. If I go, I will write you one of those National Geographic Articles of mine and tell you my impression. Before I leave this theater of operations (and goodness knows when that is going to happen, but I hope soon), I want to go to Rome and the Isle of Capri.

Tonight Mahon and I went over and saw the picture show at Group. It was "As Thousands Cheer." It was the third time I have seen it, I think. I know I have seen it once before in Italy, but it's fairly entertaining and it helps to tick off some of these awfully long minutes.

Young and Curtis, plus others, have been pretty sick today. Both have that old Army ailment — the G.I.s. I don't know where we get it, but I have sporadic epidemics. I have looked and looked, and I can't find out why. Guess we will have it sporadically until we hit the States again. Since I'm doing practically nothing but "talking shop" guess I'd better shut-up and go to bed. As it is, I know that I gotta arise with dawn. Maybe I'll get energetic tomorrow and clean up my tent — who knows? Tell Kay and Carol that Daddy still "yoves" them, as Kay says, but remember that he "yoves" Mama best of all.

Goodnight,
Joe.

July 19, 1944
Saturday, 9:30 p.m.

My Honey,

I received the longest best letter from you today — it did my heart good. I know that while you were writing it you were "putting out" just exactly what you felt. I believe you were really missing me that night.

Well we moved into our new house today. Boy it's a honey. Large screen windows (4 of 'em), three more small screen windows, double screen doors, concrete floor, large floor space, running water of course, stone, high ceiling making it very cool, and room galore. It's really nice and if we live in it for over a month, and I believe we will do that and then some, we will get our money's worth out of it. Tomorrow we are having some clothes closets built. This is as good as I have ever lived on an army base. Really it's nice. Zinn, Reese, and I are the only occupants. Even though Zinn is a very peculiar boy and rather hard to get along with sometimes, I think we will manage nicely.

Kay must be a knock-out. I never got such a kick out of anything as her astride Carol saying "Giddy-yup hosie." Poor Carol — but I'm not too worried about her. According to her last picture, in a couple of years if Kay says something to her that doesn't set too well, Carol will wipe the floor with her.

There is a BBC news Bulletin coming over the radio. Sounds like we are still doing O.K. in France. Surely hope it continues. Every day is an eternity.

I love you,
Joe.

From Joe in Rome to Babe in Memphis

July 21, 1944
10:05 p.m.

My dearest Honey,

I left the Foggia area with Zraick a couple of days ago. We started out at about 7 p.m. for Naples in a jeep. The jeep had no top on it, and we figured it was going to rain but we started out anyway. Rain it did. In fact, it rained over half the way from Foggia to Naples — half of the trip is over mountains, so

247

of course that was the part where it rained. We finally arrived at Naples about 11 p.m. and went directly to the Transient Officers Hotel and got us a bed. Ate some semi-boiled eggs that Zraick had cooked for us and then went to bed.

Yesterday morning we arose early, about 6:30, and started out for Rome via Cassino. Rome is about 150-175 miles from Naples. It was a very interesting trip. I took many pictures of Cassino. The road was very good but is not too good now due to bomb craters, bullet-holes where airplanes have strafed, etc. Even many of the larger trees that were planted in rows by the highway (the old Appian Way) have been sheared off by bullets. You can't possibly realize how devastating modern warfare is.

Crosses on the road outside Cassino, Italy.

Cassino is pitiful. You remember the Battle of Cassino where the Jerries and the Allies fought a grim battle practically all winter until the Allies started their present offensive in Italy. Well there is not a single house, building, barn, shack, or anything that has not been leveled to the ground; and Cassino was a fair sized town. For many square miles not a tree has a leaf on it and the trunks of the trees are pock-marked by bullet holes every few inches. There are bomb craters, wrecked buildings, and white crosses galore. The crosses are both German and Allied. Many, many of them unknown. It really is devastating. Old Monte Cassino Abbey which did sit high up on the nearby mountain is also leveled to the ground. From there on up, you see many towns flattened out, but none left the impression on me that Cassino did.

To complete my narrative I had better leave Cassino and come on up here to Rome, which Zraick and I did. We arrived here about 2 p.m. And rode around a bit trying to find a place to spend the night. Finally we did find a place out in the very elite part of town — we couldn't move in a hotel until today cause we were here one day before our orders stated that we should be. Today we moved into one of the many nice hotels that the Army controls and uses as rest hotels.

This morning we arose rather early and went to Vatican City. The Pope holds a special audience for military personnel at 11:30 a.m. You should see the audience — very cosmopolitan. Every nationality was present — Hindus, Arabs, South African, Slovenes, Poles, French — everything and everybody. The officers are given seats up by his chair — the enlisted personnel stand. He gives his blessings in three different languages — Italian, English and French. When he is giving his benediction, some kneel, some bow their heads, and some just stand. Many kiss his robe and kneel before him, and he touches their rosary. It's very interesting.

The Vatican is surrounded by a high concrete wall. There are about 2000 rooms in Vatican City. The part where we visited the Pope is very elaborate — practically everything is gold. There are thousands of large murals which were done by Michaelangelo and Leonardo da Vinci — sculptures, inlaid mosaic work also. It's all very beautiful and impressive. Also visited St. Peter's Cathedral which is inside Vatican City. Again much gold, statues, inlaid mosaic pictures (where they used over 20,000 different colors of tile). Most of the art in the cathedral was also done by the same two guys. There are many tombs of kings, queens, popes, etc — all catholic "big-wigs." There is no telling how much the place cost nor how many years it took to build it. All of this was built in the 1600's (I think). I got a lot of pictures — hope they turn out. I took two rolls of the Pope even though taking pictures in there is frowned upon, particularly by the Catholics present I presume.

Inside Vatican City

Tomorrow I will go around and see some more sights — the Coliseum, catacombs, Mussolini's stadium (recently changed to the Stadium d'Italia) — the balcony where Musso made all his braggadocio speeches, etc. What I see of interest I will tell you about. Rome has not experienced the real ravages of war. Few bombs were dropped here, and the Germans cleared out so rapidly that they were unable to really loot the city. The people are well dressed, there is little begging on the street, the people are lighter in color, and if you weren't acquainted with the rest of Italy, one may temporarily think that Italy was a nice place. There are many fine buildings (none tall) here, beautiful parks, etc. I've written enough "Cook's Tour" tonight — more later.

I love you,
Joe.

Dearest Babe,

July 22, 1944
Saturday night, 9:30 p.m.

My dearest Honey,

Well there hasn't been a whole lot happening today. We arose rather early again. Went down and had breakfast and then out and got a shave. We then rode around town for a while just looking in the store windows.

By chance while riding around we came upon the tomb of Victor Emanuale II which is one magnificent monument. It is really something to behold — very pretty. From there you could see the Coliseum, so we went down there and looked around. The Coliseum is just exactly what you expect to see from the pictures in your geography or history book — nothing more. There are also many old Roman ruins in the same vicinity which we looked around. Took plenty of pictures of all.

We then came back to the hotel and sat around until lunch time. Had lunch, and then split forces. Zraick and Peterson went out "gallovantin" and I walked around the shops. They took the jeep and I walked — bet I walked 10 miles this afternoon just looking in store windows. Haven't seen Zraick and Peterson since, but they can take care of themselves. As I said I went window shopping. Most of the windows are empty, but some are quite full of merchandise. When you go inside the store however, most of the time you find that their complete stock is in the window. True to my prediction, the prices are now sky-high. The first thing I priced was a linen tablecloth and 6 napkins with some lace work. They merely wanted $450.00 for it. You would give $20.00 top price in the U.S. Artificial silk is 7 to 10 dollars per yard. Sandals that you would not put on your feet 15 and 18 dollars. Since I only had about $30 in my pocket, I didn't buy very much. I did buy a pair of black earrings (mosaic inlay) for 8 bucks that were made about 50 years ago by the people in the Vatican. I dare say they are worth about 50 cents, but it was about the only thing I saw that was small, within my price range, and that I thought you may like to have. You can add these to your "earring collection."

Victor Emanuel Monument

Roman Colosseum

Hey — yesterday morning I was sitting in the Red Cross having a cup of coffee when a Lt. Colonel came and sat down beside me. We began talking, and presently I found out it was Sibley Moore I was talking to, and he found out it was Joe King. Quite funny — I didn't know him nor did he know me. He is doing fine — asked about you. I suppose I was with him 15 or 30 minutes. He is stationed here now.

Remember how I've told you about when I was a kid I was afraid something might happen and I'd never get to drive a car — but I did. When I was a little older, I wondered whether I would ever complete a long road to an M.D. — I did. Now I wonder if I will ever get back to my Honey with just the same anxiety. Don't worry — I will.

I love you,
Joe.

July 23, 1944
Sunday, 10:15 p.m.

My dearest,

Tonight my Cook's tour is going to be short. I haven't done much touring today. This morning I slept very late — finally got up about 8:30 and went down and had breakfast. Peterson and Zraick finally came in while I was at the breakfast table. After breakfast, we got in the jeep and rode around town some. Drove out the "Antique Via Appian" which as you know was the first highway ever built — goes from Rome to Naples.

Also looked at the old wall that extended at one time completely around Rome. From there we went to the King's Palace but couldn't get close enough to it to see it because of guards. We then drove out to the other end of town — crossed the Tiber River and went to Mussolini's Stadium which isn't anything to brag about at all. I dare say with the exception of a few statues and a monument to Mussolini himself the stadium is no more outstanding than our own dictator's stadium — the Hon, E.H. Crump. See, all Crump needs is a few more ignorant people and a few statues and he could be considered as big a

Appian Way, Rome

Joe at the Roman Forum

"blow-hard" as Musso himself. Of course, that's just my opinion.

We then went down to a restaurant a couple of blocks from here and bought three quarts of "Spumante." Spumante is an Italian champagne. By five o'clock, all of that was gone so we went down stairs to the bar, which was then open, had a couple of cognacs, and then had dinner. After dinner we walked down to the Red Cross theatre, but they were showing an old picture, which we had all seen back in the States, so we left and returned to the hotel. I came on up to the room to write but Zraick and Peterson stayed down at the bar to try to drink up their book of chits, so they won't have to be bothered about getting a refund tomorrow. So that's the happenings of the day.

Tomorrow we have to get back to the base. In the morning we will drive out to the airport, and if Bernstein comes up here, we will hand him over his jeep and go back by air — if he doesn't I'll have a long rough drive in a jeep.

You know the Americans have certainly left their "brand" on North Africa and Italy. Besides the devastations of war, they also have many other "idiosyncrasies" of ours to overcome which will take at least a generation to accomplish. Every kid from 18 down can speak a few words of English. All can say "Hey, Joe, gimme cigaretto," or "Hey, Joe, — shoe-shine" or "Joe — gimmie chew-gum," or "Joe — gimme chocolata." However the funniest thing I've heard in a long time was tonight walking back from the picture show, there were two hack-drivers arguing and one of them kept saying "No, no capito — you son-of-a-beetch," and then he would bop the other one. "No capito" means "I don't understand." I'm sure you understand what the other part of his sentence meant to convey. It was very funny, no kidding. I'll bet he said that 10 times, and each time he got through he would slap the other guy. There's no doubt in my mind where the Italian picked up such language — whether he knew what he was saying or not.

Well tomorrow night I should be reading a couple of letters from you. I'm looking forward to it like a kid looks forward to his next ice cream cone.

I love you,
Joe.

From Joe in Castelluccia to Babe in Memphis

July 26, 1944
Wednesday, 9 a.m.

My dearest Honey,

I haven't been able to write you a letter since I returned from Rome. I will at least attempt a note and get it in this morning's mail. Rome was nice. Picturesque, historical, and educational. The most interesting thing there to me was the audience with the Pope and seeing all religions present. Educational to say the least and although it is a very nice place, I don't care to return any time soon without you.

Anyway we started out from Rome about 2 p.m. Monday and drove down to Naples. Went through Velletri and a few other wrecked places. We intended to go via Anzio, but we got such a late start that we didn't. Arrived in Naples about 7 p.m. and had supper and then drove on to camp.

Yesterday morning I was pretty busy around here doing a little back work. Also cleaning up my tent — while I was gone we had two days of strong wind, and it had blown so much dust and dirt in my tent that I had to change bedding before I could go to bed. Everything I own is covered. Well, I got some of it out (but there's plenty more) yesterday morning. Yesterday afternoon Young was going into town for his rations and I needed some too, so I went in with him. He promised me that he was coming right back — sure enough, we finally arrived back at camp at midnight.

This morning I arose at 5 a.m. for take-off, and as soon as I got back from the runway, I started cleaning up again. I suppose I am about half through now. I gathered up all of my winter flying clothes and my winter uniforms and packed them away. Gosh, but they are filthy. You know I packed that blue wool bathrobe that you gave me for my birthday last year. I couldn't help but think about when you gave it to me. How we laughed about how I had been freezing all winter and you had been carrying the bathrobe around — remember? We surely have had a lot of fun together. It's been so long since I have really enjoyed myself that I wonder whether I can anymore or not.

Looks as though the papers are trying to make us believe that this phase of the war is going to soon be over. Wish they knew what they were talking about. Perhaps they do, but I'm not going to hold my breath until it is. I don't see how it can last but a few months longer with the Russians going to town, with inside rebellion and so forth, but months seem like years to me.

Well, honey, don't let the cherubs get you down. I know they are an awful job sometimes but they will turn out O.K., and it will be due to your raising. There's nothing to worry about. I'll bet you and I were just as much a problem when we were the same age. Carol sounds precious. It's certainly difficult for me to realize that she is scooting around, saying "bye-bye," "pat-a-caking" and the like. Think how small she was the last time I saw her. Bye Honey.

I love you,
Joe.

July 27, 1944
11 p.m.

My dearest Honey,

This morning I got up at daylight for take-off — came back and lit my furnace so I could take a bath, which I needed badly. Later in the morning, I took off with Mahon and we flew for a couple of hours or so. When we returned to the field, the boys were coming in from their mission, so we had to circle around and watch them land for a while, much to both of our disgust for both of us should have been on the runway. That's one of the very few take-offs or landings I've missed unless I was completely away from camp.

I spent the biggest portion of the afternoon cleaning up my tent. Just about finished what I started the day I returned from Rome. It really was a mess too. It had dirt and dust in here that has been collecting for months. I hope it's a long time before I get another notion to clean it up. If you saw all the dirt and filth I let collect in one tent and allow myself to inhabit the same place, you would divorce me.

Tonight I've been over talking with Young and Mahon. I wish you knew this guy Mahon — he's pretty nice and very likable. I enjoy talking with him. He tells me about his wife and expected child, and I tell him about my wife and present children. In that way we have a lot of fun. His brother was killed over here on the Italian Front — yesterday he took a jeep and went up to see where he was buried.

Hey, I'm glad to know you put up some blackberry jam. I like blackberry jam better than anything in the canned sweets. You were certainly smart to do all that canning — is there nothing my honey can't do? You sure are swell to be married to. If all those other guys — Roger, Stewart, Fraser and the boys who proposed or wanted to before I did just knew what they missed they would never speak to me again. In fact, I believe some of them realized it. Even if I had gone about it as they did, you might not have married me — I don't know how I went about it, but I sure am glad that it worked. Remember the Sunday dinner when your Mother said "Well, Joe, I hear you are going to marry my daughter." I almost sunk through the floor. I guess I'm luckier than anybody and have more to live for than anybody.

Tell Kay and Carol that Daddy said hello and tell Kay that if she worries her Mommy too much that I'm gonna spank her little bottom when I get home. What I'm really gonna do to her though when I return is buy her a tricycle — hope I get home before it's time to buy two — our finances might not stand it. I love you so much.

Joe.

July 28, 1944
Friday, 10 p.m.

My Honey,

After I cleaned up so doggone good a few days ago, my place is all dirty again. Today my little tent got caught in the middle of a nice sized whirlwind, and boy it surely did bring in a lot more dirt. Just got it to do over again. Dust around here now is like mud was this winter — I'll take the dust in preference though.

Wagner has gone to Rome and left me in the "driver's seat" again. As my luck would have it, the Air Inspector came around today and did he inspect. He inspected the narcotic book and alcohol register primarily and seemed more interested in whether a comma was missing in a sentence than whether your records were straight. Some guys are like that, and I've been in this racket just long enough now to where it doesn't bother me a bit. I say "Yes — yes — yes" and go ahead and do whatever I was doing in the first place. They don't bother me like they once would have. I'm not a very good soldier.

Tomorrow we aren't flying so all the boys are up at the club "throwing one." I went up for a while, but I can't stand it for very long. The noise gets an "old" man down. They have a very poor Italian orchestra up there tonight. Like all Italian orchestras they play much too loud. Every piece, regardless of what it is, is always played at "full blast." I've just grown into an old grumpy and just can't really enjoy myself anymore.

Glad to know you are going down to Helena (or have gone). I hope you saw about getting all of my office furniture together and moving it to Memphis. I'll bet Jim is tired of it being in the hotel, and I surely would like to get what the Helena Hospital has. Don't know whether you intend to change bankers or not. It makes no difference to me. In that theatre of operations, you are completely the boss and whatever move you decide to make will be O.K. by me.

Young is sick with the G.I.'s again. I'll be glad when he finishes up and goes home — he's been sick with one thing or another ever since this winter, and he is not a good patient at all. I've had two violent cases of malaria in the past few days — first malaria I've had since being over here — gonna have a lot more though between now and the 1st of October.

Ran across the enclosed little snapshot of "the guy who loves you so much" today. It was taken this winter while I was at the first place we were in Italy. I might have sent you one like it before. Looks like I was mad but don't suppose I was any more than usual. Too, I think I will send you the picture of the Pope that they give you in the Vatican — good politics, I guess. Got small 35 mm prints of most of the pictures that Peterson took while in Rome. As yet I haven't been able to get the ones I took developed.

On closing, I will predict that if the Russians keep up their good work

— and I certainly hope they do — perhaps we will again be together before we are too old to see one another. Even so it would be nice to know that you were just there whether I was too old to see or not. I love you so very much.

> Goodnight honey,
> Joe.

> July 29, 1944
> Saturday, 7:30 p.m.

My Own Honey,

You know that old song "I Get the Blues when It Rains?" Well, it's raining and I got the blues, so the guy that wrote that wasn't too far off. You know I sang that song once in an audition at Loew's State when I was the ripe old age of 13. The following week I sang a song there that goes "One Kiss — One girl, etc." It was a good song. Never did I think then that 16 years later whenever I caught myself singing that song, or humming it, or whistling it that a handsome blonde would appear before my eyes in all the glory that only a wife and mother can have. It's swell being married to a woman like that — nothing could take its place.

I suppose you are in Helena by now — if not back to Memphis. Hope you enjoy your trip. I do hope you get the furniture detail straightened out. Also get your piano sent to Memphis regardless of its cost.

Do you like Helena now, or is it now putrefied? It has been dead so long that I'm sure putrefaction has certainly set in by now. It had rigor mortis the first time I saw it. For goodness sake, never let anyone know I talk of it thusly. I may have to go back there to make a living someday. By the way, why don't you and Thea go down to Charlie's and have a great big steak for me with a big bowl of combination salad and a lot of pepper and sauce on it? I see in the paper that you can now get steaks and roasts in the states.

July 30, 1944

Honey, I intended to finish this up this morning but was so busy I didn't have time. Since it was the end of the month, there was a good deal of paperwork that had to be attended to and besides getting my stuff out, I also had to get Wagner's out. So I didn't get to finish.

This afternoon I called the squadron together, and we had a talk on malaria again. We have had 11 cases and 8 probables in the group in the past four days. So far I have had only one known case, but I will have more soon.

Tonight Mahon and I went to the picture show. We saw Football Flashes of the 1941 games (we get the very latest in pictures), and also saw Kay Kyser in "Swing Fever." I still think the same thing about him as an actor that I always have. He still definitely stinks. This one might have been a little better that some of the others, but he still has far to go before one can class it as entertainment.

Honey, I received another package from you today. It was only mailed June 20, which is by far the fastest I have ever received a package. I have never yet received the one from Sis. I'm afraid I won't now. It has been so long. Thanks again for the pillow slip. Now that I have two, that's plenty. The peanut candy was still good too. It didn't even taste stale. It was sweet of you to go to all the trouble, and I appreciate it. I also received a V-mail from your mother, but no mail from you. She told me that some dolls came through for the children, but I think I have sent them dolls twice. Perhaps when the war is over, another doll or so may come through. It surely is funny how messed up a lot of things are. Some of these days real soon I'm gonna get a whole bagful of mail from you — I know it's on the way somewhere.

Honey I'm going to close and go "nite-nite." I'm tired tonight for a change — my back in particular. If you were only here to give me a little rub down! Guess I'll just put that off for a while. I miss you so doggone much — I just don't see how I can go on much longer without you. If I have to I will though, but this is by far the most bitter dose I've ever had to take in my entire life so far. Goodnight.

> I love you,
> Joe.

> July 31, 1944
> Monday, 2 a.m.

My dearest Honey,

Probably won't finish this tonight, but I want to talk to you a little while before I go to bed. Today has more or less been the usual one — take-off at daylight this morning, landing this afternoon, etc. However this afternoon late, I really enjoyed myself. The staff officers of the squadron played one of the crews a game of softball. Zraick and I declared ourselves the "stars" at the outset of the game and really had a lot of fun — if anyone else made an error or didn't get a hit or something we "bitched" like the devil, but to he and I, we

Boys playing baseball in Italy

were always right — of course we raised Cain with the umpire too. Did have a heck of a lot of fun though and everyone enjoyed getting out and getting a good sweat up.

Tonight Young came over, and we talked over some squadron policies and then went up to the club. There were three English officers up there that some of our boys had invited up from a nearby tank corps, and I enjoyed talking to them very much. They were rather broad-minded and I learned much about how they feel about things in general, and I'm sure they learned the same thing from me. They were very interesting — in fact that's why I'm writing to my Honey so late — I've been up there shooting the bull with them. One of them is going to bring up a tank in the morning and take me for a ride. They are also going to drive their tank around our area for us a few times and beat down the old dry wheat for us so as to reduce the fire hazard. They are very nice fellows (if you can understand them) but you really have to watch your step or pretty soon you will be telling them why the States is the best place on Earth to live and why you think it's better than England. Of course all Americans are damn cocky about the States, and since England is their home, they feel England is a pretty good place also so that subject is best to avoid.

Well I'm starting to send more men home. I'm trying to send 25 this time — you don't know any of them, they are "new" boys that joined us as replacement crews after we had been over quite a while. Just means that I've got to sit down and write a lot of letters like I did on the old boys.

Got paid today — I will try to get in town tomorrow and see if I can help out the finance situation of the Joe King's back in the States. Gonna quit 'cause it's late. Take care of yourself for there's someone over here that loves you with all his heart.

Goodnight,
Joe.

August 2, 1944
10:30 p.m.

My dearest Honey,

Got a V-mail from you today written the night before you left for Helena. I'll have to admit that I don't envy you that trip, nor do I envy you the stay in that three room apartment. To me also, it would bring back even more acutely too many things that I hold so close to me — the things that make this nasty business worthwhile. Which I cherish to the utmost and am yearning to experience again like I've never yearned for anything prior to this. Make the best of it though, and have as enjoyable a trip as you can. It might do you good to get away from the families, but I hope they aren't causing you too much trouble. Even if they were, I am aware that I would probably never know of it.

Don't worry about the amount of money I send to you each month. It takes practically nothing for me to live on over here. The only time I've ever spent any money was in Cairo. I didn't spend but $14 in Rome — $8 of that was for a pair of earrings and $4.50 for room and board. Where I wasted the $1.50 was, I dare say, at the bar. I am sending you all I can — it's for a good purpose though. When this mess is over, we are going to need every penny we can lay our hand on and then some. Firstly, for a while I don't want to lay my hands on anything until I've spent quite a while with you and the kids — that's going to take some money. Secondly, I don't know what the hell I'm going to do and feel sure that I won't know until I can get back to the States and look around and see what the heck is going on. It would be foolish for me to try to make up my mind at this stage, and I have quit trying. I've got to think that step out with your assistance verbally — it can't be done by correspondence, and I can't do it justice in this environment — it's too important to slip on. The one thing that I have definitely decided is that I'm never going to work as hard as I did in 1940, '41 and '42 again if I can help it. I will if it's necessary to make us a living, but I'm not if it's just for the money involved. Yes, I suppose I have a different perspective, but I'm gonna live with my family regardless of how sorry people might think of me. I mean it!

Speaking of money, I went into town today to send you some money but found that the only way I can send money henceforth is through the squadron. You will still get the same kind of checks — it only means that I can only send you money on a certain day of the month instead of being able to send it anytime through the Finance Office.

Well, I think I will go to bed. I'm still very sore from playing too much baseball for an 'old' man. These kids can run me ragged, and I have to play hard as the devil to keep up with them — and even then I can't, but I can 'talk a better game' than all of them combined.

I love you so terribly much that it seems foolish for me to try to put it down on paper. It seems even more foolish for me to try to express myself verbally on said subject to you. I guess I'm the world's sorriest lover. But I do love you with all my heart.

Joe.

August 3, 1944
7:15 p.m.

My dearest Honey,

Finished writing the letters and certificates today attempting to send twenty five more boys home that have completed their missions. Guess they should get out of here within the next two weeks. As you suggested, I would like very much to be on that list but firstly, I'm afraid it wouldn't go through, and secondly, I'd be ashamed of myself for those boys have really been through hell and really deserve it. Boys are made into men in a period of three months.

Tonight I have a bellyache. I've taken too much Atabrine in the past two days, and I have never really developed a tolerance for the stuff. I can take ½ tablet a day and never notice it but the minute I take a whole one it "does things" to me. I think we now have more than 20 cases of malaria in the hospital, and they are raising Cain with us from the War Department.

Well, I've rambled enough. Tomorrow or soon I've got to write to the folks and Sis. I'll bet I haven't written them in over a month. Signing off for tonight. Take care of yourself for me. I love you with all my heart

Your Joe.

August 4, 1944
10:30 p.m.

My dearest Honey,

Well I didn't go to Corsica today — slept too late. Didn't get up until 7 — think how you used to have to work to get me out of the house by nine, and how I would never schedule 8 o'clock operations if I could get out of it.

Honey, I sent you $200 this morning. Please let me know when you get it. Just thought you ought to know. This afternoon Wagner and I drove into town and got our weekly rations. It's surely gonna be funny when you can walk into a store and get matches, cigarettes, candy, razor blades, etc without a ration card. See, over here everything you buy, except quartermaster clothes, is rationed. Rations are now much freer than when I first got here. We get enough — now we even get a can or so of beer per week. This week we got a Coca-cola; however that is rather rare.

Tonight I went up to the club and had a can of beer and a drink. Came back pretty hungry, so I ate one of your cans of sardines that you sent me, and opened a can of pineapple juice to help wash it down — surely was good even though I'm not too fond of pineapple juice.

Hey, I think pretty soon I'll bundle up a batch of clothes, towels, etc. and send them home. I'm getting too much junk. I'll bet I have 25 towels and I don't need four. When a guy comes overseas, he should bring just as little as possible. I surely wish I knew exactly what I need and what I don't. From now on, I want to travel as light as possible.

Enclosed I'm putting a copy (my own typing) of some orders that came through today. Of course it's not as bad as it might sound and I don't deserve it, but I guess the guys just wanted to give me something for my efforts — after all, that's my job. I'm sending this to you 'cause I know it will make you feel good now — not for publicity. You can show it to our Mothers and Dads, but I'd just soon that it went no further. There's so many guys doing so damn much more.

It's time for me to go to bed. I love you with all my heart. Tonight the moon is out in full and it's so pretty — wish I was with you.

All my love,
Joe.

AWARD OF SOLDIERS MEDAL

JOE W. KING, 0-496213, Captain, Medical Corps, 727th Bomb Sq. 451st Bomb Gp. For heroism at great risk of life. On 29 May 1944 this gallant officer rushed to the scene of a crashed aircraft at an air base in Italy. Although one (1) engine appeared to be on fire, many gasoline leaks and a full load of ammunition and five-hundred (500) pound bombs made an explosion imminent, Capt. King, with complete disregard for his own safety, entered the wreckage and extracted one (1) member of the crew who had been critically injured. By his heroic and voluntary action under extremely dangerous circumstances, Captain King has upheld the highest traditions of the Military Services, thereby reflecting great credit upon himself and the Armed Forces of the United States of America. Residence at appointment: Memphis, Tennessee.

Report of the rescue that led to Joe's Soldier's Award

August 5, 1944

My dearest Honey,

I'm tired tonight. I've done a lot of traveling today. Went over to Corsica via Rome the earlier part of the day. This afternoon we flew back by way of Anzio. However, you can't tell much about a battlefield by airplane — you've got to get down on the ground before it really means anything to you.

Boy, for about 1 hour late this afternoon it really rained. We landed in it, and as I said before, it really rained. Then it cleared off as quickly as it came up. Tonight it is beautiful. The moon is out in its full glory. The outside is as bright as it can be — what a wonderful night it would be to put the top down on the car and go driving with my Honey. What a wonderful night any night would be with her whether we went driving or not. I have been with Kremers tonight — he brought a case of malaria that he wanted me to see — I made him stay, and we went up to the club and talked about what little medicine we remember, also about our experiences when we were practicing. Helps you out to talk to someone like Kremers. Run out of paper again. But you know I love you with all my heart.

Joe.

August 7, 1944
Sunday, 11:10 pm.

My Honey,

Haven't done one cockeyed thing today except go to take-off and landing. Even this afternoon, I spent a couple of hours with another bored friend betting a dollar who could hit a tin can the most with rocks. Sounds rugged, doesn't it? Some days really are long — this has been just another one of them.

Boy, I wish this thing would hurry and get over. I want to get back to you so bad it is difficult for me to think about it without becoming depressed. The luxuries, the lack of ants, grasshoppers in your bed, fleas, lizards, tents, C-rations, etc. don't bother me one iota. I could live with those things for the rest of my life. But, being away from you is really getting me down. It would be worth more than I could ever afford to pay just to be in your presence so I could feel happy again.

I love you,
Joe.

August 7, 1944
7 p.m.

My Honey,

I think I dated the letter I wrote to you last night the 7th also — don't know for sure. This is Monday, I'm quite sure though. It's very difficult to keep up with the days — every day is exactly alike. Every time that you find out the date you are surprised that it isn't a few days past that — time passes very slowly.

Got a letter from Mother today. It was the first that I had received from her in quite a long while. Received a V-mail from you too written the night that all the electricity was out on your street. By gosh, I'm losing my mind! I can't even think of the name of that street! I know the number is 1227, but for the life of me I can't think of the name. Guess I am going nuts after all.

Played baseball again today. Had to quit after 3 innings however for the planes were coming in. Didn't hurt myself today — believe it or not. Maybe I'm getting "younger."

Well, today the boys in France cut off the Brest Peninsula. Don't know how important that is from a military point of view but it must be something, and they surely did it quickly. Surely would like to see this part of the war over by winter. I've spent all the winters I want to spend in Italy — one's sufficient.

Well, my honey, think I will go up to the club and have a couple of drinks with the boys — also got a New York to read. Kiss the kids for their Daddy but save a bigger and better one for yourself.

I love you,
Joe.

The 451st completed their 100th mission in August. On the bombers' return, they were greeted by a giant "100" on the ground. About 300 men were standing in formation with white cards over their heads to celebrate.

* * *

August 10, 1944
Thursday night, 10:45 p.m.

My dearest Honey,

My new house is just about finished. Think I will wire it and put in the waterworks tomorrow — or, at least, the other two guys and myself will. Boy, it's nice and cool and roomy. We will have so much room that we will probably get lost. There's room for 12 men to live in it if we live as crowded as we do in tents, and only three of us are going to occupy it. I think we even have hired a little Italian boy to keep it clean for us and make our beds. Perhaps he can shine our shoes once every month or so for us too. I'm rather anxious to move in.

I sent 25 more boys home today and wrote out papers trying to get some more home. Certainly is nice getting them back — funny you don't hate to see them go at all even though they are your friends and you will probably never see them again. Nor do you envy them for you know how much they deserve it.

Well, honey, this is the last sheet of stationary that you sent me. I will have to dig up some more tomorrow or use V-mail. Tell the kids hello for Daddy.

I love you,
Joe.

August 11, 1944
9:35 p.m.

My dearest Honey,

For the past two days this army life has almost been like private practice. Night before last, I had to get up four times to go see someone at the dispensary. Last night four of the officers had a jeep wreck and crippled up all of them — 2 rather badly. One had a broken back and one a severe fracture of one elbow. So I was up all night getting them into the hospital. Consequently, I am pretty damn tired tonight and am going to bed early. I don't like night work any better now than I ever did either.

While I was in town this afternoon, I went to see a pretty fair show — "The Angels Sing." That Hutton gal is pretty good. I like to watch those expressions of hers even though I'm not too sold on Lamour. It was entertaining though, and that is something for the shows that we get over here. Have the shows in the States deteriorated until practically all of them stink?

Guess what I bought today? A Colliers and two Saturday Evening Posts. Got them at the PX, and they aren't too old either — all printed in June. That's the first time we have ever been able to buy reading material other than the "Stars and Stripes" (our daily news paper) since we have been overseas. I hope we are able to continue to get magazines. In a few minutes, I'm going to pile into the sack and start one of them. Surely hope I don't have to get up tonight.

Honey this piece of discourse may be a bit shorter tonight, but I'm tired-er than usual. My back hurts right between my shoulders and in the small of my back. Goodnight, my Honey, and goodnight to the kids. I love you all.

All my love,
Joe.

August 13, 1944
Sunday, 10:20 p.m.

My dearest Honey,

I'm trying something that I have never tried before. I'm trying to write a letter while up at the club with all the distractions about. I bet that I will never be able to complete it up here.

Got a nice long letter from you today. In it you were telling me of how you may hurt feelings, etc. Go ahead and hurt feelings — something tells me if I were down there and Bill Butts were around, I'd probably make a few sore at me. Just do the best you can, Honey, but get it all cleared out — plus your piano. (The furniture, of course!) I want everything that we own out of Helena. Perhaps I will have to go back down there someday to practice, but I certainly don't want to.

Last night, I got the hiccups and bad. Went up to the club tonight, had 2 bottles of beer and one drink, and it has given me the hiccups like I have never had before. Saw one of my English friends up there who is a Lt. in a nearby tank corps and had an enjoyable chat with him — talking about the difference of the States and England — amiably, of course. He is very interesting — I enjoy talking with him. He invited me over to his place tonight for Scotch, but you know how I am — I don't like to go to someone else's place — if they want to come here O.K., but I like to hang around here.

Hey, Honey, how much have we got in bonds now? I really haven't the slightest idea how much I left in the safety deposit box. When you write and tell me, be sure to tell me whether you are telling me how much in actual value or maturity value — that is whether you are counting a $100 Bond worth 75 bucks which we paid for it, or $100 which it will be worth at maturity. Also let me know what your bank balance is. I'd just sorta like to know how things stand now. We should be getting a little better off all along. When I get home, we will need it for we are going to really enjoy ourselves. Of course, we will enjoy ourselves whether we have any money or not.

Got a letter from Dan yesterday. I think he is hurt 'cause I haven't been over to see him. I suppose I should have gone over, but you know how it is and how I am. I'll try to get over to see him soon but I never have any business over that way, and there is nothing we can do together due to that Officer and Enlisted Man stuff. If we go into town, he has to go to one club and I to another.

Not long ago when all the commotion occurred in Germany, I thought this damn war might terminate pretty soon. As it turned out, as far as I'm concerned, it might have just strengthened Germany. Don't like to hear of you weighing only 110. That isn't enough — eat and quit worrying. You are doing a swell job. Take care of yourself for me 'cause remember how much I love you..

Joe.

August 14, 1944
12:45 a.m.

My Honey,

Got the longest nicest letter from you today. The first of that length that I have received in a long time. It surely was good too. I feel much better tonight just because of it. Your letters, and yours alone, are just about the only things that keep me plodding along. They are all important. The contents are read, reread, ciphered, deciphered, and studied. I can tell how close you feel to me at that given moment, your mood, almost your thoughts sometimes when they aren't down on paper. I realize you can do the same by mine.

Terribly glad that you got the children inoculated for diphtheria. It is something that I should have done for Kay years ago. If she had gotten diphtheria, I certainly wouldn't have forgiven myself for such gross neglect. You know, diphtheria inoculations are practically 100% preventative. It should be a must for all children at 6 months. By the way, did Storm use the "one shot kind" or the "two shot kind"? It doesn't matter a lot, but the two shot variety is probably more effective in little children and babies. I know it hurt you to watch them have to take it, but think how much it would have hurt us to sweat out a real illness where life was at stake. I'm certainly glad you had it done.

Don't worry about me groaning at how much you eat at 110 lbs. If I were there, I would probably poke as much food down you as you use to do to me. I'm doggone glad you have the newly found appetite, so go to it. I certainly won't even make cracks until you are well past the 120 mark. What do you want people to think — that I won't give you enough money to keep you from starving? From Kay's and Carol's pictures, I see no reason for me to worry about them on the eating score — they look like butter balls.

It's time for me to turn in, I'm sure. It's almost 2 a.m. Lately I've been reading quite a bit at night, so I've been staying up later. It surely does feel funny to have to get up the following morning still sleepy — I'm just not used to it. These relatively new magazines surely do help out. The articles in them are stuff that you have not yet heard about through other sources. I hope they keep coming in — it's nice to sit down and have something to do to pass away a few minutes. Goodnight, honey. I surely do love you. Tell Kay and Carol "hi."

All my love,
Joe.

August 15, 1944

My dearest Honey,

Today has been a real scorcher. Boy, it's been hot, and not a ripple of a wind, which is very, very unusual for this part of Italy. We usually have a rather brisk breeze all the time. Tonight I'm practically burning up — the first time I've ever had this sensation at night in Italy. I certainly dread getting under that mosquito bar for I know how hot it's going to be under there. I'm writing by candlelight again. We can't get the generator started — it's just about worn out.

I ate that last can of sardines today that you sent me. They surely did taste good too. Also drank the last can of juice. Honey, I wouldn't bother about sending me much stuff. I can get practically anything now. Of course we are still rationed, but our rations are plentiful. I probably can get as much of the stuff that I need as you can what you need. A soldier's needs are not much. I've run out of things to send you though. Can't send you anything else without repeating myself.

The picture I'm enclosing is of me congratulating one of my pilots on making a safe crash landing of the ship we're standing by. That's why the nose of the ship is so low — I really haven't grown to such an extent that my head is nearly to the level of one of those airplanes when it is parked! It was taken by our Public Relations personnel. This kid in the picture with me has since finished his tour of duty, and I wrote letters on him only today attempting to get him back to the States — nice boy. Notice the "Flying Boxcar" that is so heavily armed on his jacket — that is our squadron insignia.

Your mother mentioned in a V-mail that she hadn't heard from Hugh for quite a while. Don't worry too much — it's going to take a long time to get the mail situation straightened up in that sector — especially as fast as they are

Joe' shaking pilot's hand after a successful crash landing.

moving. Give Hugh my best when you next write to him, and tell him I think of him often and sincerely hope the going is not too tough. I know it must be pretty rugged at times though. Well, honey, goodnight.

> I love you,
> Joe.

> August 16, 1944
> Thursday, 11:10 p.m.

My Own Honey,

    I'm a little later than usual in writing to you tonight. We have had a big party up at the club tonight — Italian band and five signorinas — the most that the boys have seen in quite a while apparently. They are all still up there making "whoopee," jitterbugging, etc. Me, I'm much too old. I did "try myself though." I had two beers tonight instead of one. Capt. Bell (the doctor from the service squadron attached to us) came over and we went up and had a couple of beers.

    The C.O. of the wing, while presenting the medals yesterday, told us that if the war was over tomorrow it would probably be 8 months before we got home even though we weren't shipped to some other theatre. That didn't sound so good. Firstly, the war isn't going to be over tomorrow, and, secondly, I've been without my honey too long already. When he said that, had it not been for the toughness of my feet, my heart would have dropped completely out of my body — even though I have known that it was the truth, I still don't like to hear it.

    My little "valet" (more or less), Di Domenico Leonardo, has been tugging at my coattail all day. He is 12 years old and smart as a tack and cute as the devil. He does everything for me now — makes my bed, shines my shoes, gets me water, my mail (when there is any, doggonit) and so forth. He is an enjoyable little cuss. Speaks English and Italian and acts as my interpreter. This afternoon he was gone for a while and came back with a watermelon for me. The Italian watermelon is no larger than a cantaloupe in the States.

    Honey, I'm going to bed. I miss you, I want to be with you, I just love you.

> All my love,
> Joe.

THE STARS AND STRIPES
MEDITERRANEAN

Vol. 1 No. 58. Thursday, August 17, 1944   Printed In Italy   TWO LIRE

RIVIERA COAST SECURE

August 17, 1944

My dearest Honey,

Today has been another scorcher, and I have been standing up almost all day long. Today we didn't fly, so the group took the day off to present medals. This morning we were presented with the blue ribbon you wear over the right pocket for good showing, more or less, known as the "Presidential Citation," which means just what it says. This afternoon the commander of this wing came down and presented medals to everyone but me. I got just the ribbon — the group didn't have any soldiers medals on hand. So even though it has been officially presented by the general, I got the ribbon but not the medal as yet.

We haven't got lights yet. I'm writing by candlelight again tonight, and the candle is just about gone. In fact, I'd better get up and scrape for another right now. Got a chance to go to Naples and Corsica tomorrow, but think I will stay home and work on my house instead. Might move in tomorrow. Take care of yourself and the kiddies. 'Cause I love you with all my heart.

Joe.

Joe's Selective Service Medal and Soldier's Award

August 18, 1944
10:40 p.m.

My dearest Honey,

I've had a lot of fun the last couple of days out of this little boy working for me. He is one of the smartest youngsters I have ever seen — and to hear him talk will knock you out. Every time I leave the tent, he immediately demands to know where I'm going — how long I'm gonna be there — when will I get back etc. It's the first time I've had to say where I was going at every move since I was about 8 years old. Today he asked me "who did your laundry before I work for you?" Of course, he just assumed that since he worked for me, his mother was naturally going to do my laundry. When I come around, he runs out and grabs my hand and walks back with me — he won't go through a door before me for nothing. Today I made him take a bath — after the first attempt I found out that he didn't know how to adjust the hot and cold water and was only getting hot water, so he only washed his legs. Later I adjusted it for him and he took a "whole" bath.

He has impetigo over one side of his face and of course the itch, but we are clearing all this up nicely. He sits in my tent and plays the harmonica when he's not busy at some chore — his favorite song "Lay that Pistol Down Babe" and "Red Wing." I have gotten a kick out of him really, he's one of the brightest kids I've ever known. His dad has been a prisoner of war of the English down in Abyssinia for the past 8 years. Enough about the kid — I wish you could see and hear him in his typical Italian accent.

This afternoon Wagner and I went into town and got my rations. Came back and sat around talking. We were having a dust storm, so tonight there is dirt a half inch thick on everything. After taking Wagner home, Kremers came over and we went up to the club for a while. Otherwise the day has been exactly the same as usual. Time to say goodnight to the best woman there is.

Your Joe.

August 20, 1944
Sunday, 10:45 p.m.

My dearest Honey,

Would just write a V-mail tonight for I haven't much to say. However, I was given the enclosed picture today by the Public Relations officer and I thought you may be interested. As you can see the scar even shows up on my neck. That is Wagner standing beside me, but his face is hidden.

Today has been like most of the rest. The only thing different about it was that my enlisted man on duty last night let me sleep through take-off. So I didn't arise as early as usual. It's one of the very few I have missed.

Went to town today and got a shave, haircut and shampoo. When I sat down in the barber's chair, I suddenly realized that I didn't have one penny on me when he was about half through cutting my hair. I was supposed to meet Zraick downstairs when I was finished but since I was broke I had to keep the barber busy on me until Zraick got tired of waiting and came up for me, so I could borrow some money from him — hence the shampoo. Yes, tonight I guess my hair is red, but if I can find enough oil it won't be tomorrow.

Kremers came over tonight — he is as "down in the dumps" as I am. He was no help at all. We had a couple of drinks and then I carried him back to his area, so I could come write before it got too late. Everyone else is in bed, so I guess it's time for me to turn in too. I love you with all of me.

Joe.

August 21, 1944

10 p.m.

Dearest Honey,

Tonight 9 months ago I went to sleep with Kay playing on my tummy — remember? You had to roll me into bed. I was so tired and sleepy. Only nine months ago but really centuries.

Got a letter from you today. In it, you had just found out that your Dad was going to have to rent a truck to come get our furniture. Hope you made it alright. Sorry it was all so doggone much trouble. Too, in today's letter my Honey was pretty blue. That is probably reflected however, for the past few letters that I have written to her have been pretty terrible. Sorry that Janie's gone for your sake but don't have the audacity to compare her cooking with yours. You cooked better the day we got married then Janie did the day she quit working. I'm eager to taste some of your scrambled eggs, and biscuits again — plus some good cold milk.

I love you,

Joe.

Mission #108: 22 August 1944
Target: Lubau underground oil storage, Vienna, Austria.
Results: Main concentration S/E of plant on canal dockage, partly destroyed. 1/3 bombs directly in the target area, probably destroying and damaging the underground receiving tanks and section of the blending plant. 1 726th ship bailed out just off Italian coast, picked up by ASR. 6 aircraft lost to fighters, 1 missing.

This was the first of two missions, and the repair crews had to work through the night to prepare for the next day. The holes in the planes were patched with masking tape and painted because the repair crews didn't have enough time to fix all the damage. The "patch job" was more to set crews' minds at ease than to actually repair the planes.

Mission #109: 23 August 1944
Target: Markersdorf, A/D, Austria.
Results: Bombs start at the S/W corner of A/D, across to the hanger line on N side. Of 20 enemy aircraft visible on ground, 5 probably destroyed, 2 damaged. 9 aircraft lost to fighters, 8 aircraft damaged.

On the way to Markersdorf Airdrome, the 451st was met with resistance from the Storm Group, elite fighters of the German Air Force. The 451st was successful in spite of the enemy fighters. Bombers flying in a tight formation destroyed buildings and supplies at the crucial airfield. However, this was one of the worst days for the 451st. They lost 90 men. Once again the 451st was awarded the Distinguished Unit Citation. They were one of only a handful of Army Air Force units to earn three DUC's.

August 23, 1944
11:05 p.m.

Dearest Honey,

This is the third V-mail I have written to you. The first time I have ever pulled such a trick on you before. I'm sorry but for the past few days it just hasn't been in me to sit down and write a decent letter.

I've been with Wagner quite a bit today, and tonight he and Marco have been down here. We couldn't shake Marco — I still have as little use for him as I have ever had. But it is very enjoyable to be around Wagner. He surprises me sometimes with some of his observations. Didn't hear from you today again — suppose I will tomorrow. Did just get through hearing over the radio that Romania is asking for peace terms with Russia. Don't know if that means very much to us, but at least it might be a step in the right direction. Perhaps it won't be too long before all of this is over with, and I can be home with my Honey again. Oh, what a happy day. Until then just realize that I love you with all my heart.

Your Joe.

THE STARS AND STRIPES
MEDITERRANEAN

Vol. 1, No. 242, Thursday, August 24, 1944   ITALY EDITION   TWO LIRE

Patriots Report Paris Freed

August 24, 1944
Midnight

Dearest,

Today was a red letter day. I got two Air-mails and a V-mail from you. The first time I've received more than one letter from you in ages. They surely were a sight for sore eyes, and they helped my morale out no end — it has been needing a little help recently too.

Tonight they are having a big party up at the club — much too big. I went up and told Young off about it more or less — maybe he will put an end to such. I've tried — but without backing, it's no good. In the morning, I'm going to call the whole bunch together and give them all a piece of my mind and then some. Gonna really step on a lot of their toes and pull no punches regardless of who it is. These boys are going to think I'm really the chaplain and not the understanding flight surgeon when I get through with them. They really set a beautiful example tonight for the enlisted men to follow, and I'm pretty griped at them. I don't know whether the answer is lack of character, lack of morals, ignorance, or just don't give a damn. I strongly suspect the latter two though, and I should be able to change some of that. Some of these boys are fatalists and are pretty hard to change.

You asked a lot of questions in one of your letters today that I will attempt to answer but for the most part your opinion is as good or better than mine. 1) When the boys come back, I don't know what will happen to them. They will all rest for 3 or 4 months I presume — then some will go back to combat and some won't. 2) When the war is over, will being a Flight Surgeon help me to get back quicker than the other doctors? — that I don't know. I can only say I hope so, but I'm afraid that is going to be left up to the "breaks." Some of us, I believe, will be left over to be just plain doctors in reconstruction, some will return. Again, that is merely my opinion, and the army never does what I think it will do so that is bound to be erroneous. 3) Kremers is still with us, but he is still having trouble — I wouldn't be surprised if another winter in Italy didn't just about finish him up though. Until cold weather I believe he will be O.K. 4) The nights are still cool here — sometimes you only need a mattress cover for warmth and sometimes a blanket — but you always need something. I can't remember but one or two night that it has been "stuffy" here. The altitude is not very high here though.

By the way the blue earrings you mentioned did not come from where you thought they did. They are a product of Porto Rico — the other (the silver ones) British Guinea. By the way, in the papers that I've seen recently from the States, they seem very jubilant and apparently feel the war is won. I even read where the "Luftwaffe was dead." Don't be misled. Certainly, I think much progress is being made, but the boys are pretty sure the Luftwaffe isn't dead. I still say the Hun soldier is strong, courageous, brave and a damn good soldier. Don't sell him short. Seems that they are worrying more about what they are going to do about peace when it gets here than anything else. If they don't get off their fannies and do something about winning it, there will be no use to worry about peace. The war is not over — many lives are yet to be lost, in my opinion.

To get off of that subject and finish up with something really pleasant to talk about. It's almost impossible to think of Carol as being 11 months old. Isn't the little tyke even trying to pull up yet? If she doesn't start pretty soon we are going to have to start her in school in a wheelchair. Kay was pulling up and going everywhere (and getting into everything) at 11 months, wasn't she? Guess you will have to have a heart to heart talk with our youngest daughter.

I've saved up about 15 packs of gum for you and Kay. I am under the impression that it is even more difficult to get the popular brands back in the States than it used to be. I'll send them as soon as I get a chance. Did the other batch along with the rest of the stuff ever arrive? Honey, I must say goodnight. It's pretty late in the morning. I still and will always love you with all my heart.

Joe.

August 26, 1944
10:10 p.m.

My Honey,

Oh! Today has been O.K. Got two nice letters from my Honey today, one from Sis and one from Mother. Not only that, yours were written the 15th and 17th so it sorta brought me up to date. I'm glad that you are again back home — see, 804 Spring St. means something to me too — even though I have never had the pleasure to really live there. I know I would be happier there than in Helena so, naturally, I guess I presume you are too.

In your letters today you were spraying me with pride about the Soldier's Medal. Thanks, although again, it wasn't so much. You just feel that it was because it was me involved and that you knew about it. Most of the boys do more, and many of them get no recognition for it whatsoever. Thanks too for not putting it in the papers — I appreciate it.

I guess I have a mild surprise for you. You mentioned seeing a picture at Mother's of me when I had curly hair. I don't suppose I have sent you a hatless picture in quite a while 'cause your old man has now got curly hair again. Why? I don't know, but it is just as wavy now as it has ever been. Sometime I'll take a snapshot and send it to you, so you can see my hair. If it has turned straight by the time I get home again, then you won't think I'm a liar. The peculiar thing about it is that the right side was curly for 3 months before the left side even had a wave — it gradually spread across and I started pulling the part nearer the left side in order to separate it. Finally I started parting my hair just to the left of the mid-line which I have been doing ever since — not in the middle anymore but not really on the side. As the Italians say "Come si, come sa," which means 'just in between" or "on the fence" or something like that. So you see if it means so much to you that I have curly hair, then I'll grow you curly hair! I aim to please — and I aim to please you for the remainder of my life — and I'm not intending to fail! You'll have to admit that damn few men would change their hair from straight to curly naturally just 'cause their wife likes it better that way.

Thanks for all the letters — they really help me an awful lot. I feel better tonight than I've felt in a long time. You are a sweet, good, wonderful woman. I'm just as proud to be your husband as you are my wife. Kiss the kids for Daddy. Goodnight and I love you.

Joe.

August 27, 1944
10:45 p.m.

My own Honey,

Today has passed a little faster than usual. I have made myself busy all day. I've been painting. All the furniture, or at least lots of the furniture, that is in our house now is either a bright blue or a bright red. Tomorrow I'm going to get some bright yellow and paint some more. Somehow everything reminds me of our fiesta set of dishes — but it surely makes the place look nicer. The steel crates that 2,000 lb. bomb fins come in make quite attractive side tables by the chairs when they are painted. Fuse cans make excellent ash trays. You would certainly be shocked to see how nice everything looks. Tomorrow I'm going to get me some plexi-glass (airplane window glass) and make glass tops for all the tables.

We got a Lt. Colonel assigned to our squadron today. He is going to be squadron C.O. as soon as Young finishes up and goes home, which shouldn't be over a couple of weeks. The new future C.O. seems to be a fairly nice guy, but I will tell you more about him after he actually takes the squadron over. It seems he has not had a lot of heavy bombardment experience, but soon as he catches on that shouldn't make too much difference. I hope he turns out O.K. 'cause he will probably will be with us a long time.

Mahon and I went into town this afternoon for his rations. I still enjoy being with him more than anyone else in the outfit. He, too, should finish up very soon and be on his way homeward. I sorta hate to see him go — for selfish reasons of course 'cause I enjoy his company very much. His wife is supposed to be having a baby just about now, however, and it will be nice for him to get to go home soon. His wife is an ex-school teacher also. I have seen her picture, and she resembles you very much, believe it or not. She is blonde, tall, skinny, and fixes her hair like you used to with the "horns" on the side.

Gosh, Honey, I would like to be with you tonight. Any night for that matter — better still would be every night from now on. May it be soon. I love you with all of me.

Joe.

August 29, 1944
Tuesday, 10:35 p.m.

My Honey,

Charlie Bently went to the hospital today. He was taking a routine exam-ination to come back to the States and take some specialized work. On his chest plate, he was found to have active pulmonary tuberculosis. I haven't yet seen the X-rays so don't know how bad it is but it's definitely bad anyway you take it. I certainly was sorry to hear it and am also sorry for his wife.

Major Marshall returned from the States today. I suppose he will be given over Bentley's job now as Group Adjutant. I was sorry to see him return to this outfit for he has always been much disliked by everyone — they call him "Tojo — the Black Major," which isn't too complimentary.

I was supposed to have had a wing inspection today. Had everything in pretty good shape just waiting for him. The inspector came while I was having lunch. He just came in and had his lunch and asked me how everything was. I told him O.K. so he didn't inspect. It sorta griped me for I was "cocked and primed" for him. If things had been in a sorry shape, he probably would have had a minute inspection.

Tonight Young, Mahon, the new Lt. Colonel (Col. Hoppock) and I got in my ambulance and went into town. We went in to see a show, but when we got in there was a line over two blocks long, so we decided to just go by the Red Cross and then come home — which we did. However, the lights got shorted out on my ambulance, and it was really a rough ride. I couldn't see the ruts in the road and it was surely terrible.

I hear that the Germans evacuated Ploesti today. That's the best war news I've heard in a long time. To me it ranks with the Normandy Beachhead — this is a selfish point of view however. It has always been a rough target — so they say. Maybe this business will end someday after all. I just hope they send us to the States for a while before we go to the Japanese theater. Course one can never tell. I surely would like to know whether I've spent half of my "sentence" yet away from you but realize there's no way to find that out. Can't stand it too much longer.

Young is about finished up. He should be on his way home soon. It seems that McManus may get to come home soon for some special training. He is going in Bentley's place, I hear. If that happens, Zraick will probably take his place and that will just about leave me out here by myself as far as old personnel is concerned. Guess I can make some new friends. In fact, I think the new C.O. and I will get along pretty good — he seems to be a likable sort. Already, in two days, he is calling me "Quack," so he's at least friendly to get so familiar. I'll hate to see Mahon go more than any. He has so much more nature than Young or any of the rest of the guys around.

I've "writ" enough (by hand). I wish I could kiss those lips that made those lip prints on the letter I got today. I shall never be happy until I can kiss them again. Tell Kay that Daddy has received all those kisses that Mama 'wrote' to him and answer them for me to you and her.

I love you,
Joe.

August 31, 1944
10:45 p.m.

My dearest Honey,

Today has been nothing out of the ordinary. We didn't fly today which adds a little to the boredom. Mahon and I went into town this afternoon and got our rations. Tonight I went up to the club with our new Colonel and drank and chatted with him — still believe he is going to be alright. I wouldn't be too surprised if he didn't turn out to be better than Young. However, the squadron is getting worse daily — it doesn't feel like "home" anymore. I'll bet I don't know 25 guys in it and I used to know everyone.

I have been spending the majority of my time lately working in our house. It's the finest thing in this Air Force. The carpenter has finished private wardrobes for each of us and each of us a bedside table. I put a plexi-glass door on my wardrobe, and we have plexi-glass tops on all our tables, and most of our front doors are plexi-glass — quite modernistic. We get something new put in or built daily and each of us paint something daily. It is really artistic — and gobs of tables — one at each chair about 2 x 2 feet in size and even more ashtrays. There is certainly no excuse for putting ashes on the floor. It's the nicest thing I've lived in since I've been in the Army — either over here or in the States. We are starting a built in tile bath as soon as we can get our stonemason back. We have already obtained all the fixtures and fixed the boiler for hot water and got a 600 gallon tank for a reservoir. We, of course, already have running water in the lavatory. Plastered walls, blue shutters — everything. Everyone envies us (instead of getting out and fixing up something for themselves).

Honey, I sent you $150 today. Got paid $183 and owed Zraick $25 that I've spent this month for "chit-books" at the club and the mess hall and helping to pay some of my part in the house. I still have about 15 dollars left though, which is plenty — it doesn't take much for us — unless you are buying stuff. Let me know when you get it.

I like the idea of the 3 piece gabardine suit — get a good gabardine now. Poor gabardine is nothing like as good looking as good gabardine. I still want whatever you buy to be something you really want and especially nice. In fact, I think it is much cheaper in the long run to get nice stuff you really like and admire rather than something that will "just do" — cause then you take pride in wearing it rather than hoping it would hurry up and wear out.

Honey, this is my last piece of paper: must close. When I get with you again nothing is going to separate us, — mothers, dads, children, wars, or anything — I'm not worrying about ourselves at all.

I love you,
Joe.

Fall 1944

At the end of the summer of 1944, the Allies rescued almost 1200 POW's from Romania. The multi-day mission, called Operation Reunion, sent B-17's to Romania to pick up POW airmen. Those prisoners were taken to Bari, the 15th Air Force headquarters. Then they were returned to their units before they were transferred to the U.S. The surviving crewmen from the Shilay Lee B-24, shot down over Ploesti in May 1944, stayed with Joe for a few days before leaving Italy.

* * *

September 1, 1944
10:15 p.m.

My dearest,

Got an Air-mail and a V-mail from you today. The mail situation has been much better the past few days. Surely hope it continues. Glad you got the jewelry, pictures, etc. I will send Mother something the next time I send something to you. That is except some gum which I already have packed and intend to send the next time I go into town. I hear you don't get good gum in the States. I don't care so much for it, so you and Kay might as well enjoy it. Both of you can have plenty. So she can get plenty in her hair, and you can sit down in plenty in the chairs (as if you don't have enough to worry about).

Honey, I would write and ask you to send me stuff, but it is no use for you to go to the trouble. My needs are small and I can get everything I need and then some. I don't care about the eating side of the question so much, as you know, so why send me food. Just tonight I sat here and ate a K-ration simply because I was too lazy to walk over to the mess hall, I guess. Eating is not such a problem. However, one does get tired of C-rations I'll admit. I'll tell you what to do — every time it enters your mind to send me a package, send me a picture of you and the kids instead (if you can get any film at all). I much prefer that.

Yes, we now have a good boy to do the cleaning up for us and making the beds. We got rid of my little friend and got another, and this one is really good. He remakes my cot daily layer by layer, and uses insect spray between every layer, sheet, blanket, or mattress. He has even gotten rid of the fleas that I have been sleeping with ever since I've been overseas. I'll bet you didn't know that fleas didn't bite me — see, you learned something about me that you never knew before. As I just said, I've slept with them for many months and they haven't bitten me yet. They don't bother me except sometimes get in the center of my back or under my undershirt crawling. Anyhow, I haven't noticed them since the new boy has been working for us.

Today we started building the walls to our shower. It's going to be a knock-out too — we are quite comfortable — you would certainly be surprised to see how we are living now — We are really living in class plus.

Tonight I went over to Group to the picture show. Saw Betty Grable in "Coney Island" — not too good, nor too bad. Just served to pass another hour or so, and the hours can't pass too quickly until I'm back with you — then they can put 100 minutes in an hour if they want to instead of the regulation 60. Love is a wonderful thing — something so vital, so damn deep, so sacred. I have never missed anything in my life before — relative speaking. Nor have I had such a great anticipation that I have had to control before. Sometimes I don't control it and it gets the best of me for a few days. I hope I am back with you before I have many more "bouts" with myself. I love you with all my heart.

Joe.

September 3, 1944
10:30 p.m.

My dearest Honey,

I did my honey bad last night. I didn't write her. So many things happened around here that I didn't have a chance. One of the old fellows that came over-seas with us who was shot down 5 or 6 months ago over Romania turned up and spent the night with me. He is still with me tonight. I have to take a boy down to Bari tomorrow, and he is going with me down there. I will be in Bari tomorrow night and will return here the next afternoon. It surely is good to see an old face and find out that some of the old boys that went down are safe. We were together so doggone long and I knew them all pretty well.

The last few days and nights have been so hot it's almost been unbearable.

This afternoon a little rain blew up, and tonight we have a very brisk wind, which is not only cooling things off considerably but also stirring up so much dust that one can hardly breathe. However, it's surely welcome. But I haven't been in as much heat as you have this summer.

Honey, Young finished up his missions yesterday. As soon as I return from Bari, I will write up the necessary papers to get him started homeward. Don't think it has quite dawned on him yet that he is through with this grind. I am glad for him and will be glad to see him go.

Honey we received another presidential citation today. We are about as much decorated as any outfit in the army. We got our second citation today for our 1st raid over Ploesti. Honey, I'm gonna cut this one short on you. Take care of yourself. Tell the babies hello for their Daddy.

I love you,
Joe.

September 5, 1944

Dearest Honey,

Should write you a long letter tonight, but if I get to complete this I will be lucky. In our house, there are nine of the old boys besides us sleeping on cots. It's surely good to see them. They have been prisoners of war for the past 5 months, and we didn't know who was alive or dead. I suppose they will be around for a few days and then may be sent home.

I went down to Bari yesterday to get rid of a guy. Did my little business and yesterday afternoon went to a picture show. Last night I went to 15th A.F. Headquarters and listened to the orchestra that plays there. They are really good. I enjoy them every time I hear them. Honey, I will write you a long letter the first chance I get. The boys should be outta here in a couple of days. It surely is swell to see them again and listen to their tales.

I love you,
Joe.

September 7, 1944
9:30 p.m.

Dearest Honey,

I haven't been writing to you very regularly for the past few days. All these boys in our house with us are just too distracting. They are all leaving tomorrow, for which I'm glad. One reason is that they will all be going home, secondly they will move out and we will have some peace and quiet and room enough to move around. It certainly has been nice to be with them for a few days again.

Young got relieved from the command of the squadron yesterday, and Col. Hoppock took over. Young was rather surprised about it and a little disappointed, I think. However, he had no reason to be 'cause we need a C.O. who is on combat flying status and, of course, Young had finished up. He and Zraick are now in Rome for 3 or 4 days. When they return Young will go to the Group until he ships out — which should happen in about one week. The old squadron is certainly not the same anymore. Everybody is a stranger — even Mahon has only 3 more missions to fly and then he goes. Pretty soon I'll be just about the only one left.

Got a letter from you yesterday — also a V-mail from your Mother. Your Mother has been and still is mighty nice about writing to me. I must write her within the next couple of days and your Dad too, also my Mother and Dad and Sis. I'm so terribly far behind in my correspondence — it's a wonder they ever take time out to write me. I want you to make them know how much I appreciate hearing from them even though I am always so dilatory in answering. They have made me feel as though I'm just about as much in their family as you or Hugh, which I certainly do appreciate. You have a good Mom and Dad.

Perhaps the Good Lord will see fit to let us get back together real soon — I hope so — for I love you with all my heart.

Joe.

September 8, 1944
9:15 p.m.

Dearest Honey,

Not writing a long one tonight for I have no lights. As it is, I can hardly see for the wind is upsetting my candle too much. The old boys who returned here from being Prisoners of War left today for the States. They all borrowed money and clothes from me so if you get any anonymous packages or checks in the next couple of months you will know why. A couple of them said they would drop you a card. I was glad to see them get out of here. They had been through so much.

Went over to Wing Headquarters today with our new C.O. He still seems to be an alright sort of guy. Found out tonight that we are having a surgeons' meeting over there in the morning at 11 a.m. Hope something is brought up slightly different and a wee bit interesting for a change.

Sent you and Kay some more "chew gum" today. Since I don't smoke cigars, I trade the cigars for chewing gum for you two. Tried to buy Mother something in town today but couldn't find anything suitable. Here's a good night kiss for you.

I love you,
Joe.

September 9, 1944
Saturday, 10:15 p.m.

My dearest Honey,

You know 1 year ago tonight I slipped away from Fairmont, Nebraska and went to Omaha, then caught a plane to Chicago and from there to Memphis via St. Louis. About 11:30 a.m. I hit Memphis — boy, wouldn't that be swell if that was happening now!

I went to a meeting today and my opinion is that after this phase is over, even if we don't go to the other theater, it would be 8 to 10 months before we get home. Notice that it has increased a couple of months since I last heard. It sounds damn discouraging especially for the medics regardless of whether they are on flying status or not. Since hearing all of this my morale is way below zero. As far as the demobilization plan is concerned, it does not apply to officers nor personnel with specialties. Since I come under the heading of both of the above, I'm sure there will not be a chance there.

Last night when I started to crawl in bed, there were two letters under my pillow from you. Don't know yet how they got there. In one of them, there were pictures of Kay and Carol with Mattie and Laura. Gosh you don't realize how fast they grow. Kay is certainly a young lady. In the picture Carol looks like she has a "mature" face rather than the face of a baby. She is pretty large, isn't she. Gosh, but I would like to see both of them.

You were asking about my roommates — yes, Zinn is a very peculiar guy — besides being peculiar he is also one of the most selfish guys there is too. However, he has some good qualities — after you know fellows for a long time, you know their faults and just sorta look over them. Reese is one hellova swell boy, and you would probably remember him if you saw him.

Honey, I've written enough for one sitting. How I wish I could be with you instead of write you but I had better repress such thoughts and not let them come forward. Our day will come! Take care of yourself for me. By the time you get this, Carol will be a year old — tell her she better quit being so lazy.

I love you,
Joe.

September 11, 1944
1 p.m.

My dearest Honey,

Had an inspection this morning. Wing inspector came around and said hello and that was the inspection. The last time he did the same thing. One of these days he's gonna really inspect, and then I'll probably be caught short. Colonel Hoppock has finally taken over the outfit — still seems to be an extraordinarily nice guy. I believe he's going to turn into a better C.O. than Evans or Young.

Young is still around but apparently feels very much out of place. He has been transferred up to Group but only goes up there to sleep. In the daytime, he is lounging around the squadron area here and occasionally makes a decision for the Colonel — which the Colonel doesn't appreciate. Young should be leaving in a couple of days.

At 1:30 we are having a meeting of all the section heads, and Hoppock will officially tell us what he wants done and how he wants it done. We have been the best outfit in this group for a long time, but something tells me we are going to be an even better outfit pretty soon. This guy hasn't had too much experience in flying B-24's, but he has had plenty of executive experience and I think everything will work out alright.

Zraick and Reese just returned from another 3 day pass to Rome. Bernstein is up there now. I suppose I could go now, but I don't have the desire. In fact there is no place to go around this area now that is interesting. Perhaps later I can wiggle me a trip up into France for a couple of days. I really don't have much of a desire to do that either, but I get so tired of this place. Wish to hell we would move or something just to keep me busy for a while. I'd like to get transferred to a job that I didn't know one single thing about, so I'd have to work like the devil for a while to catch on to what was going on.

10:30 p.m.

News flash!!! Something different is going to happen tomorrow. I am going to fly up into France. Wagner is going also. He is going to remain up there for about 10 days, but I am probably going to come back tomorrow and act as Group Surgeon until he returns. I will be in France only four or five hours but at least will get an idea of what the country looks like. It can't be worse than Italy. I have been trying to get Wagner to let me stay up there for the 10 day period, but he won't budge. So I suppose I will have to learn "all about France" in a relatively short time. Wish I spoke the language like you do, and then I could get around. I will write you one of those "descriptive" letters of mine when I return.

On this news I will say goodnight. And will again tell you that I love you with all my heart as I have and always will.

Your Joe.

September 14, 1944
Midnight

My dearest Honey,

Day before yesterday when I wrote you last — yes, I skipped last night on you — I said I was going to France, the next day. Well, it didn't materialize. I didn't go to France then, but I may be going over there almost any day now.

I have been working on our shower for the past two or three days. Think tomorrow we will get it finished and then we can take a shower right in our house, which will be a boon to cleanliness.

Today I received two letters from you — the first I had gotten in four days. In one of them you sent a very good picture of Carol and a mediocre one of Kay — none of my Honey who I had rather received pictures of than anybody. Hey, I can tell that Carol's hair is dark but what color is it? I'm guessing that it is a pretty dark brown. She looks like a pretty serious child — but looks very attractive. Of course, I probably have her figured out all wrong by her pictures, and will find out she is completely different when I see her. Gosh, how I would love too. Here I have a daughter 1 yr old, and I haven't the slightest idea what kind of a personality she has — I know she has been a good baby but with a definite mind of her own. For one who hasn't seen her, those two characteristics almost make a conflicting personality — I would like to see her and see what I thought of her. I will never know her as well as you do — probably I won't know Kay that well either. But I wish I had been there to grow up with them — even though it is a lot of trouble perhaps, it certainly is a privilege.

Young just came by to say goodbye. He is leaving in the morning. Says he will call you when he gets home. He really does hate to leave this outfit. I almost believe he had just as soon stick around. Saying goodbye to everybody really was a job for him. Well, honey, again I will say goodnight to you. Kiss the kids for me.

I love you,
Joe.

September 15, 1944
10 p.m.

Dearest Honey,

There's not much news tonight. I'm sitting up here listening to the news, and not much has changed in the past few days. A stalemate is occurring on the more important fronts. Finished our shower today — also put us in the stove — which we will probably be using very soon. The weather is gradually getting cooler especially at night.

I was sewing up a couple of cuts tonight on a kid, and while washing my hands in alcohol, I caught the alcohol on fire. Burnt both of my hands some — the right one worse but it's not very bad. I can still write with it. Carelessness, of course. I'm tired tonight — worked pretty hard all day — since I'm so unaccustomed to it, it gets me down when I work for a couple of hours. Well, honey, guess I'll get in the "sack." Tell the kiddies hello for Daddy.

All my love,
Joe.

Dearest Babe,

September 16, 1944
7:45 p.m.

My Honey,

Haven't done much today. Just went into town with Kremers — went to the hospital for a short time and to the Red Cross. Mahon finished his missions today. Certainly am glad for him but selfishly sort of hate to see him go. He just got word that he's the father of a male offspring born September 10th. Guess I will write up his papers for him to go home within the next couple of days.

Hey, we are installing a small hospital at the Group Headquarters for the winter. Microscope, other laboratory equipment, portable X-ray equipment and so forth. Apparently it will be a very nice set up with some exceptions. The exceptions being laundering facilities, technical personnel, etc. We will have to do our lab work ourselves, which won't be too boring at first but which will soon get that way. Looks as though they really want us to get set up for the winter in a big way. I still hate to think about another winter in Italy and find it very difficult to reconcile myself to it, but suppose I will in time. It's one sure cinch — as long as we are here I will spend a much more pleasant winter this time than last winter on account of our house. All day today I have had a low fire going in our stove. It's not very cold — just cool but the small fire takes the chill off the air.

Looks as though the war has settled from a blitzkrieg in reverse to a more stalemate affair. I hope soon that it again breaks into the fast variety. Certainly would like to see it ended here before the end of the year! I'm listening now to the broadcast from London of over 1,000 men returning to England — some of them being prisoners of war for as long as 4½ years — boy, don't you know they are happy. Well, honey, I'll quit for tonight. Take care of yourself. Remember that I love you with all my heart.

Your Joe.

September 17, 1944
Sunday, 11:45 p.m.

My dearest Honey,

I'm a little late in writing you tonight, but I've been in town to see "Barret's of Wimpole St." Katharine Cornell, Briana Aherne and there troupe were here. She is just as good as you have always heard, and Briana Aherne playing Robert Browning is better than anything I have ever seen him do on the screen. The first act is fairly slow, but the last two acts are fast, exciting, tense, and very entertaining. I enjoyed it as much as "This Is the Army," however both were extraordinary. The fellow that plays the tough, hard, villainous father, Mr. Barret, almost steals the show from Cornell and Aherne. It was really good — I wish you could have seen it with me.

Saw Koenig while at the theatre. He looks and acts the same. He is as fed up with war, separation from his wife and child, and Italy as I am. After the show went by the Red Cross and had some ice cream and coffee — also saw Koenig there and got to talk to him quite a while.

All day today I have been working on our water reservoir. It has had a leak ever since we put it up, and we can't stop it. The tank is aluminum, and aluminum doesn't weld very well. I did manage to get the holes dripping in a little trough that I made however and piped off — so at least you don't see the leaks now, nor the effects of them. It was a pretty tough job even though it doesn't sound so, and it took the whole afternoon to do it. Now if I could think up something to try to help pass the time away tomorrow.

10:45 p.m., next day —

Honey, I didn't finish this up last night — so will tonight. Nothing has happened today — we didn't fly so I didn't have a whole lot to do. Went to town this afternoon with Col. Hoppock and Mahon for awhile and just messed around the Red Cross club.

Tonight, we went up and saw some shorts at the Group movie and then went over to a "shin-dig" that they were throwing as a farewell party for Col. Eaton. He is finished up with this outfit and I think he is going to England from here. We have already gotten a new group C.O. in — who apparently also has been pulled from the depths of oblivion by that three letter word WAR. Funny how so many of the so-called "big shots" have been — from West Point graduates to running a war in short order — or, better still, from CCC camps to running a war.

I could write to you for hours — I could write about you for longer, but I could never express the feeling within me concerning you — I do know, and can express, that I love you with all my heart.

Your Joe.

September 20, 1944
11 p.m.

My dearest Honey,

I took a kid into the hospital today that had gone "nutty as a bedbug" suddenly — when I got into the hospital I didn't know whether to enter myself in the nutty ward or just the boy. The war situation looks very good, doesn't it? Don't know how much longer Germany will last, but it appears to me that it shouldn't be too long. The sooner it is over, the better I will like it. However, while we are waiting around to either go to the Pacific or to the States, it's going to be as difficult to keep up our morale as you can possibly imagine.

Went to town to get my rations today. I've been needing hair oil for quite a time and the PX has been out. Today I bought a $1.59 kit with toothpaste, soap, razor blades, etc. just so I could get the 4 oz bottle of hair oil in it. Aren't I extravagant? Mineral oil makes my scalp break out.

Went up to the club tonight and played rummy for about three hours. Enjoyed it very much even though the boys took me for almost $2.00. Anything to pass away some time. First time I have played any cards in a long time. Well, honey, I'm quitting early on you tonight. Somehow my "talking" has given out. Lately there hasn't been enough happening to even make interesting conversation. Take care of yourself and the children. I wish with all my heart that I could be with you.

All my love,
Joe.

September 21, 1944
10:30 p.m.

My dearest Honey,

Didn't get any letters from you today, but did get a package which you mailed me July 20. The jam had leaked a little bit but not much. Thanks for all the jam, jellies, and sandwich spread — plus the Kodak film. The nights are getting cool enough now where we will be sitting around the fire sipping coffee and eating late snacks. The food will come in handy.

This morning I took off with Mahon and Col. Hoppock to fly to Naples. We were going under a layer of clouds through the pass in the mountains when the weather closed in on us. We got out O.K. but by scaring all of us half to death. Hoppock is nothing like the pilot that the old boys were — Mahon is, but he was flying co-pilot. When we got to Naples area, the city and the field were "socked in tight" by weather, and we couldn't land so there was nothing to do but turn around and come back in the "soup" — instruments all the way. Then the Colonel tried a 3 engine landing, which he finally accomplished but not before he scared the devil outta me — I guess I'm just leery of airplanes lately.

Heard some bad news a few minutes ago. Kremers was going to Capri yesterday and they crashed on landing at the Naples airport. I just heard a few minutes ago that Kremers broke his leg, and Pearson, the executive officer in 726th, broke his back. I don't know whether that's right or not. Tomorrow we will definitely find out.

I also hear tonight that Reese, one of my roommates, is being transferred to another squadron. Certainly I hope this isn't the truth for Reese is one heck of a swell guy and is doing a swell job, and I know he doesn't want to be transferred. Too, we have spent quite a bit of time and money getting our house in shape for the winter, and if he is transferred, he will just go over there in a tent.

This afternoon Mahon and I went into the hospital for me to see the boy that I took in yesterday — he is still as "batty" as he was. Then we went on into town for doughnuts and coffee. In the morning Mahon is going up to Rome. He is supposed to stay for three days up there, but I'm pretty sure he will come back tomorrow night.

Went to see Jean Arthur and John Wayne tonight in "Lady Takes a Chance" — Jean Arthur was just as silly and entertaining in it as she usually is. There was hardly any plot at all but it was entertaining, light, and enjoyable.

Well honey, I think this is all the news. Thanks again for the package. It's mighty nice and thoughtful of you. I know it's quite a bit of trouble to get things wrapped and such. Tell the kiddies hello — and tell Kay that her Daddy "writes her a kiss." No one could love his wife as I love you.

Your Joe.

September 23, 1944
10:15 p.m.

My dearest Honey,

I'm surely glad to find out you have taken up the piano again. It'll give you some diversion and also is a very nice thing to be able to do. I wish I could play one. Also I'm glad to know you went to town and bought that new outfit — only wish I could see you in it. Hope you got something that is snappy. However, when you say you are going to "meet me in it" it makes my heart bleed. Honey, Honey don't be building yourself up for a disappointment. In your letters I can read between the lines and see that you don't think it will be very long before we are together again — I'm afraid that won't be so. No one would like to believe that more than I, I assure you. Most of the time I can hardly bear it much longer, but, we will if we have to.

Yesterday I flew over to Naples to see about Kremers. If he has a broken leg, I couldn't find any record of it anywhere — and I'm pretty sure that is all just a latrine rumor without any basis. Stayed in Naples all day, looking in shop windows and walking over the town. Didn't land here at the field until nightfall. Was pretty tired and went to bed real early.

This afternoon I went down to the nearest river bed and loaded two tons of boulders by hand and brought them up here to fill up my soakage pit that our shower empties into — and then unloaded them after I got here. Good exercise, and makes you feel like you have done something. More importantly it passes the time away. About another two tons and I will have enough to finish the job.

Well, in the morning I have got to arise real early, so I had better say good-night to you, my Honey. Don't let it all get you down. I love you now, and I always will. Tell the daughters goodnight too.

All my love,
Joe.

September 24, 1944
8 p.m.

Dearest Honey,

Do you remember that awful dose of Castor Oil you took a year ago tonight. I know you do, but I too know you remember the wee hours of tomorrow morning more vividly. Bless your heart — you have gone through a lot for us. I appreciate it too. Not only do I appreciate it, but I'm also proud of you.

Since writing the above paragraph I have been interrupted. We have been playing rummy for a couple of hours. Mahon came in and got up the game. After playing all that time, I still came out exactly even. In fact, the big and only winner of the game won 30¢. Don't know whether we can afford such stakes.

Looks like the war fronts have reached another stalemate. These "stalemates" are the tough parts of the war. More damage is done then than any other time to personnel and property. Maybe this one will break soon in our favor, I hope. The Italian Front stalemate looks to be breaking up now.

I was informed today that we are having an Air Force inspection tomorrow or the next day. I dare say they have caught me with my "pants down." My area is in the worst shape that it has been in an awfully long time. We have been digging new soakage pits and new latrines, and everything is a mess. I'll bet they really look around this time too.

Honey, I gotta go. I'll write you a better and longer letter tomorrow night.

I love you,
Joe.

September 25, 1944
10 p.m.

Dearest Honey,

Just a "shortie" tonight. Today is Carol's Birthday. Hope next one we are together. Seems to me we separated on the last one rather abruptly with rather unusual conditions. That was one of the hardest things I have ever done — to go off and leave you that day one year ago today. I realized about 8 a.m. that day just what a plight I was leaving you in — and there wasn't a damn thing I could do about it. Does all of that seem centuries ago to you too, or does it just seem that way to me? Wish Carol many happy returns for me.

Had part of my big inspection today and so far everything is O.K. Have some more tomorrow on the administration end, and I probably won't fare too well. Goodnight.

I love you,
Joe.

September 26, 1944
9:05 p.m.

My dearest Honey,

Just got back from the picture show at Group. Saw Irene Dunn and Spencer Tracy in "A Guy Named Joe." It's screwy. Did you happen to see it? Why they send such pictures to combat I don't know, but they seem to use very little brains in their selections. There was good acting in it however.

They finally finished up the damn Air Force inspection today. It has been a ditty. They want us to be more "spit and polish" over here than in the States. Somehow it gripes you no end. There is nothing sillier than the Army. I think this is just a prelude to what it really is going to be like when this affair is over. Better that the "big dogs" wait and see that it is over soon first. We have got plenty of time for this sort of stuff when this is over and done with.

In one of your letters that I received from you (written September 9), you mentioned that Bryan had called you up. Apparently he had called you a night or two previously. He's a nice guy and was a good morale factor in this outfit — on the pudgy side, serious in a funny sort of way, good ideals, and just a nice fellow all around. I'm sorry that you heard about the boys that went down. Only Williams returned out of the first pilots. I'll tell you all about the rest sometime when we are together and I'm in a talkative mood. Did any of the other boys call you up or has Bryan been the only one?

Sounds like Carol has a luscious appetite. Tell the little devil that her Daddy envies her all of that milk she is drinking. Even though I haven't had a drink of Scotch in a long time, I would trade a good big drink of Scotch for a good, big, ice-cold bottle of pure old cow's milk.

Next night, 10 p.m.
My Honey,

Never did finish writing to you last night, so will finish up tonight and send you both at once. Today has been another dullard. I worked for a little while this morning and about noon Mahon and I went into town to get our rations. They didn't have anything but Raleigh cigarettes, so we just went by the Red Cross and got some coffee and cookies. Guess we will go back tomorrow and see if we can do better — too it will pass off that much time. Mahon's orders came through today, but I think he is going to wait around four more days and get paid before he leaves. However, he may change his mind. Don't know who I will run around with when he is gone.

We have been sitting around playing rummy tonight. It's a sorry way to pass away the time but it is better than nothing. Hey, I bought me a new winter hat today. The only two pieces of apparel I have purchased since being overseas have been a summer hat this spring and a winter hat now. A total of a little better than $6. Doin' alright, aren't I? And I still have plenty of clothes, except socks. Pretty soon I'll probably need some socks — no, I won't either. I've got some winter heavy socks stored in my trunk that should last me until next spring easy. Surely am wearing out some clothes that I've been wanting to wear out for years — especially underwear.

We have had a cold wind for the past three days. Run the stove all day long. We are still thinking about putting in a fireplace — got the stone and everything but haven't started yet. We are going to need another stove whether we put in a fireplace or not 'cause when it gets really cold, one stove isn't going to be enough. Honey, I think I will take a bath and turn in. I do want you so terribly much and love you so much it hurts.

Goodnight,
Joe.

September 29, 1944

My dearest Honey,

Skipped writing to you last night. There was a party up at the club last night, and I stayed up there until about midnight. Drank too much too for today I feel loggy — no, I didn't get "tight." I haven't been that way in a long, long time — no desire to do so. They got about 10 English WAACS last night and the boys had a very nice time. It was a "clean" party too — not like when they had the "Italians" over.

Hey, I thought I would write and ask you for something. How about sending me a pair of bedroom slippers (like the blue ones Kay gave me once) — size 8? I lost my bedroom slippers while I was in the hospital last winter and haven't been able to get any since. I don't know of another thing that you can buy that I want. I wouldn't go to the trouble of sending me something fancy for Christmas cause really I don't know what you could send me that I want or need. Too, I already have so much that the next time we move it's going to be a job packing all the stuff in what they will allow one to carry.

Today has been about the same. The weather here has held us in quite a bit lately. It seems we rarely fly anymore. It's O.K. with me, but the days seem even longer when we stay on the ground. However, Mahon has been just as bored, so he and I have been going to town almost every day. Yesterday we went in and saw Bette Davis and Paul Lucas in "Watch on the Rhine." I had seen it once before, can't remember whether it was with you or not, but it was just as good this time.

Enclosed is a picture of Bernstein, Mahon and me. (Mahon in the center, of course). Bernstein is sticking his tongue out for some reason and I have a "peculiar" grin on my face. Too, it's quite evident that I have on that small pair of pants that I have been trying to wear out for ages.

We haven't been getting mail through lately. Pretty soon I'm going to get a bushel basket full from you. It will come through pretty soon — it always does. Honey, I get paid tomorrow, but I don't think I'm going to send you any money. Hope you aren't planning on getting it. You will find out later what is happening to it — it'll be for a good cause. Don't want to just raise your curiosity but didn't want you to think I was throwing our "livelihood" away.

Got a report from the Air Force Inspection that was held the other day. For my squadron, it just said "No major discrepancies noted." Listed a whole lot for the other squadrons and headquarters — sorta made me feel good even though it doesn't mean anything. Well goodnight, my love. Kiss the kiddies goodnight also. I love you with all my heart.

Joe.

September 30, 1944

My dearest Honey,

Got two air-mails from you today — written the 6th and 7th of September. In one, you told me of Duncan Sheats' death, which I am very sorry about. I don't know the Sheats' very well, but if it won't be "opening an old sore" by the time this reaches you — and I'm afraid it will — convey to them my sympathies. It always hurts when you see a youngster meet an untimely death — regardless of how many you have seen before or how "hardened" you have become. I believe the actual hardness is only skin deep. Did Duncan get killed in a plane he was flying or was he a passenger on a transport?

The other letter you wrote was written the night you talked to Judge Bryan. I wish you could have met him instead of just talked to him. He had one of the most beautiful philosophies of life I have ever seen to be as intelligent as he was. A nice guy — not because I'm sure he said a lot of nice things about me to you, but a nice guy anyway, and he helped the morale of the old boys around here more than any other one person, too.

I don't know whether Young has called you up yet, but I'm sure he will the first chance he gets — maybe even before he gets wed. I'm sure that will be as soon as Weezie can make the arrangements for a big affair after he gets home (God help her — and I mean it). Mahon is leaving the first thing in the morning. He has promised me if he goes through Memphis, he will look you up. He will call you anyhow. Again, he is one of the nicer fellows I have ever had the pleasure to meet. I value his friendship with almost anyone's, and I will miss him very much — more than anyone who has left heretofore. I don't want to sound like a man's friendship means so much to me, but over here, in this womanless (or wifeless, is better) place, a good friend is very hard to beat, and I have had more in common with Mahon than anyone around.

I have had a little business tonight. One of our trucks had a wreck, and four of the boys got hurt — one had a very severe fracture of the right upper arm. I fixed them all up and sent them into the hospital for some jerk in there to do the interesting part of it, which still gripes me very much.

There's a fierce wind blowing tonight, and it's raining like the devil. And it's biting cold outside. Typical Italian winter weather, which I had hoped so much that I wouldn't have to endure again. This is just the beginning — it will get worse soon. "Sunny Italy" they say — that's even a much worse exaggeration than "Sunny California" — remember? I started getting out my heavy clothes. Took quite a few things into the cleaners this afternoon. Another reason I hate to see winter come is that over here the cleaning situation is very difficult when you can get just any Italian to do your washing.

It should be fairly good sleeping tonight. The wind whistling outside and the rain beating down on the tin roof. Living in luxury in comparison. Think how tough it must be for the boys up a little further in the fox-holes. I hope and pray that it isn't too tough for Hugh. Maybe it isn't. I think of him a lot — hope his outfit isn't having it too rough.

Yes, it should be good sleeping tonight. That is as good as one can expect to sleep when sleeping solo. It might have taken me quite a while to get used to sleeping double, but, since I have been broken in, I shall never be happy solo again. I love you with all my heart.

Your Joe.

October 2, 1944
Midnight

My dearest Honey,

You deserve better than a V-mail tonight, but Kremers has been here all afternoon and I haven't had a chance to write. Even so, I would write a longer letter if I were not sitting up here freezing to death. It is very cold and the wind is whipping around very hard.

Got two V-mails from you today and your dress sounds wonderful. I'll bet you look real good in it — gold and black go good together and look swell on a good-looking blonde.

Mahon finally pulled out this morning — I miss him already — he had such a swell sense of humor. Am not getting along too well with the new squadron C.O., but so far no one is. He hasn't as yet learned that he is supposed to work for the squadron instead of the squadron working for him — which is of the utmost importance if you are going to do the job right.

All my love,
Joe.

October 4, 1944
6:30 p.m.

My dearest Honey,

Got a V-mail from you today — also a letter from Judge Bryan telling me about calling you up among other things. He said, by the way, that you had one of the most pleasing Southern accents that he had ever had the pleasure of listening to. I could have told him many more things that made you a remarkable woman but after so long a time it may have become boresome to him.

Today we finally got started on building us a fireplace. By tomorrow night it should be finished. I am eagerly looking forward to its completion. A nice big cozy fire somehow does the heart some good. Today passed a little quicker than usual. To begin with, this morning we took off for a mission, which we haven't been doing too much of lately because of bad weather. Then later in the morning, I went and got a load of bricks to help build our fireplace. This afternoon I attended landings and did odd jobs.

Tonight I am eagerly awaiting for 7:45 to come around, so I can listen to the first game of the World Series. Thank goodness that will give me something to do for the next few nights. I hope the St. Louis Browns beat the Cardinals, but I have little hopes for them. Remember the ball games we used to go to in Helena? The right fielder that played for Helena then — Zarilla — is now playing right field for the St. Louis Browns. This is probably pretty nonsensical to you — cause you aren't a very ardent baseball fan. I'll turn you into one though when we have a good chance to go to some good games together.

My house boy, Giovanni, brought a little puppy to work with him this morning. And when I say little, I mean little. It hasn't got its eyes open yet. Don't know where he got him. All today he has been trying to feed him condensed milk with little success. This afternoon I got an eye-dropper, and I think he has had a little more luck since then. It's a cute little thing — brown with white nose and toes. Looks as though it's going to be a long haired pooch but as yet I suppose it's a little early to tell. Of course it will probably die in a few days due to starvation. Giovanni is spending the night with us tonight — he missed his truck again. He is a good clean kid though and we don't mind. He really keeps this house spotless. He has been leaning on the table watching me write. I think it intrigues him to watch someone that writes rather rapidly.

Today has been a pretty day. The first even mild day that we have had in quite a long time. I hope it stays like this for a while — winter will get here in its severity soon enough.

So Carol is trying to talk? Bless her heart, when is she going to try to walk? Seems funny to talk before walking. It certainly is difficult for me to realize that she is that large. I have certainly got a lot to get "used to " and learn when I get home, but I'm ready to start that part of my education immediately. Of course I think Uncle Sammy has some different ideas.

10:15 p.m.

Just took time out to listen to the World Series. Sure enough, the Browns came through with a thrilling 2 to 1 victory.

Col. Hoppock tells me that he and I are going to Rome for three days in the near future. Personally I have no burning desire to go at all. We will wait and see. I don't care to leave here for I get too lonely when I have absolutely nothing to do. It's bad enough as it is. At least I can do something around here during the day when I get too bored. Well, it will be over someday. When I don't know, but I'm eagerly awaiting that day.

All my love,
Joe.

October 5, 1944
11 p.m.

My dearest Honey,

Due to the fact that I have no envelopes of any variety it will have to be a V-mail tonight. For some unknown reason, I'm tired, so it's probably just as well. My back is really tired — why, I don't know for I haven't turned my hands to a thing today that I can remember.

Finished our fireplace today. Have had a nice roaring fire in it since noon. (Even before the chimney was finished). A fire in a fireplace provides food for thought for some reason. I have been looking in it tonight and just thinking about you, the children, and many more pleasant things. Also I have been listening to the second World Series game, which I enjoy very much.

Went in and got my rations today. It's been raining hard all day, so there was nothing else to do. Looks as though this winter will be identical to the last one. Don't tell anybody, but I don't care very much for Italy. Too, I still need my honey very bad.

I love you,
Joe.

October 7, 1944
9 a.m.

My dearest Honey,

Didn't write to you last night on account of listening again to the World Series. I'll bet you will be glad when these ball games are over, so I will be more apt to keep up with my correspondence.

It is still bad weather around here. Not as cold as it has been (probably 'cause of my new fireplace) but it has been rainy and real muddy. Early this morning it appeared that today was going to be a little different and the sun might shine. Now it is turning true to form and I can hear a few drops of rain now and then on the top of the house.

Since we have put on winter uniforms (October 1), we have been having to wear ties. Last winter the wearing of a tie was optional, and this summer they were definitely out. Now, since I haven't had one on for a whole year, it feels like I'm trying to strangle or hang myself. They are very, very uncomfortable when one isn't used to them. In fact, my neck must be quite tender for this wool collar close against it has gotten sore. Don't worry, in certain ways (ties, "spit-and-polish," etc.) this has gotten to be a sissy war and pretty soon we may again become civilized in some respects. The new Group C.O. makes all the Headquarters personnel wear blouses to dinner three nights a week. Too, they have to wait until he sits down to eat before they can. That is certainly a far cry from the war of last winter, I assure you. Then everyone just kept warm with anything he could dig up that was dry — and why even go to the mess hall for K-rations. The "upstairs" that run this war are amazing. I would certainly like for many of them to walk into my office during private practice, so I could tell them to get the hell out! Only yesterday I heard a bright remark. My C.O. (Hoppock) reminded me that he gave up a good job when he got in the Army. Said he went from $220 per month to only $150 as a Second Lt. I suppose I should have wept in sympathy for him, but, honestly, I didn't feel like it. He is the damndest guy! Thought he was going to be O.K, but day by day I dislike him more. Lately I find it very difficult to speak civilly to him — and usually don't.

I was so bored with my own company yesterday that I jumped into the ambulance just after noon and decided I would drive over to Dan's outfit. So after driving the approximate thirty miles over there, he wasn't there. He had gone into town earlier in the day.

Just had to leave you for an hour or so to go out to take-off. They don't worry me like they used to for if anything happens I don't know the fellows involved so well. I have only one man flying now that came over here with us, and he should be finishing up pretty soon.

This has been a rather sorry letter. Apparently all I have been doing is griping. I'm sorry — there's not a lot you can do about it nor I either. Perhaps when the mail comes in this afternoon there will be a letter from the sweetest and best woman there is. That will, no doubt, have a profound effect upon my morale in general. Haven't gotten a letter from you now in three or four days. The mail used to come through rather steadily — now when it comes, it's in bunches. I think I like the former better. Well, if I want this to get in this morning's outgoing mail I had better make it snappy. I'll be writing later on tonight.

All my love,
Joe.

October 8, 1944
Sunday, 10:10 p.m.

Dearest Honey,

There is nothing much to say again today. Everyone has been just sitting around waiting all day for it to quit raining. Sure enough — it never did. Things are surely dull around here when it rains — not to mention the mud — quite depressing weather.

Flew a mission today — one of those kinds that makes you realize we are still fighting a war — made us think of "the old times." I sound like a well-seasoned veteran, don't I? Tonight there's a heavy fog with sporadic rain. The moon tries to come out occasionally, but I think it is fighting a losing battle.

I've been listening to the World Series again tonight. My underdog team got beat again — doggonit. If they get beat once more, it's all over. Can't let the Series be over a minute before it has to for it certainly helps to pass the evenings. If it wasn't for the baseball games and "rummy," we would go nuts sometimes.

Reese, one of my roommates, is getting transferred to another group soon. That will leave only Zinn and me in this place, which is plenty large for a couple of more men easily. We are going to get the Assistant Operations Officer to move in with us. He is the only guy that we can think of who we would like to have. There are too many people around the squadron that are undesirable as roommates (the Colonel particularly). Selfishly, I hate to see Reese go. He is such a nice likable person. He will probably get his captaincy by transferring though, so I'm glad for him. This outfit has not done him fair, and it's time for him to get a break.

You should know this guy Col. Hoppock honey. I'd like to hear your opinion on him. Very selfish and stingy person. Always asking for something — drinks, cigars — everything. Anything you get for yourself he sees and remarks, "You know, I could have used that" — quite seriously. He has been with us over a month now and has never bought even a $2.00 "chit-book" (coupons to get trade with) at our club, but drinks off others' consistently. It's getting where none of us can stand the guy, but of course, there's nothing you can do about it. It strikes me mighty funny that apparently it's the guys of that calibre who definitely get ahead in the Army — someone who is presumptuous, selfish in every respect, egotists, and aggressive. Of course a person of that type is usually very aggressive — the only admirable trait they possess. I'll just be so damn glad when I can sever my associations with fellows of this particular calibre. I could get along with them but only by living a life of the hypocrite, which I refuse to do for more than a short time. At present Hoppock and I are really having personality clashes too. In fact everyone on his staff is — well this is not important enough to make an issue of.

A lonely night, when rain is coming down on your roof, is only good for one thing, which is going to bed. Goodnight, my Honey.

I love you,
Joe.

October 9, 1944
Monday, 6:30 p.m.

Dearest Honey,

I received a letter from you today written September 29. There is such a large gap in between your letters that some of the things you say don't fit. You mentioned Mother and Dad back down to Shannon and being able to see Aunt Maud while she was still able to recognize them. I gather that she died — and since she had cancer, it is just as well — particularly since it was apparently so far advanced. Too, you mentioned keeping Kay's bandage on her arm wet, but as yet I haven't heard why. Again I imagine that one of her diphtheria-whooping cough shots abscessed and had to be incised and drained. Possibly it ruptured on its own hook and didn't have to be incised which, for your sake, I hope was the case. Sounds like Kay is really getting big going to the toidy all by herself. I'm eager to be with her and Carol and learn them all over again — Carol, of course, for the first time. I have a lot of future happiness to look forward to — and I'm looking forward to it too.

I have been in town practically all of today. Went in at 11:00 and didn't come back until 5 p.m. — just messing around. Went to a picture show and waited for 45 minutes for it to start and then they announced that there would be no show due to the fact that the movie operator had disappeared. Went by the PX and bought me another brocaded 15th Air Force shoulder patch for my blouse. Guess I will sew that on in the next couple of days and get my fingers sore from sewing on that heavy stuff again. My blouse will look good if I ever get back to the States to wear it. I have a pair of gold brocaded Flight Surgeons' wings that I had made while in Cairo and now this brocaded shoulder patch. Now when I get all the ribbons, decorations, stars, overseas stripes, oak leaf clusters, etc. attached, I'll look as decorated as that gal used to look that J.C. married — remember? Gosh, how that girl could adorn herself with distasteful dress and ornaments. I'll never forget how hideous she sometimes looked.

Well I just took time out to listen to the St. Louis Cardinals win the World Series. Now, that the Series is over, there will be nothing to do to pass the evening hours away. I believe the happiest day of my life will be when I am again with my Babe, and I won't have to worry about how I will have to pass my time. Tell the children goodnight for me, Mama — and goodnight to you. I love you all with all my heart.

Your Joe.

P.S. Sent you some more "chew-gum" today. Kay should have enough to stick up the whole house for you — more worries!

October 11, 1944
Wednesday, 9:30 p.m.

Dearest Honey,

The letter I received from you today sounded as if you were down in the dumps. What's wrong — was Kay giving my Honey a hard time? So she called for me when you spanked her, eh? Sounds sorta sad in a way but somehow cute at the same time. It's probably a good thing Daddy wasn't there for if your patience gave out on you, think what mine would probably do. I surely would like to see that mischievous little devil — she sounds like she is an education within herself. Carol, of course, still sounds vaguely foreign to me — even though she "scoots" on the floor and "sticks her tongue out of her mouth." I'm really sorry to hear that her teeth are separated in the front. Sounds like the dentist will get some business on our children at a later date. We can't have two nice looking young ladies on our hands who appear that someone has knocked one of their central incisors out. Perhaps their permanent teeth won't be separated.

I've been into town twice today. The first time I have ever had enough courage to tackle that road between here and there twice in a 24 hour period. This morning I went in with Kremers just to pass the time away more than anything else. Tonight I had a boy that was pretty sick and I took him in myself thinking that I may go to the picture show afterwards. However, by the time I was through with him it was too late to go to the show.

The weather here is still bad for flying. However, the coldness has let up somewhat. The stars are out tonight but it is still pitch dark outside — either that or I'm developing "night blindness." I just walked down to the latrine and got lost on the way even though I have been walking that same path at least once every five days (remember) since April — missed the latrine 50 yards. It's a wonder I hadn't broken my neck falling in some g.i.'s private fox-hole (which should be filled in now anyway). That's how dark it is here even when the stars are out. Last night I was coming from playing rummy at Zraick's "casa," and I couldn't find my own house until I was near enough to touch it even though the two houses aren't 50 yards apart. Now you see it is not only dark over here figuratively but literally also.

Now, Honey, you get outta the dumps. They are no good — I'm afraid either of us being down in the dumps won't get us together any sooner. When you and I get together again everything will be rosy.

All my love,
Joe.

Mission #134: 13 October 1944
Target: Osterreichische Motor works, Vienna
Results: Excellent bombing with concentration on rail-way workshops, locomotive and car repair sheds. Numerous direct hits on main M/Y. 4 ships lost to flak, 6 missing. Some crews bailed out over Yugoslavia. 3 727th ships lost.

October 13, 1944
9:20 a.m.

My dearest Honey,

I didn't write last night so I will make an attempt to write a "shortie" before the mail goes out this morning. I'm already tired today. Got up early this morn and made take-off. Then came back here and we started putting large stone blocks up on the top of our house to hold the roof down. It's a common prac-tice around here by the Italians, but we haven't done it cause it looks funny at first. The other night our roof did everything but leave us, so we thought we had better "do as the Romans do." In a few minutes, I am going to get a truck and load up a couple loads of dirt and haul off from the side of the house and bring some gravel back to finish filling up our soakage pit. That should take just about all day, so tonight maybe I'll be tired and sore. Then I'll let everything go for another couple of months until I feel ambitious again. Tell the kids hello from Daddy.

> I love you,
> Joe.

> October 13, 1944
> Friday, 10:30 p.m.

My dearest Honey,

Tonight I'm sitting up here in that nice, blue, woolen, gabardine bathrobe that you bought for me at Sweet's in Riverside and gave me for my birthday in 1943. It still looks just as nice as it always did. In fact, right now it looks nicer for I have just had it cleaned and they did a nice job on it. The sleeves are still a couple of inches too long, but that is easily overcome by folding the cuffs back like I do on the striped job too.

I wrote you a V-mail this morning telling you of all the things I was going to do today — (one of my ambitious days). Well I worked until I completely gave out this afternoon, and I mean I really gave out too. I couldn't have shoveled another shovelful of dirt if my life depended on it. I cleaned up all around the house. Picked up every rock above 3/4" in diameter from around the place, all the trash, and every little splinter of wood (three big boxes full). Outside really looks nice now. Wish I could get some rye — I would plant a winter rye crop around the place and fence it off. After doing this I shoveled three trailers full of dirt and hauled it away. Tomorrow, if I'm not too sore, I have about three more loads to haul away and I want to bring in about 2 loads of gravel. Just finished taking a bath though, and it revived me somewhat.

We have had a rather rough day otherwise too. Looked as though there may be something to that Friday the thirteenth stuff. It has gone on into the night — I have just returned from the enlisted men's club where they have just finished a good fight, and I had to take down the ambulance and pick up what was left of one of the boys. He really got the hell beat outta him. I think tomorrow I will close up their club for them for a few weeks until they decide that they can enforce a little order among themselves. I have had scraps down there for the last few nights, and it's gotta be stopped before someone finally gets hurt.

Oh, I also put in a connection to the fireplace today where we can burn fuel oil (half gasoline and ½ Diesel oil) instead of wood or coal. It works real well too. I know it must sound bad to you as little gas as you all get for driving for us to be using so much to heat with, but if there wasn't gasoline, I don't know how we would keep warm. They put out an order earlier that we couldn't use gas for heat this winter but have unofficially rescinded it for there is no other fuel. We have three 55 gallon barrels to keep full now. Wish I could send you some.

Well honey I'm going to make me a cup of K-ration bullion, drink it, and go to bed. That's the only thing in K-rations that's good. The way I feel I'm not going to have much trouble going to sleep either.

All my love,
Joe.

October 15, 1944
Sunday, 9:20 p.m.

My dearest Honey,

Today has been as usual. I stayed in the sack until a little after lunch. First time I've done something like that for a long time. But my tummy was upset, and I didn't feel like getting up and around. It was my own fault though, and good enough for me. I think it was upset because I drank too much Scotch that Bernstein beat some Englishman out of last night.

I'm without a roommate tonight and will be for the next few days. Early this morning Zinn and Reese took off on a four or five day pass for Naples, Rome, and probably Florence. Florence is still out-of-bounds for passes, but I expect they will go up there anyway. I think I mentioned to you that Reese was going to transfer — well that fell through. I think he was glad for he didn't much want to move out of here — this is such a far cry from living in a tent.

Don't have as many of my boys as I did have. I transferred four of them up to Wagner — which left me with four. Of course I kept Mason and Rosie and Wilson. With those three I can run just as well as with the original eight. At present, Mason is sick though — and he is, of course, my right hand man.

It's one of those deathly dark nights outside. I just walked right into a jeep parked between my and Zraick's house. It doesn't really matter though for when the moon is out in all its glory, it just goes to waste. It's said that this war can't last forever, but sometimes I wonder. If they aren't able to break that stalemate around the Siegfried line, this war is going well into '45. That's just my opinion of course. I surely have been watching what goes on up there with much interest. You have to really hand it to the Germans for their profound tenacity. I surely would like to see this phase over soon.

Early next morning —

Got up early this morning for take-off. And I feel much better. It's a beautiful sun-shiny day out. Today I'm going to go get that load of gravel and answer some letters if it's the last thing I do. Perhaps I will go to town and get my rations. So far I haven't got my this week's rations and tomorrow is the last day.

I'm gonna sign off early and get this on its way. I'll do better tonight perhaps. I love you with all my heart.

Joe.

October 16, 1944
Tuesday, 10:05 p.m.

My dearest Honey,

Here I sit all by myself. Listening to music and the news over the radio, but just primarily thinking of you. Just finished fixing me a snack. K-ration ham and eggs (which are pretty good) and coffee which I made far too strong. I'll bet I won't get to sleep for hours. I wasn't particularly hungry, but it gave me something to do.

I have kept myself pretty busy today. I wrote to a lady in Tucson, Arizona. I received a letter from her a few days ago asking about her son, who is missing in action. Don't know how she found out about me — possibly her son mentioned me sometime. Those kinds of letters are particularly difficult to answer.

Did something else today too. I finally went and got two more loads of gravel, which I have been trying to talk myself into doing for a couple of weeks. I need one more load now, and I will have that fixed like I want it. The trouble is I want a special kind of gravel for this last load, and I don't know where to get it.

The news programs are usually pretty fair. We get American news once daily at 6 p.m., and we can get BBC twice daily — 7 and 9:45 p.m. They say again tonight that Aachen is encircled. That's the sixth or seventh time that it has been said that it was encircled — hope it really is this time. Today in the "Stars and Stripes," it said Hungary was asking for peace. On the radio tonight, they say that instead it has gone under a complete Nazi regime. You never know what to believe — but it does give you something to think about. Sounds like the tempo is stepping up somewhat in the Pacific. I hope they are able to get plenty of forces over there, so we can come home for a while when this is over, but I don't see how they can. That just about covers the current events.

Honey, I have shot the bull enough for one night. Think I will take a hot shower and go to bed and see if I can go to sleep. I know that I'm going to have to arise early in the morning. Kiss the kiddies for Daddy — I love you with all my being.

Your Joe.

October 17, 1944
Tuesday night, 10 p.m.

My dearest Honey,

Got two swell long air-mails from you today written October 9 and 10. That's the best service I have had in a long time. It surely leaves a long gap in between that I haven't heard from you at all.

The first thing I want to mention is that fellow, Durham, who lives behind us. I think fixing up the little swing for Kay was about the nicest thing I ever heard of anyone doing. Will you deliver my personal thanks to him?

So you have paid the furniture notes completely off now? We own some furniture for the first time in our married life. That's good! You had to wait a long time for some, but it's good to know that we have some furniture that we can call our own at least. I'm proud of that, and it cuts your monthly expenses down some. I've lost completely out on all of our business. Approximately how much do we owe on the house now? You are doing swell with everything apparently — I knew you would though. Really, I had no worries.

Kay and Carol sound more precious to me every time you write about them. I certainly am glad that they are "talking together" more or less, and that Kay, in her own childish way, is taking up for Carol — even though apparently she slams her around a bit. I'd give almost anything to watch them and listen to them for just a little while. I know they are sometimes a lot of trouble, but I doubt if even you realize what a pleasure and joy they are to you. They are bound to tax your wisdom, courage, and patience to almost the limit sometime, but gosh are they worth it. Just stick it out until I get home, and honestly I'll help a bit.

Today I went into town and finally got my this week's rations — today was the last day. Stayed in town for about 3 or 4 hours. Also picked up my summer woolens that I had cleaned. I surely am short of winter clothes. When the boys were here that were Prisoners of War from Romania they swapped a lot of clothes with me. None of them had but one suit, so when theirs got dirty, they just gave their dirty clothes to me and took clean ones. Consequently, now I have nothing but an assortment of clothes of different sizes. This morning I put on a pair of trousers that were a good three inches too long and just about that much too large in the waist — the sleeves of the shirt were a couple of inches too long and the neck was too small. Tomorrow I will take them over to supply and trade them for some clothes that come nearer to fitting me. The only woolen clothes that I have left that I came over here with as far as shirts and trousers are concerned is one pair of greens and one pink shirt that's much too small. All the others have been swapped in some of these Italian cleaning joints, and I just don't have any good woolen clothes — but I'll be darned if I buy any over here. I'm just going to keep on wearing O.D.'s.

We are supposed to have another inspection tomorrow, and I have made no preparation for it whatsoever. By the time the inspector gets here, I can at least have some excuses formulated in my mind.

So you think you married me 'cause it was leap year? Don't fool yourself. I had my mind made up about you two years prior to you ever giving me a thought. I loved you then — but more now.

All my love,
Joe.

October 18, 1944
11:15 p.m.

My dearest Honey,

Another shortie tonight. There has been nothing happening today. I didn't get any mail either so there is practically nothing for me to write about. Had my inspection today, everything was in pretty good order however this one was done by the Wing Surgeon who is a pretty good friend of mine.

So Hugh is in Germany — keep me posted on him, will you? I think of him very frequently, and wish for his safety and comfort.

Don't worry about my Christmas — I know you are there waiting for me, wishing for me, and taking care of ours — that's all the Christmas I need. Next Christmas perhaps I'll be home — then we will celebrate for all three. I love you with all my heart.

Your Joe.

October 19, 1944
10:20 p.m.

Dearest Honey,

Well we had a cold front to blow in on us today. It is quite a bit colder and the wind has really been blowing today. The wind blows across Italy practically all the time though. It has rained off and on all day and we have done no flying.

Because of the weather and nothing particularly to do, Zraick and I went into town around lunch time to get our rations and go to the picture show if there was anything worth seeing. When we got in, we found that Bing Crosby was on at the Ensa Garrison theater (English Theater) — neither of us caring too much about Crosby that we almost didn't go. However, we did finally go and saw absolutely the best picture I have seen since being overseas. The name of it was "Going My Way" and it was good. Highly emotional, good acting, little singing, and very entertaining. Did you see it? If not and you get a chance to make an effort — there is one scene in it — the last scene — that will almost tear your heart out. I enjoyed it very much.

Reese and Zinn got in tonight from the front lines. They brought Reese's brother back with them who is an Infantry man on the front. I have been here listening to him tell his stories. He has been on the front for quite a while and the poor boy (34 years old) is just about shot. This Italian campaign, i.e. on the front, is rough. It doesn't get much play on the headlines, but probably the toughest battles of this war, so far, have been fought in Italy. This fellow has to go back to it by Saturday night, and he really does dread it.

Next morning —

Looks as though I'm writing most of my letters in two stages now — sorry.

Did it rain last night! I believe it rained harder than I have ever seen it before. This morning it's really cold. When I went out for take-off this morning, I almost froze. I was so cold that I came back and jumped in bed just to get my feet warm.

Well Honey, I'd better give this to Giovanni and let him get it in this morning's mail. Will do a better job tonight. I love you with all my heart.

Joe.

From Joe in Rome to Babe in Memphis

October 23, 1944

My dearest Honey,

Here I am in Rome again — without any warning or mentioning it to you. Yesterday morning I got fed up with it all suddenly, so I decided I would get away for a couple of days, so I picked up and left — with permission, of course. I think what I need to get away from is myself more than anything else. So far, I haven't found out how to do that.

Last night, I drove into Naples. Called up Harvey Carter and went out to spend the night with him. Enjoyed talking and being with him very much. He lives with four other doctors in a very nice apartment. This morning he took me over to a hospital only a couple of blocks from where he works to see the Hughes Twins, Phil Bleecher, John Wilson, and Dr. Harwell Wilson. I didn't know the latter Wilson too well. He came to Memphis after my time, but of course I know the rest very well. I was very glad to see them all and had a very enjoyable chat with them for about an hour. By the way, Jimmie Hughes is a full colonel now, which is pretty doggone good. Anyway, after talking with them for an hour or so, I pulled out for Rome with Bernstein in his jeep. Tomorrow or the next day, I think I will go back down there and stay a day before going back to camp. All of them look the same — Harvey appears to have gained a couple of pounds or so.

Got to Rome about 2:30 this afternoon. Bernstien immediately went to see his Italian gal friend, and since then I have been lying around the Red Cross, the Hotel, sleeping and reading. It's now about 11 p.m. and soon as I finish writing, I'm going to bed. Personally I don't expect Bernstein back, but we were put in a room with another lad, and though his clothes are here, I haven't met him yet. Perhaps he won't be in either. I am staying at the Hotel Regina this time. It is larger and slightly finer than the Savoia, where I stayed the last time I was in Rome. Have a private tub bath and three half-beds in an average sized hotel room. The hotel is moderately old — at one time very popular and cosmopolitan and still nice. The walls of the room are "upholstered" rather than papered and the tapestries are very nice.

I have been sitting, or laying, here tonight and going over our life together. From the beginning I supposed it has been about as perfect as any couple could have. I have given particular thought to our short but happy premarital romance. I say "premarital" for it didn't stop then. Never, if I were to live to be a thousand, would I forget the conflicting expression on your face at the top of the stairs the first night our date was over without Dan. All the time I rode back to Helena I had two prime emotions, which were certainly conflicting. Firstly, I felt guilty — that I had kissed you with your guard momentarily down thereby taking an unfair advantage. Secondly, I was glad that I had 'cause I had wanted to. Funny, I don't think I've ever talked to you about this before! I loved you so very much then but even then you nor I yet knew what the word really meant. It surely has been swell.

Well, my honey, I will again say goodnight to you. May God see fit that it won't be too much longer until we say goodnight without having to pick up a pen.

I love you so much,

Joe.

From Joe in Castelluccia to Babe in Memphis

October 25, 1944

11 p.m.

My dearest Honey,

I owe you a nice long letter for two reasons. Firstly, I haven't written in two nights — which you'll have to admit is unusual; secondly, when I returned home today I had three long Air-mails waiting for me. The reason I have missed two nights is just because I haven't had the chance. Night before last in Rome, I couldn't write because of my roommates in the hotel — last night I was again staying with Harvey Carter and didn't have a chance. Tonight I will attempt to make up in part.

Yes, spent the night with Harvey again last night. We went to see an English Characterization play, which was sorry. English wit is so stupid and dumb. They are certainly far behind us in comedy. Got up early this morning and went out to the hospital with Harvey and saw some pretty interesting cases. Then got in the jeep and came home. Everything is just about as it was when I left. I did enjoy the trip, however — we didn't stay long enough in one place for me to get bored. I enjoyed seeing Harvey and the other boys — and some sick people (first ones I had seen in many a day).

When I opened one of these letters that I had waiting for me today, I almost got my head bitten off for saying that you could have been happily wed to any decent chap — because any decent one would make it his business to see that you were happy. I was "raked" over the coals so mercilessly that I am considering going back to Rome and taking back the most gorgeous thing I have been able to buy for my wife since I have been overseas. I haven't completely made up my mind yet, but if you give me much trouble I will consider these drastic actions in a rather strong frame of mind!

Too, there was mention in a letter today something about twin beds. Whatcha' tryin' to do, get a divorce from me! Twin beds — of all the useless household equipment ever made, that's the most useless. They were made for twins. If we get twin beds, what are we going to do with the one that's left over?

I love you,
Joe.

October 26, 1944
2:45 p.m.

My dearest Honey,

This is a rather unusual time of the day for me to be writing to you — but I have been thinking about you all day today (living up in the clouds), so I thought I would sit down and talk a little while.

General Mud has taken over again! Last night it rained all night and the camp area is just a mire now. It'll be that way the majority of the time from now to next spring, no doubt. Even so, I don't imagine it will be as bad as it was down near Constantine, North Africa, so I imagine we can stand it — and living conditions are so much better now.

By the way, you mentioned Young in one of your last letters. He is married now. We haven't heard from him but someone did send some clippings concerning his wedding to us. Apparently it was quite an affair.

In talking about your coat in your letter, you mentioned a black one with a silver-fox collar. Have I ever seen it, or did I just pull a "faux pas par excellence"? I just don't remember it — how my memory concerning some things slips. I must be getting old!

On this writing business — I wish you would find out if somehow I offended your folks. I don't think I did, but sometimes in letters, things don't sound to the reader as the writer intended. I haven't heard from them in a long, long time. Are they O.K. in every respect? Perhaps they are "paying me back" for not writing for such a long time, and I certainly have it coming to me. That's O.K. — but I certainly don't want either one of them to harbor any ill feelings toward me whatsoever, 'cause I like them both quite a lot as you know. Don't say anything — just find out for sure that I haven't written something to them that perhaps they didn't understand my true meaning, and if so, fix things up for me. That's putting you in a new role for me — my "good-will" ambassador. You are sorta infringing on a priority of sisters' in that role. Sis' has covered over more rough spots for me, especially if something was up between the folks and me, than anyone. She's pretty good at such too. She's been a good Sis to me — but somehow she feels like she is my little Sis, rather than quite a few years older — perhaps it's cause she's so short.

So you are trying to get Carol to walk, eh? Perhaps with the combined efforts of you and Kay, your folks, and my folks, finally she will decide that those big legs are made to walk on. After all of you trying so hard some day, she will probably get mad or provoked and get up and run completely across the house. I never will forget how quickly Kay learned — I mean, how sudden.

Now, I'll hush but not before I tell you that I love you with all my heart.

Your Joe.

October 28, 1944
8 a.m.

My dearest Honey,

We worked all day yesterday trying to tar up the roof so the soot and rain won't come in. Personally, I don't mind the rain, but the soot gets me down. We got about 2/3 of it done and will finish it the first pretty day we have — which are rarities to us now. You should have seen us when we finished yesterday, we were covered in tar.

Honey, I've got to run over and do some 64 examinations on some neophytes who just joined us. It's a pity they don't finish these guys up back in the states where it is convenient. Kiss the kids for me.

I love you,
Joe.

October 29, 1944
10:10 p.m.

Dearest Honey,

Wrote you an Air Mail this morning — at least I made an attempt to — I was interrupted so much that it probably wasn't coherent.

Today has been rather dull. The weather closed in again this afternoon and the ships had to return early. That's the way it was much of the time last winter. I haven't done a thing today — just sitting around.

Heard Kremers had been down in the dumps, so I called him up and asked him to come over. He came over about 6:30 tonight and stayed about 2 hours. He was very restless. He is down physically and psychologically. I think his leg is giving him quite a bit of trouble, but perhaps most of his trouble is psychological — which is certainly explainable. It's pretty hard to keep from getting really depressed sometimes — particularly when you can't see anything like an end to this sort of stuff. I think Kremers will give up within the next month and go back to the hospital. Perhaps they will send him home this time — I believe they will and hope they do for his sake. He's a nice, quiet, level-headed guy who is getting too old for this sort of stuff.

Honey I've been thinking about writing to Dr. Boyd at Campbells for the past couple of weeks, and at least letting them know that I am interested in getting that fellowship there, but don't know whether to or not. I have got to make up my mind what I'm going to do after this war. If I'm going to try to do more studying, I have got to start making some sort of connections — otherwise, I won't be able to get into anything for years after I'm out of the Army (if that day ever comes).

However, as I see it, we won't have enough money to see us through the 3½ years that it would take — in fact I'm damn sure we wouldn't. Guess we could borrow enough to see us through from one source or another, but at the end we would be destitute. I don't know whether to enter a venture of this sort. You know if I got it, I would be making the total sum of $50 per month — room, board, and laundry. The latter of course, wouldn't be worth very much in my case. Next time you write me, express your honest opinion. And when I say honest opinion, I mean honest! Many things have to be considered — our age, our happiness, the children, finances, what we can do, etc. Think about it and let me hear from you — I don't guess there's any particular hurry — looks like it's gonna be a long war.

Well, honey, I think I will hit the sack. I didn't get enough sleep last night, and I have been feeling the results of it all day long. (Don't tell anybody, but I think I'm getting to be a sissy). Don't know what I'll do when I have to start working for a living again. Goodnight my sweet.

All my love,
Joe.

P.S. How were the cherubs today? Was Kay in a mischievous mood or was she a little lady? Bless 'em — they must be little devils sometimes, but bless 'em anyway. I love 'em.

October 30, 1944
8 a.m.

Dearest Honey,

Honey, I fixed up a box to send home yesterday. However, I got to town too late to mail it. In it I put that Palm Beach summer coat that I should have left home in the first place. Also a bunch of towels that you might be able to use and that I don't need — they look soiled, but that's the way they are when the Italians return them from laundering. I believe after they have been sent to an American laundry that they will be usable at least. Also I'm sending some do-dads. By the way, the Purple Heart medal in there is not mine. I have just picked it up as a souvenir. I definitely have not been wounded in action and that is what the Purple Heart is given for. While I was in town yesterday, I looked around to try to find something to send Mother as you suggested that I do the next package I sent home. I couldn't find a thing that was suitable. I will find something sooner or later though — I have not forgotten.

I hear Read is in Naples. He is the fellow (pilot) that returned to the States and was slated to come back to this outfit. I will certainly be glad to see him. I certainly hope they put him in this squadron for we certainly need him. We have long since gone to the dogs in the hands of inexperienced fellows at the wheel.

Today looks as though it is going to be sunny. I hope so — we have had so very little of it lately. We had a mission today for the first time in ages. When we don't fly, the time just drags and it seems that there is no reason to be here.

Went to see a show yesterday. It was a Universal Western — "Frontier Badsman," I think it was called. The funny part of it was that I enjoyed it. The cast was very good.

Well Honey looks as though I'm gonna be bothered all morning by one thing or another. I'll bet I've had 10 interruptions since I started this letter. Now I've got to hurry so my trusted aide, Giovanni, can get it in the mail before it pulls out this morning. Tell the babies that Daddy says hello.

I love you,
Joe.

November 1, 1944
10:30 p.m.

Dearest Honey,

I sent you two boxes today. I put an article made out of some parachute nylon in one of them — I think it's pretty. Every stitch is nylon, so don't wash it in anything real hot or iron it with a hot iron cause nylon will just melt away. Even the thread used in it is nylon from the shroud lines of a parachute. That's a good 'chute — it saved a guys life once. Also in that box (the smaller one) is a "doile" set for you to give Mother for me. Too there are some Life-Savers that were given to me (a whole carton of them) for Kay. Pete Massare, who finished his 50th mission gave them to me to send to Kay. He knows I send gum all the time, so this morning he sent the Life Savers up. Hope you will like the things.

Got paid yesterday. After I paid all of my debts, I didn't have much left. Around the 10th of the month I think I will be able to send some though — perhaps 50 or 60 bucks. From then on I will do better unless I see something that I think perhaps you may want. It's silly to buy anything over here though with such inflation. The American dollar isn't worth 20¢ when it comes to buying something. It would be far wiser for me to send every penny I could to you — which I'm gonna do henceforth. It's probably difficult for you to realize how you can get to feel about money. Money just isn't worth anything. You can get one egg for 20 or 25¢ now, but I can trade one pack of American cigarettes for 2 eggs — the cigarettes cost me 5¢. To get a few pieces of laundry done it costs $1.50 plus a bar of soap — but the bar of soap means more to them than the $1.50.

Thank goodness we have a more or less stable economy. The Italians are the worst people in the world at stealing. Every time you leave your car, or ambulance or truck for 15 minutes it has been ransacked when you return, but you can understand it — the laborer makes 65¢ per day but a pair of shoes (wooden soles) would cost him at least $25.00, and it would be impossible for him to get a suit of clothes and every family has more than 10 kids — poor, ragged, ignorant, barefooted kids that plague you for cigarette butts and that you soon lose the feeling of pity for and get to almost despising. Well, I could write volumes on Italy and Italians but it is uninteresting — and according to the old Sigmund Freud Theory something that will be forgotten because it isn't pleasant to think about. Let's just be damn thankful that our children won't be brought up in the same environment. But, again, I want to get off this subject.

Got a short note from Burr Craycroft today. He is still living very comfortably in the same place. He is still under the delusion that I'm happily riding around Italy on my motorcycle. I will drop him a note in the next day or two. Went into town twice today. My back is almost breaking from going over all of those bumps 4 times in one day. Don't think I will try it again — at least until I convalesce for a while. I went in this morning to send the packages and get some cleaning, and tonight Zraick and I decided to go in and see the show. Ray Milland in "Uninvited" was on and was reputed to be good. On the contrary, I thought it was lousy.

I haven't heard from you in two or three days. Perhaps tomorrow there will be much mail. Since there's no use trying to write everything in one night, I'm gonna stop — but not until I tell you that you are undoubtedly the most wonderful woman on earth and the only one for me.

I love you,
Joe.

November 3, 1944

My Dearest Honey,

Went into town last night to see a picture show and really enjoyed it. It was Jean Arthur in "Impatient Years." She is really good. I think I have gotten to where I enjoy her more consistently than anyone else.

The weather here has been bad so long that things around here have become pretty boring. I think all of us are getting tired of looking at one another. From the morale point of view, that isn't good. The monotony is going to be broken somewhat for me however for either this afternoon or in the morning, Wagner, Quinn, and I are leaving for Bari to go to a medical meeting. Don't know how good it will be, but at least it will beat sitting around here doing practically nothing.

Hey, I've been wondering this morning — do you suppose that I might come under the heading of the G.I. Bill of Rights that they have been making such a fuss over back in the States? If so, we could do alright. I could get some more work off, and we could still live and eat. I just wonder. You know on 50 bucks per month that I would get if I did some more work, and $50 or $75 from the government we could see our way through for 3 years, couldn't we? I don't relish the idea of being on the government "dole" though, do you? That doesn't sound like I'm upholding one of those promises I made to you one time. I wish I could think things out for myself, but being away from it all and not knowing exactly your idea, it's sorta difficult for me to make up my mind. I just can't quite visualize the set-up back in the States — but it would be far better if I could get a line on something now — if I'm gonna do it at all. Send me Dr. Boyd's name and home address. I think it's Harold G. Boyd, but I'm not sure. — I might write to him.

It's pretty difficult for me to think of education, practicing medicine and things concerning our happiness in the future. All those things seem at the moment to be so secondary. But, at some future date after we are together, they will again become important. The important thing now always is the same — that is, get back together so we can all be happy. Look — after I go crazy, what are you going to do — how are you going to get along? Something you had better start thinking about. Honey, It's high time that I go to lunch if I want to get any. Tell Kay Daddy said "to be a sweet girl and that he loves her." I love her mother most though but don't tell Kay that!

All my love,
Joe.

November 5, 1944
Sunday night, 8 p.m.

My dearest Honey,

Well I have a lot of writing to do. Friday afternoon I pulled out of here for Bari to go to the Medical Meeting. As I was leaving, I was brought in 8 letters. Four from you, 1 from your Mother, two from some of the old boys who are back in the States, and another from University of Tennessee. This afternoon when I returned to camp, I had about six letters from you and one from Mother.

We had quite a meeting down in Bari. Wagner and I took my ambulance and went down Friday afternoon. We stayed in a tent out at the hospital. I suppose there were 150 or 200 doctors down there. Among them were the Hughes twins, Phil Bleecher, Emil Koenig, Priver, Tripi and a number of others that I know. (Priver is losing his job, by the way, he has a new C.O. and they aren't getting along. I hear Priver is trying to get out of the Army completely, but don't know whether he will make it or not.) Anyhow to go on with the meeting — I enjoyed seeing the guys. I went around to some of the clinics, but I was so far behind time that I didn't learn too much. One morning of it was all we could stand, so yesterday afternoon Wagner and I pulled out and went to some sorry picture show. Then last night, we again went out to the hospital for dinner, cocktails, and a putrid bit of entertainment that they put on for us after the dinner. After all of this was over, we started out to our tent, but as we were passing through the club towards the door, we passed a table where 35 to 40 doctors were standing around playing a game called dice or "African dominoes" or something like that. I thought I would be sociable, so I decided to donate 10 bucks. I picked up the "toys" and, as beginner's luck would have it, in 5 minutes I won exactly 200 dollars — so I quit. Consequently, due only to the fact that I was trying to be sociable, I was able to send a P.T.A. check to you today for $250 rather than the fifty dollars that I had intended to send this week. The moral to that story, if there is any moral shown, is that it sometimes pays to be sociable.

I'm not going to try to answer all of those letters that I have received from you since I have written tonight. However, I want to relieve your mind concerning one point. Apparently in one of my letters I led you to believe that I had just flown a mission, which is erroneous. I'm certain I didn't intend to lead you to think such. I have never flown a mission, and at present, don't intend to. I get all the thrills I need in other ways, and I'm not the least bit curious. In fact, I would just be plain scared. I'm only telling you this not to just admit how scared I would be, but so you won't be sitting there thinking that I might be shot down somewhere. I repeat — I have not flown a combat mission, nor, at the present time, do I intend to. So don't worry about me.

Before I sign off tonight I want to say something about the pictures you sent me too. The ones of Carol are particularly good — she looks like she is getting much prettier and has very sensitive features. Kay looks like she just didn't feel like posing and used that mind of her own, so didn't pose (as one would expect from her). Honey, every day you seem to get more and more beautiful. Pictures of you sorta knock me back on my feet and make me realize more and more just how good looking my wife really is. I'm an extremely fortunate man if for no other reason than I have you for my wife.

I love you,
Joe.

November 6, 1944
Monday, 11 p.m.

Dearest Honey,

Guess who just walked in? Read, who has just returned from the States. I haven't talked to him much as yet — he has gone to get his stuff to bring over here to live with us. He was Bryan's first pilot. He is supposed to be returning for a second tour of duty with us, but I think he will probably be turned around and sent home. Of course, I hope so for his sake, but the squadron could certainly use him.

Yes honey, we are pretty well separated out here, and I don't see the other doctors very often. Kremers, McFarland, Quinn and Group Headquarters are all away over on one side of the runway, and I am down near one end on the other side. I am nearer Kremers than any of the rest but almost 3/4 of a mile from him. This suits us fine though.

I was so glad to hear that Johnson came out to see you. He is a good boy. Did you notice how nervous he is? He doesn't smoke, drink coffee, drink or anything for fear that it would increase his "shakes." I don't imagine they will let him have pilot training, but you can never tell what they will do back in the States. I am so glad you got to talk to him though.

No honey, don't worry about sending me any pajamas. The shorts do wonderfully. They might be a little chilly getting into bed sometimes but I know I won't hang myself or feel like I have a rope tied around me during the night. Shorts are safe, and as yet I'm not so sure about pajamas. Don't guess I'll ever turn into a gentleman concerning those things. Another thing, I don't like to try to sleep on a lot of knots.

Now about having the house insulated. To begin with, I'll tell you I don't know much about it, nor do I know, or rather remember, about the price of last winter's gas bills. You know much more about the situation back there than I do, Honey, so do what you think best. I think it would probably be a good idea to wait until the house needs to be papered though and then have it insulated, cause of course when you have it insulated you will have to have it papered too. Glad to know you got the windows put in and it seems that you certainly got it done reasonably — you are a smart girl.

Now I've answered enough questions for one night. I will finish other things tomorrow. I gotta get busy sometime soon and answer a lot of other letters. Tell the children that Daddy loves them.

All my love,
Joe.

November 7, 1944
Tuesday, 11:00 p.m.

My Honey,

No mail for the past two days — that is letters. Today I received a package from you containing candy, fruit cake and a couple of books. Thanks very much — it's mighty nice of you. The cake looks good, but we haven't cut it yet. The package was marked "Christmas," but I figured you just did that so I could see it. Thanks again.

Glad you got to go to Cape Girardeau — I don't know whether you enjoyed the trip or not, but I'm sure Kay enjoyed the ride both ways. Bless her heart — she loves to ride so. You were back in Memphis before I knew you had gone.

In one of your letters you told me how "colorless life was with me." Aren't you ashamed! And I thought we had been so happy together — of course, I guess it was a "typographical" error. Well, honey, it's late and I must get to bed. Today has been so usual there is nothing much to write about. Just wanted to tell you that I still as always love you with all my heart.

Joe.

November 8, 1944
Wednesday, 10:30 p.m.

Dearest Honey,

Honey, of course I didn't mind you joining the women's Medical Auxiliary. I'm honored that you did — sorry, I never said anything about it. You know, I'm glad for you to do anything that you get any pleasure out of. I have the utmost faith and confidence in whatever you do, or whatever steps you decide to take concerning anything. I don't worry about how you are running things back there one iota. It's even difficult for me to remember that I ever did any of our business. In fact it's hard to realize or remember anything but war and Italy!

I missed the news tonight, but I hear that Roosevelt is still president. I was torn between personal selfishness, and what I thought best in this election. Before Roosevelt gets out, during this term especially, he is going to socialize medicine just as much as he can. You know my opinion on that. On the other hand, at the peace table if that time ever comes, I want Mr. Churchill and Roosevelt to be sitting there with all the prestige that they can muster up for the occasion. I do believe if Dewey had put his campaign on a little higher level of politics, he would have ran a better race, and from what I hear he didn't do badly at all. For years the Republicans will win even if they ran Kay for president.

We have been eating all evening. Zinn got a package containing some anchovies and crackers that were delicious. He is a peculiar cuss — I believe he varies his love for his wife daily according to what she sends him. Tomorrow he is going to Cairo. The colonel is taking off in a couple of days for a 4 or 5 day stay in England which I would love to make, but can't for a number of reasons. One of them being that Wagner is also going to Cairo for a week. He too is leaving tomorrow. I hope the Inspector General doesn't expect me to know too much about the Group Medical situation cause I don't know as much about everybody's business as I used to when Wagner left. Somehow I'm always Acting Group Surgeon when we have big inspections, and Wagner doesn't do it on purpose either — it just works that way.

Well honey I think I will hit the sack — read a little bit, and then go to sleep. Hey, I dreamed of you nearly all night last night. It would wake me up and I would go back to sleep and dream of you some more. Nice pleasant dreams too. It was the most delightful and refreshing thing that's happened to me in a long time. Tell Kay and Carol hello.

I love you,
Joe.

November 12, 1944
Sunday, 9:30 a.m.

Dearest Honey,

Figured perhaps I had better drop you a line this morning and get it in the mail since I didn't write last night. Last night Kremers came over for a visit and he, Read, Reese and I sat around and talked and drank until about midnight. Kremers is leaving in the morning for the hospital — he thinks they will send him home this time and for his sake I hope they do. Wagner is down in Cairo now so until he gets back, we are going to be a little shorthanded.

It's been icy cold here lately. Surely does make it difficult to get up early in the mornings. Snow has covered all the mountains around here and this morning all the puddles were frozen over pretty solidly. So it looks like ole' man winter is really here. Don't imagine the snow will be off those mountains until next May, but I might be wrong.

I've been sticking pretty close to camp lately, but I think that I will go into town today and get my rations. Probably will take Read along and maybe go to a picture show. You know Read just returned from the States — well, apparently the Air Force is going to reverse their decision and send him back to the States now for more rest and recuperation. That's the Army for you.

Hey honey, if you want to send me anything — I think canned stuff (soup, sardines, fish or anything) would be about the best thing. When it's cold, we sit around a lot and eat. Even soda crackers are good to send for we never see any — cheese would also be O.K. That's just some ideas. For gosh sakes don't go to a lot of trouble nor don't use up a lot of ration points that you need for Carol's baby food. I get plenty to eat otherwise. You asked me how much I weighed — Honey, I haven't the slightest idea. I don't think I've seen a pair of scales since the States. However, I'd guess I weigh around 155 or 157 — about my maximum. No, I'm doing alright in every way except missing you which is practically saying that "I'm not doing worth a damn." But I am doing as well as could be expected in the physical set-up.

I love you,
Joe.

November 14, 1944
Tuesday, 9:30 p.m.

My dearest Honey,

This is going to be a short one tonight — I'm so tired I can hardly move. I've been working hard all day long, which is a bit unusual. We had a severe crash today and it has kept me busy all day with Kremers and Wagner gone — it wasn't one of our ships however — nothing for you to worry about — just a lot of work.

Now, Honey, about the Realty Tax on our property — I too, was under the impression that all of the taxes and insurance was taken care of by F.H.A. However, I certainly could be wrong. Guess finding out about that bit of business will have to be taken care of by you also. One of these days, I promise to take most of that stuff off your shoulders.

It touches me pretty deeply to think that Kay still asks about me — you'd think that she would have forgotten by now. Bless her heart. Even so, I'll bet she will steer clear of me for a few days when I see her next. Well, Honey, I'm going to bed. Bet I'll be asleep in 5 minutes after I get in. Goodnight.

All my love,
Joe.

November 15, 1944
Wednesday, 10:30 p.m.

My dearest Honey,

We had a big inspection today — the biggest yet (made by the Inspector General). It wasn't bad though — I think they must have been rushed for time. Anyhow I think we came through with flying colors. In fact, I'm sure we did. This morning I had to trudge around with the inspectors. This afternoon I have been hanging around my casa taking it easy.

Sounds as though Miss Carol is finally going to make a stab at standing and walking — the 'fraidy cat! It's about time. There's not but one thing bad about it — she will now probably be a lot more trouble to her Mama, and I'm afraid Mama has her hands full as it is. I wish I could be of some help to you, but I'm afraid all I can do is stand on the sidelines and wish for you (literally and figuratively). No doubt Kay is a nuisance with her 1000 questions per hour, but that's the only way a child has of learning. I expect we should be thankful that she has the intelligence to wonder about everything (perhaps she is going to be scientifically inclined too). I do know that becomes awfully boresome at times though. You know it seems to me that Kay is entering the "question stage" a little early too. We must have a pretty smart youngster there.

Just opened me another can of tomato juice — they weren't rationed the other day and I bought a case. Since I have had this cold, I've been drinking a can just as often as possible, and now I've damn near finished the case. Of course, I use the cold for an excuse for my gluttony, but frankly I like the stuff. Honey, I'm going to quit early tonight — kiss the kiddies.

All my love,
Joe.

November 17, 1944
Friday, 1:15 p.m.

My dearest Honey,

I probably won't be able to finish this epistle in one sitting for the planes are due back in about 30 minutes, but at least I will get a start. That is more than I did yesterday — and I had no good excuse either unless it was laziness. Last night the boys collected around for a card game, so we played cards until about midnight. I did do a little more work than usual yesterday. I worked up four cases of Kremers' to go to the Medical Disposition Board. I may take them down Monday, so I can go by and see an Ear, Nose and Throat man in Bari.

I have been doing a little better with my job as Flight Surgeon the past few days. I finally got disgusted with myself and decided I would make an honest effort to learn these new boys at least fairly good — my conscience already is partially salved. Perhaps with a little more effort, I shall again be doing the job as it should be done.

Midnight —

I have been playing poker for hours, and my mind is so dull. I don't know whether I can write a decent letter or not. However, I'll try. After playing poker all this time, I think I lost about $2.00. We play a small game of 25¢ limit, which isn't bad. I guess you could lose $15.00 or so in one evening, but no one ever does. We either play poker of that type or rummy. In rummy, it is almost impossible to lose over $3.00 per evening. I was the big winner in rummy last month, and I won about $12.00 for the whole month.

Colonel Nelson (Wing Surgeon) spent the morning with me. He told me that he had been trying to get me a Group, so I could get a majority but hadn't been able to find an opening and didn't know when he could. I told him I didn't care whether I got to be a Major but that I wish to hell he could get me some job somewhere that was different from this one, that I was bored. Don't worry, nothing will ever come of it — I'm stuck right here for the duration plus. He predicted that we would get out of the Army in 1949 — he'd better be wrong! Goodnight and tell the kiddies hello.

All my love,
Joe.

November 21, 1944

Dearest Babe,

I have been down to Bari for the past couple of days. I left yesterday morning and just got back tonight. Went by to see Kremers — they have not decided whether he will be sent home or not yet. They were supposed to hold another Board on him today.

Upon returning tonight I had a letter from your Dad, Sally and Ruby Mae, and Mahon. Mahon is still on his leave and is due in Miami now, I guess. He was going to drive down and take his wife with him — perhaps he got to come through Memphis, and if he did, I certainly hope they looked you up. I'd like to know your opinion of his wife — bet she's a nice likable sort.

In your last letter, you said you wrote me the night before about your "honest opinion" about more education. As yet I haven't gotten it. I'm eager to get it but don't know what exactly to do 'cause there's no telling how long this is going to last — tonight I'm a little down and a little more homesick perhaps, so it's not a good idea to think much about it. I'll wait until I get your letter and feel more chipper.

Looks as though the west front is becoming more active. I surely hope they are able to do something surprising up there before the weather closes in on them too tight. I don't anticipate too much — fighting the elements is sometimes a big job within itself.

Well, honey, I'm a wee bit weary tonight. I think I will shut up and thus bring to a close a very rambling and disjointed letter. It's raining like everything. It gets dark now around 4:45 every evening. Certainly makes the nights seem long. Goodnight.

I love you,
Joe.

November 22, 1944
7 p.m.

My dearest Honey,

Since they have opened up on the Western Front again, perhaps a miracle may happen and this war will be over soon. Gosh, I hope so, but am not going to allow myself to become too optimistic 'cause I don't see how it can be over until next summer. The longer the European War lasts, the higher the possibility that we shall come home before going to the other parts of the world.

Just got through reading about 20 of your letters. I had been saving them for just such an emergency. I haven't heard from you in three or four days, so I reread them. There are a couple that I have saved a long time. They always serve the purpose of giving me a little extra boost even though I have reread them so often that I practically know them by heart.

My cold is doing nicely as are my sinuses. I'm only wondering when the process will reverse itself and I begin to do nicely instead of the cold. Honestly, today I have felt considerably better. Everything will be fine in a couple of days.

Got out this afternoon and kicked and passed a football around for quite a while. Tonight my leg is sore for having kicked so futilely for so many times. It does make one feel better to get a little exercise occasionally even though it does make him sore.

We are planning on having a big dinner tomorrow night (Thanksgiving). Even though it is cold, we are going to have ice cream. We, of course, don't have a freezer now, so tomorrow we are going to send an airplane up to 20,000 feet with 40 gallons of ice cream mixture. It's minus 50 up there, so when it comes down, it will be frozen. We tried it out today when we sent a small amount on a mission with the boys — it was good when they got back. That's the way we got beer and cokes cold this summer, but sometimes it freezes so hard that it breaks the bottles. I guess there's a lot of things that's nice about being in the Air Corps. I realize that I'm very fortunate in being where I am, and I've thought many times how fortunate I was that I was not accepted by the Navy. It's a cinch that I fight a very easy war in comparison to the average boy.

Well I've rambled enough for one night. My letter writing has been sorry for the past few nights. Sometimes I sit down and it's so easy to write a good letter, and then sometimes I sit down and write one that isn't too good and I realize it. Tell the Kiddies hello for Daddy.

I love you,
Joe.

* * *

On Thanksgiving 1944, the officers played the enlisted men in a Thanksgiving Day football game. The loser had to serve the winner Thanksgiving dinner. The enlisted men served dinner that year. They had planned ice cream for dessert with Col. Hoppock and Charles Thomas piloting the B-24 for "Operation Ice Cream." However, the mixture would not freeze. As Charles tells the story, to make up for the lack of ice cream, Col. Hoppock added whiskey to the mix to make a sort of egg nog.

* * *

November 23, 1944
Thanksgiving, 5:30 p.m.

Dearest Honey,

Even though it has been prettier than usual today, we did not fly, so this afternoon some of the boys and I went into town to get our rations. We intended to go to a show but it was so poor that we decided against it and came home.

We had a very good meal tonight. Turkey, even — so can't gripe about the eats. The ice cream that I told you about last night didn't freeze after all — we carried it up to 30,000 feet but no ice cream. I think we had too much of it in one container.

This morning the deputy surgeon of the Air Force and a couple of other fellows, including Major Bassett who was a friend of mine at March Field, came by to see me. Don't know what they wanted — just said he thought they would drop by — didn't inspect or anything and only stayed for a few minutes. He said he didn't know what would happen to Kremers yet. Wish they would make up their minds, so we would get a replacement.

Well I think I will play cards with the boys and then hit the sack. Hope I get a lot of letters from you tomorrow. Kiss the kiddies and always keep in mind that I love you with all my heart.

Joe.

November 25, 1944

My dearest Honey,

No mail again today — 6 days now. My morale is lower than the proverbial snails belly. I started to write you this afternoon but decided that I would put it off until tonight after the mail came in so I would have more to write about. Uncle Sam didn't come through. The mail will be coming through in a day or two, I'm sure.

Went into the Wing today to have a chat with the Wing Surgeon about one of my boys that I'm having a little trouble with. He wasn't there, so the trip was made in vain. Took Colonel Hoppock along with me — we all got in a discussion sitting around the table on who was "trifling" the most — soldiers overseas or their wives back home. What a hellova thing for a bunch of men to sit around arguing about. The consensus of opinion was that it was about even. I couldn't believe that, nor could the colonel that was there, who must also be very happily married. We took up for the women. However, after knowing some of the things that has happened to some of the boys (I think about five of them have wives that are trying to divorce them or who are now pregnant) either way things have gotten to an appalling state. More reason why I thank my Heavens for you every day.

For the past three days all I've been doing is writing letters. I have been trying to answer all the letters from the guys that have gone home that write back to get me to do something for them. This is getting to be quite a problem and a hellova lot of trouble. Sounds like all I'm doing tonight is griping. Maybe I'd better stop and tomorrow night I'll probably be in a better frame of mind — at least, I hope so. Living by myself is bad enough at best but when I'm in a bad frame of mind it's horrible. How do you put up with it?

I love you,
Joe.

November 26, 1944

Dearest Honey,

Finally, after waiting for 6 days, one lone airmail letter arrived from you today. It has been since last Monday since hearing from you. That letter is the only thing that connects me between reality and the ones I love — between the present and memories of the past. Almost daily, our life gets dimmer and dimmer in the past and somehow seems to get farther and farther out of my grasp. I still remember you, Kay, Carol and "us" and look forward with all the anticipation possible of being together again. But sometimes it seems like a dream — the dream of getting back together sometimes seems too good to be true. Sometimes you seem so far away — so many miles that cannot be spanned just because I have the desire to do so.

I got something for Kay today. A little nightgown made by an Italian Senora from Parachute Nylon. I believe it is much too large for her but she can grow into it. These Italians do beautiful handwork. I'm having another Italian working on gowns for both Kay and Carol. All I tell them is that I want them made for a child four years old and one two years old (I realize that isn't their respective ages, of course). The one I sent you was just made for my wife. Guess the Italians look at me and just guess the size of my wife and go to work. What we give for this sort of work is funny. If we paid with money, it would cost a small fortune, but for this one for Kay, I gave two tubes of toothpaste and about 4 bars of Palmolive soap — all costing me about 50 cents. It's really funny. I have a little more nylon left but it's in small pieces — if I can get something made for you out of it I will. I'll send this of Kay's (also some more gum for you and Kay) the first time I go into town. Wish I knew of something I could get little ole' "Cawol" that she would enjoy but that hits a blank note with me. I'll have to make-up to Carol later. I can't let Kay believe that I don't "yove Cawol" too.

Hey, did you happen to read that book you sent me "God is my Co-Pilot"? I happen to know some of the people that were involved in the story. I had read a synopsis of it in Reader's Digest but only today got down to read the book. It is interesting to me cause I've been to some of the same places, and it is written in the flyers vernacular, but I don't guess it would be too interesting otherwise cause the author is definitely no writer. The style is strictly amateurish.

Well, honey, I certainly was glad to hear from you today. I haven't gotten the "honest opinion" letter yet, even though the one I received today was written one week later. However, knowing you, and your loyalty and love, I am sure that I already know just about what it's contents are. In today's letter, you sent me Dr. Boyd's address. I think however, I may postpone writing him for a while. I somehow can't reach a definite, logical conclusion for naturally I can't tell when this is going to end.

Dearest Babe,

We got a beer ration today — I am drinking my third bottle tonight even though I don't care too much for beer. Gluttonous, that's me. It's only 9:45, but I believe I will stop and go to bed. The last few mornings I have allowed myself to get lazy and not get outta the sack until about 8:15 or 8:30. This has gotta cease 'cause it means that I miss those delicious dehydrated eggs. If I ever see something that's supposed to be food dehydrated back in the States, something bad is gonna happen. I wish you could have the experience of tasting a chip of dehydrated carrot — that's the nastiest thing there is — tastes 50 times worse than the fresh carrot and that's bad enough.

One of my medical boys (Wilson) has the "hives." And he is bringing me some adrenalin to give him a shot and keep my eyes on him for a while, so I'd better stop and get busy. Goodnight, my love. And tell the Kiddies to be "good gurls" 'til Daddy gets home.

I love you,
Joe.

November 30, 1944
Thursday, 7:15 p.m.

Dearest Honey,

I'm tired — and you couldn't guess why. I have been playing ping-pong all day long — imagine. Dew, Brien, and I rigged us up a ping-pong table and really have been playing all day. We certainly have enjoyed ourselves. I have to play so hard 'cause the guys had rather beat me than eat. Naturally I have to do everything in my power to keep them from doing it.

Another Italian winter day. It has rained constantly all day. It has been somewhat different however in that it has been light but continuous. You know, I'm always griping about the rain, but really I've gotten to where it really doesn't bother me very much. I just slush right along in it. I almost slip down many times a day and frequently do a split when both feet go in opposite directions at the same time. So far, I haven't hit the mud entirely.

We had a bit of ill fortune today. Giovanni (our house boy) was playing with another Italian, and he threw some sand and lime in Giovanni's eyes. I got all of it out as fast as I could and neutralized it with boric acid, but I'm sure his left eye is completely put out and some damage to his right eye. Poor kid, I sent him to the Italian Hospital and will try to get in to see him tomorrow. We all liked Giovanni so much. He has been staying here at night as well as day since Read went home. He's a good kid and I hope he has some vision in his eye, but I'm quite sure it's all gone for good.

No mail again today — as yet there hasn't been any to come in. I hear a whole lot has just come in up at Group, but I'm afraid it will be tomorrow before it gets distributed. All the packages certainly have upset the mail schedule — I'll be glad when Christmas is over for that reason alone.

Tonight I am again writing by candlelight. Our generator has been taken up, and we will be using candles now until we can acquire a generator from some place or figure out something. Candlelight might be picturesque and romantic, but it isn't worth a damn otherwise. No reading, writing, radio, or cards — just sit and look in the fire. Looking in the fire is only good for reminiscing, and that is good only up to a point. Well, Honey, I'm gonna call it quits for tonight. Tell the kiddies nitey-nite and I love you so very, very much.

Joe.

Boys goofing off at camp

Dearest Babe,

While they were in Italy, the 451st started a newspaper called *Ad-Lib* that covered all the happenings on base, including news, contests, game scores, and noteworthy achievements. One such achievement was The Eight Ball, an "award" presented to the squadron with the highest VD rate. An embarrassment tactic, the officers hung The Eight Ball in the mess hall to discourage the men from contracting VD. The first squadron to get this award was the 727th. In the next issue of *Ad-Lib*, there appeared a report from one of Joe's lectures about the evils of VD.

Doc King Talks on Bugs:
"Sweat, Brother, Sweat"
Capt. Joe W. King, flight surgeon and raconteur of the 727th, put over some strong VD thoughts recently. Here we publish some excerpts:

"Boys, these 'bugs' that cause the various venereal diseases are tough babies to battle. They don't care whom they tackle. Whether you are big or small, young or old, general or private, they will tackle you with the same tenacity and ferocity. Therefore, if one lays himself liable for attack by them, he must realize their capabilities."

"One must realize that it is sometimes almost impossible for the medical profession to tell whether a woman has VD or not, even with the help of microscopes and other laboratory facilities available - since that is true, it stands to reason that the G.I. can't possibly tell whether she is infect-ed or not by just looking at her. It is the so-called 'clean-looking' girl

that is responsible for much of our venereal disease, not the wholesale prostitute."

"The doctors in this outfit are so tired of hearing that age old saying - 'Well Doc, this one looked so clean!' We have heard it so many times that we can probably tell you the course of events of your escapade better than you can tell us. Most G.I.'s have sense enough to protect themselves with all precautions when they have been on a 'party' with a prostitute but somehow they do not see fit to take the same precautions with the 'clean-looking' girl. Occasionally we run across someone too dumb to use any protection whatever with either."

"I still maintain the only way to be positive that one won't contract a venereal disease over here is by abstinence from sexual contact. But, for God's sake, if you must enter the sacred sanctum of those vicious guys, the VD bugs, use every precautions that you have at your disposal - both of them, and then Sweat, Brother, Sweat!"

The Chaplain's T.S. (tough shit) card was introduced to improve morale. This card listed the common complaints of the men at camp. The boys spent their non-flying days trying to get the Chaplain to punch their cards.

December 1, 1944
7 p.m.

My dearest Honey,

Well it's still raining — it's the same rain that I've been describing for the last few days to you — not a new one. The camp area has changed considerably. First it was just a mire — then it turned into a slush, now it has turned into a sea — even though we are located up on top of a rather high plateau.

Went into town about 11 a.m. today and got my rations. While I was in, I went down and mailed Kay her little nightgown. I'm quite sure it is too large, so I enclosed a couple of packages of "chew-gum" so she wouldn't be disappointed. I'm having her and Carol another nightie made, so if this one is too large perhaps the next one will fit. I also went to the picture show. It was one of the "Falcon," series so we left after the first reel. We draw the line somewhere still.

I went around to carry my little Italian boy, Giovanni, some candy, cookies, mints, cigarettes, etc., at the hospital. What a hospital. The second I stepped inside I smelled gangrene. It was so dirty. I finally found Giovanni and learned that they had done nothing for him — not even had they given the poor kid anything to relieve the pain, so I bundled him up and brought him back to our house. (I told you in yesterday's letter that someone threw some sand and lime into his eyes and he would lose the sight in one of them.) So all afternoon I have been nursing and treating Giovanni. After looking more carefully, I am hoping that with diligent treatment we will be able to restore some little sight in the poor kid's eye. He is such a good kid for something like this to happen to. He is 15 or 16 years old and the oldest of 11 children — he needs his eyesight to help make a living. So far his parents haven't been around yet. They live about 10 miles from here. I'm hoping his dad will come tomorrow.

Got a V-mail from you today — was mighty glad to get it. You had been varnishing the new windows. Also got an Air Mail from Mother with more religious clippings — I'll be a good boy yet when I come home to you. I also received a package from Mrs. Brooks. It was a fruit cake from Habibs. I remember how good they are. I haven't finished the fruit cake that you sent me yet though. Looks like I'll have a lot of fruit for this Christmas, but I do like it very much — as do the other boys.

The squadron is planning to throw a big party December 14 — celebrating one year overseas. I guess it's going to be a big party — they assessed us enough to put it over. We are even buying a piano to put in the club. The boys are trying to get English Nurses and Red Cross girls plus Americans to attend. Due to the very amorous qualities of the Italian girls, I have forbidden them (I think the boys must think I'm a wet blanket).

After I wrote to you last night, I went over to one of the fellows house that had some electricity. Guess what I heard over the radio — part of the broadcast of the Texas-Texas A&M game. When the lights went out, the score was 6-0 in favor of Texas — don't know whether it ended up that way or not. Remember where we heard that game three years ago? That's right — riding around the streets of St. Louis in the Chrysler — "Them wuz the days." I'm eagerly awaiting for some more to come around just like them.

Well it's sign-off time again, Honey. I've had a terrific headache all evening. Think I will work on Giovanni a little bit and then go to bed. Surely will be nice when I have my only Honey to go to bed with instead of all alone. I not only love you as a bed-fellow.

I just love you,
Joe.

December 2, 1944
9:45 p.m.

My Honey,

Finally some mail broke through — today I received two nice thick Air Mails from you — one written November 8 and one the 17th. The latter is the latest I have heard from you. Both were extremely welcome.

In the letter of the 8th you sent me some pictures of Kay and Carol, Mr. Black and Mother and Dad. All of which I appreciate — but, Honey, don't be so modest. You are a very beautiful young lady, and I must have my "pin-up" girl in some of her latest poses. I like pictures of you primarily — I will admit that pictures of the children are surprising. But sometimes pictures of you are surprising too. I sorta become used to the ones I have of you, and then I get a new one and all over again I realize how good-looking you are. This isn't blarney — it's straight from the shoulder.

I just returned from Zraick's house where I have been listening to the Army-Navy football game. The Army won 23-7. Secretly, I was pulling for the Navy for they were the underdogs in the football forecasts.

Guess what happened today — it did not rain. Of course we still have the usual sea of mud, mire and muck but no rain! We even flew a mission. Hurray, we are again in the war (or should I say dammit). And tonight is beautiful. There is a nice, large, bright moon out and some stars. Makes one wish he could get romantical if the right one female was around — but, many moons are yet to come. And, when we are together again, it will just make us more sure that they do not go to waste — not that they weren't always appreciated.

Next night — 9:05 p.m.

Honey, it seems as though last night I just didn't have what it took to write a letter. Tonight I will finish this and get it off in the morning mail. My boy, Giovanni, is doing well. Yesterday his father came and took him to a nearby town to see an Italian eye specialist. I had been putting packs on Giovanni's eyes, and the specialist advised against it saying it might "give him a cold in his eyes" — of all the stupid, idiotic remarks. Just goes to prove how decrepit everything is in this sorry country. Anyhow, when they returned yesterday, I told Giovanni's dad that I didn't give a damn what the specialist said and that I was going to keep him here and treat his eye as I saw fit — which Giovanni wanted me to do. The Italian said that Giovanni would be permanently blind in the affected eye. Tonight, much to my amazement, he can tell me how many fingers I am holding up in front of his face. I feel proud and elated. He should get twice as much sight before he quits progressing. It's the first real treatment that I've done in quite a while, other than the usual trivial stuff which you get no kick out of whatsoever.

Guess what I have done today? I read Oppenheim's "Great Impersonation," which I enjoyed very much, although I have either seen a version of it on the screen or read a summary of it before. It was quite interesting. I also read another shorter book today by an ex-war correspondent called "One Damn Thing After Another." Fairly good but only because I was familiar with some of the places he wrote about. I probably will be familiar with other places he wrote about before I see you again but I hope not. (Our improvised lights just went out and now we are on candles again.)

I want to be with you and the children so bad I can taste it. The children are so large, especially little Carol, that it is almost impossible for me to realize that she is mine (or rather ours). Poor little ole' daddyless children — rather poor daddy instead. Daddy is a fantasy to them, rather than a fact. Due to your almost daily reference to me, Kay probably thinks she remembers me, but I wonder whether she really does or not.

I sometimes think that perhaps it would be better to let Kay just forget for the time being — then you and she wouldn't be reminding one another of me daily. That would relieve you of the agony of having to answer "When is Daddy coming home?" When the day comes that I do come home, I will take no offence whatsoever when the children remark "Mama, who is that man?"

In the meantime, I want you to know that I love you more than I can tell you. I have more faith in you than I ever believed possible and respect which only you deserve. And I wish that I were there, not only for my own personal happiness, but so I could try to make you and the children happier.

All my love,
Joe.

December 5, 1944
Tuesday, 6:30 p.m.

Dearest Honey,

Yesterday, I didn't do anything. It was just one of those days. I did go into town about 11:30 and stayed in until about 5 p.m. Went to the English Theatre and saw "Battle of Britain" over again, which is put out by the U.S. Army. I had seen it way back at Fairmont, I think. Perhaps I saw it even before that. It did serve its main purpose though of passing away a couple of extra hours.

Came back last night to find Giovanni's eye much worse, so I worked with him some more last night. Found another rock about the size of a match head embedded in the bottom part of the eyeball, so I had to extract the rock. Today his eye looks much much worse but he says, quite naturally, that it feels much better. Don't know what's going to happen to the poor kid's eye now — the only thing I know that I can do is sit by and bide for time. I certainly hope it starts getting better though.

Today has been a bad day. You know some days you just don't see how you can stand. Today has been one of those. I have been so bored that this morning I went up to see Wagner and just sat around and shot the bull with him. Wasted all the morning for him and me too. Then came on back to the squadron and had lunch — didn't know how I was going to pass the afternoon so finally decided I'd fly, which I did. Took off in a 45 mile an hour "gale" and flew down to one of the fields where we used to be stationed. We picked up some boys who were down there and then came back and landed in the same gale just in time for supper.

Guess what I had for supper tonight — a great, big, fluffy, brown biscuit. How's that! The first biscuit I have had in over a year, and it was delicious. Guess the mess sergeant must have been celebrating a year overseas. Yes, they left the States on December 3, that is the boys that came over by boat. We didn't (or at least the plane I came over in) leave the states until the 10th. So I have 5 days more to go before I can celebrate my 1 year mark. Two "hash marks" is enough to wear on my sleeve, don't you think so? Maybe Uncle Sam will see fit to send me back to the States before I have too many more — however, I will settle right now if I just knew I wouldn't have to wear four. In other words, I'll settle for anything under two years.

Honey, I think I will leave you for tonight. The boys over at Zraick's house are calling me to come over and play poker. They should want me — they have already taken about $6 off me this month. Maybe I'll get it back tonight. Didn't get any mail from you today but did get a very nice letter from your mother. Goodnight, honey, and tell the kiddies goodnight for Daddy.

All my love,
Joe.

December 7, 1944

My dearest Honey,

Do you notice that fateful date up in the right hand corner of the page? Boy, I remember it! Three years ago today it was Sunday. I had been out to the Hospital all morning working. I was on my way to the drug store to buy you and I some ice cream. Over the radio in my car, they announced the bombing of Pearl Harbor. I came on home, and for the remainder of the day, we listened to the radio and wondered what effect it would have on our future lives. The following December 7 we were comfortably at March Field — the following one I left Lincoln, Nebraska for "points unknown." Since then, a century no less, a lot of things have come to pass, and this December 7 I find myself in Italy. Wonder where I will be on the next one. Just hope I am with you.

Got two V-mails from you today and an Air-Mail from your Mother. In one of the V-mails, you told me that you received the $250 and was going to put it and $125 more in a War Bond, which I think is an excellent idea. I really think you've done swell. Why — now we've even got some furniture — something that I've never owned before in my lifetime.

Giovanni's eye is still not doing too well. I'm going to be fortunate even if I can save it, but I'm hoping for the best — poor kid. He is now sitting across the table from me with his head down on his arms. Giovanni has been correcting and teaching me Italian. Truthfully, I'm a pretty dull student — I won't make any effort. But we have had a lot of fun anyhow, and he gets a lot of kick out of me trying to pronounce some of the words.

My lights just went out again, and I can't get the motor to run so I am again on candlelight. Well, honey I think I'll finish and go to bed. I wish I could do it double, but I'm afraid it wouldn't be too comfortable sleeping double on an army cot. (Who said anything about sleep anyway!) Take care of yourself and tell the babies hello for Daddy. I love you with all my heart.

Joe.

December 8, 1944
Friday, 5 p.m.

My dearest Honey,

Here I am writing you with a pencil again. Sorry, I will find my pen soon, no doubt. I once had a school-teacher who preached that "if you didn't think any more of someone than to write them with a pencil to just not write." I can assure you that she didn't know what to hell she was talking about.

I'm all beat up tonight. Today the staff officers played some of the enlisted men a game of football — we beat them, but they damn near broke our necks in doing it. We had an excellent time, but I'm paying for it tonight in being skinned up and beat up. Maybe football is too "young" a game for one so "old" as I. Both my knees are bruised up, I have a "charley-horse" in my right thigh, and my hands are all skinned up. It'll be a week before I will be in shape to play another game. We have divided the squadron up into many teams, and the winners get a three day pass and 1 quart of American whiskey to divide up between them. The boys play with all their hearts in an attempt to get part of that American whiskey.

This afternoon I've been lecturing to a batch of replacement crews. Trying to get them off on the right foot. They are able to glean something from all of our experiences and shouldn't have to learn many things the hard way. I hope we do them some good but sometimes I wonder.

I have been over to check the mail situation — and even though the mail room is loaded with packages (2 trucks full), there are no letters. There is no doubt about it, the Army Post Office Department is making every conceivable effort to get Christmas packages here by Christmas — even at the expense of the regular correspondence.

I see Zinn just got us another stove made. Two G.I.'s just brought it in. That will make us two large stoves and a fireplace to keep us warm this winter — quite a far cry from last winter. The fireplace doesn't work too well any more. I think we will abandon it soon and just use the two stoves.

When I get home, we aren't going to need any stoves for a while 'cause whether it's summer or winter, we are going down on the Gulf for a while where it's warm in the winter and hot in the summer. To me, I believe we have had some of our most enjoyable times in New Orleans and down on the Gulf. I could never forget the Edgewater Gulf Hotel with the big room done in pale green, the double beds (damn 'em), the fresh frosted bowl of fruit on the table with the compliments of the management — remember? Those were the days and we've gotta have some more like them — wait and see. Goodnight honey take care of yourself for me, 'cause I love you, and you only, with all there is.

Joe.

December 9, 1944
3:30 p.m.

My dearest Honey,

I'm in a pretty sorry mood. I think what I need more than anything is to get drunk. Don't worry, I won't do that. But I feel like it would be a help — just an outlet that would soon close up — with a headache. I'm so tired of being bored I don't know what to do. I'm not the only one, plenty of the other guys are just as bored. Most of the time we are just one bored bunch.

Got a V-mail from you today. Somehow or another it sneaked through. It was written November 24. You had just finished selling $20,000 worth of War Bonds. I'll bet you led everyone in the Axillary by far. In one day, you probably did more for the war effort than I have done. I'm immensely proud of you.

Zraick got a letter from Young today. He is in South Carolina. He's still in the 4 engine jobs even though he wanted very much to get out of heavy bombardment into single-engine fighters. Guess it's not too easy a deal to pull. Apparently he is doing fine but is "p'd" off at everybody in the Army back there for not knowing that there's a war going on. This must especially be true of the Army in the States for everyone we hear from says the same thing.

December 10, 1944

Seems as though I didn't have it in me last night. I'll finish this epistle up tonight sure. By the way, in Young's letter to Zraick he said that he had tried to call you, but you were not at home. He did say that he would try again however.

You know this morning a year ago at about 1:30 a.m., I was flying over Nassau. Nassau is pretty from the air at night. It looks like a large lighted wagon wheel. It's just one big circle of lights with spokes and hub. Later in the morning I was flying over Haiti and the Dominican with its dense forests and jungles and two high mountain peaks. Everything looked real green is about all I remember.

Tomorrow I have many things to do. I gotta bathe and clean up. Take some clothes into the cleaners, go to town and get my rations, and do quite a few things in the dispensary. Today was about as usual with the exception that I had to again lecture to some new crews. Hope that job is finished now for a few days.

Wrote a note to Burr Craycraft today and also one to a boyhood friend of mine who is a Captain in the Engineers in Italy, Bud Herbert. He and Henry Jones and I used to run around a lot together when we were youngsters. Didn't get any mail today but when it does start to come through I'm going to try to get a three day pass so I can read it all. Think I'll hush, honey, and go over and play poker for a while. So far the boys are "in me" for about $7 this month, and I gotta get it back. I love you with all my heart.

Joe.

December 11, 1944
Monday, 11:20 a.m.

My dearest Honey,

Today has been a little different so that it passed a bit quicker. This morning I arose about 7 a.m. — had scrambled fresh eggs for breakfast which were delicious, then went out for take-off. After all the planes were off, I returned to our casa and decided logically and methodically that it was high time for the proprietor of my body to get said object in a more presentable, less obnoxious, and less odiferous state of being — so took a bath, changed clothes and combed my hair (with oil).

Then, since Giovanni went home yesterday and failed to show up this morning — I don't know why, but imagine his family probably took him to their Italian eye doctor friend — and the house was so damn dirty, I decided to try to get it a little more presentable. So I tackled my domestic duties with the vim, vigor and vanity of a freshly married school-teacher who has just returned from her honeymoon (no "puns" meant or intended). After sweeping out about twenty-five pounds of dirt — and I'm not given to exaggeration — bricks, tin-cans, and cigarette butts, making beds, dusting chairs, rearranging furniture, etc., I finished up just a wee bit filthier than I was before I started, but at least one could tell some difference with the house.

About this time some of the boys walked in and before they could speak what I am sure was on their minds, I informed them that my duties as a housewife could only go so far and any passes of any nature, whether with intent or accident, would not be appreciated. This of course, only threw these guys off-guard for a few minutes, so to further prove my undesirability I went out and passed the football with them. As I have remarked so many times previously, there is nothing to a man but his ego, so all I had to do to have them completely subdued was to out-pass them a little bit. With them thusly deflated, I had complete command of the otherwise possible precarious situation.

Then to lunch, which for me consisted of three bowls of thin tomato soup and some burnt coffee. After lunch I returned and tried to write Frank Posey a letter but couldn't fill up one page, so I again pushed it to the side and will try again at some later date. The rest of the early afternoon slid by somehow — don't recall now just what happened. Later, I went out to the runway for the landings and was busy for a while. Then to the mail room where I was again told, "No, Captain, no letters for you today," (a familiar sentence now). However while at the mail room, the clerk asked me what to do with "all these packages for Captain Read." Since Read has gone back to the 'land of milk and honey' where the contents of said packages are more plentiful, I instructed the clerk to give me the packages and I would make proper disposal of them. Needless to say that the 2 pounds of fruit cake, 5 pounds of chocolates, reading material including a Coronet and McCalls (of all things) are being enjoyed by all. Also there was a copy of "The Importance of Living" by Lin Yu Fang enclosed, which I will bundle up unread and send back to Read with our thanks and the enclosed Christmas Cards, which I know Read will appreciate.

Tonight I went up to the Group and saw some antiquated "shorts" on G.I. movies which proved to be of regular Army standards — dull and uninteresting. Tomorrow I think I'm going to take a jeep and drive to Naples to buy some stuff that we will need for our Christmas party, which comes this Thursday night. I volunteered for the unpleasant job — thought I'd stay overnight and maybe see Harvey. Perhaps if he can, I will bring him back with me.

I signed off writing you last night just before sitting down and playing two-bit limit poker. As I mentioned in my letter, they had won $7 off me so far this month, but last night I was a little more fortunate. I won $20 of it back, so now I can afford to lose a couple of nights — no I can't either 'cause I intend to buy me a pair of pants outta that.

Well, honey, since I'll be in Napoli tomorrow night, I probably won't write to you. Nevertheless, be not dismayed, for you may rest assured that I will be thinking of you, talking about you and wishing that I were with you as I do many times every day that passes — not because I'm a love-sick, homesick baby, but just because I love you so terribly much. The place that you have filled so completely in my life for the past few years is just so darned tired of being partially vacant. Goodnight to the kiddies.

All my love,
Joe.

December 13, 1944
Wednesday, 10 p.m.

My dearest Honey,

I arrived back to camp tonight from Naples, and I am so tired I can hardly hold this pen. Riding the distance from here to Naples is quite a job in the winter time in an open jeep. There's mountains all the way, and it is quite cold. The trip over was quite hellish for it rained practically all the way over — slowing us down considerably. We arrived there about 3 p.m. and were unable to do any shopping until today.

Last night we went to the San Carlo Opera House and heard a very good symphony concert. At least it was all good except one symphony (#3 in F, I think) by Bach, which I thought was rather dull. However, they played four more, which were very good. The opera house itself is very nice. That's the only thing that the States don't possess. That is, every city of any size, and by that I mean 20,000 or over, has at least one opera house. They certainly believe in their operas and concerts. Too, any Italian can pick up practically any instrument and play it — particularly the violin and accordion. Musically, they are certainly talented. Socially, scientifically, technically, architecturally, etc., they are so far inferior that it is very difficult to conceive. 'Nuff of that for no doubt I am extremely prejudiced and look at all Italians with a jaundiced eye.

After the opera, we went to the Transient Officers Hotel and slept on a hard cot with three blankets as mattress and cover combined and fought the ever present bugs. Today I spent trying to buy stuff, and I spent the whole day buying 300 paper plates, 6 rolls of crepe paper, 5 lbs of floor "dancing" wax, some stove polish and 1000 paper napkins. After finishing my shopping we set out for "home." Believe you me, this place looked like home tonight and that soft "sack" of mine with all the mattresses, blankets, comforter and pillow looks especially good. When I arrived here, I found three Air Mails and one V-mail from My Honey. Gosh, was I glad to have finally gotten some mail through from you. It's a good thing, for the "standby letters" — those masterpieces that I have saved which give me especial "oomph" or lift — are just about threadbare from rereading. What would I do without my wife?

I love you,
Joe.

December 15, 1944
Friday, 6:40 p.m.

My dearest Honey,

We had our party last night, and it was a rousing success. At least it was still being successful when I left at three this morning. For everyone to be as inebriated as they were, it was remarkably tame. Only one fight, which I stopped and caught a good lick in the process — no gambling, no broken glasses and few obnoxious drunks. I was rather satisfied with the boys on a whole. As I said before I got in bed about 3 a.m. and had to get up at 7 a.m. for take off. (Last night I chased all the boys home early who were to fly today).

After take-off I came back and ate and then returned to bed about 8:30. Slept until noon and this afternoon have just been "piddling" around. Even so, I'm still tired and sleepy, so I'm gonna go to bed early tonight. Another reason to turn in early is that there are no lights and reading by candlelight is no good. Well, Honey, I'm going to bed — imagine Joe King going to bed at 7:30 or 8 p.m. — I must be getting old. Hope you and I both get letters tomorrow.

I love you,
Joe.

Dearest Babe,

My dearest Honey,

Today has been a red letter day. Four big fat letters tonight and all from my Honey. Also I received five packages. They were very much appreciated but not with the same zeal as the letters. By the tone of your letters, I see the newspapers and radio as misleading you again. I'm going to disagree with you. The war certainly won't be over this year as by the time you get this letter you will be able to see. I just hope it will be over by next June or July. Yes, I'm speaking of the European war only. I haven't even begun to think much about the other one yet. It is my honest opinion that Hugh will never have to go to the other theatre. At least until he has had sufficient rest in the States. Whether I have to go directly there or not, in my opinion, will depend on when this affair is over and how soon they will get us out of here. I certainly hope and feel that Hugh will be exempt for a while though — he certainly should be.

I forgot to tell you, but the last time I was over in Naples, I looked around for a tricycle. Couldn't find one however. In Italy there is more children's stuff than anything else. Due to Mussolini's Baby Campaign, there are streets and streets of baby stores. Then when he went out of power and the women quit getting medals and money for having babies, there was a sharp decline in the birth rate leaving much stuff on the retailers hands. There are many things available that you can't get easily in the States — beds, strollers, chairs, furniture sets (baby size) etc. If and when I'm in Rome again, I'll look around. If I could find one, I could tear it down and send it through the mail piece by piece if necessary.

Today has passed rather quickly. They have had a pretty sick boy up at Group all day, and I have been up there trying to find out what's the matter with him. I've enjoyed it very much even though he isn't my patient. Whatever it is, it's screwy. I've got a hunch it's malaria or typhus fever but there are things against either — possibly typhoid. Anyhow maybe tomorrow I can come to a definite conclusion. It's a hellova thing to say, but it really is interesting to see someone that's sick with something other than a cold or diarrhea or an earache. Maybe someday it will all come back to me. Seeing someone sick really makes me "homesick." I must go to bed for it's early rising tomorrow. I love you so very, very much.

Joe.

December 19, 1944
Tuesday, 6:30 p.m.

My dearest Honey,

In mail, I hit the jackpot. I received a package from you and the babies — one from Katie and Otha, and one from Sis and Royce. Also a letter from you, one from Mom and Dad, and one from Kremers. The one from Kremers was just telling me that he was pulling out from the hospital and heading for a port to catch a boat for the U.S.A. What a hellova swell feeling that must be. I envy him but am certainly glad for him. He also told me that he would call you the first chance he got. I hope Kremers will get a discharge from the Army when he gets home and believe he has an excellent chance. Yesterday I received a letter from Mahon, who is now at a redistribution center in Miami, and he tells me they are giving discharges to medically and psychologically unfit airmen quite readily down there. If so, it's a good thing for it's tough enough for good men.

I'll take it upon myself right now to thank you for all the nice things you sent me for Christmas. Even though I just got the wooly bedroom slippers tonight, I have them on right now and they are quite comfortable. The T-shirt that was in a previous box I am also wearing at present. The cigarette lighter has already been put to use too. The ocherina — I only played a scale and one of the roommates had to ruin the melodious outburst with "God, do we have to listen to one of those damn things?" so I figured it best to put it down. Even though I have enumerated only part of the things, thanks for them all.

I have been working again today — makes about four days in a row for me, which should be some kind of a record. I did a couple of 64 exams this morning and this afternoon worked up at the Group hospital. I was supposed to see the dentist but I didn't have time. Perhaps I will get a chance tomorrow. I certainly want to get it done before Peterson leaves us (I think I told you that he was being transferred to the Air Force) 'cause I know he is a good dentist.

Hey, what do the people in the States think of the new German counterattack on the western front? Do they still think that the remainder of the campaign will be mopping up exercises? I just heard about it today, and I certainly hope it is short-lived. I'd really like to see that sector (and the Russians) get really moving. Looks as though they are making some progress in the Philippines.

I've rambled enough for one night. Think I will stop and read a short story or two and hit the sack. Glad to know you are sending me a ping-pong set — a couple of us will put it to use cause we like to play. I hope you have had a nice Christmas and New Year by the time you get this and may it be our last apart. Tell the kiddies hello for Daddy. Tell Kay that I love her and Carol very much and we will have a good time playing when I get home. I love you even more.

All my love,
Joe.

December 21, 1944
12:30 a.m.

My dearest Honey,

Well I got up this morning and thought that since we weren't flying today that I would go over and see Dan. I borrowed a jeep and got over there at about lunch time. We sat around and talked for a while, and then about 3 o'clock he and I got in the jeep and came back over here. We have been sitting around here ever since shooting the bull and talking about everything and everybody — particularly in Helena. He is the same Dan — still likes to feel his importance in an adolescent manner but still a nice guy. He weighs quite a bit more — I think he said that he weighs 197 now. Poor guy, I feel sorry for him — imagine, he has been over here 31 months and still hasn't the slightest idea when he will go back. Gosh, I surely hope I don't have to stay in this place for 31 months. If I have to stay away from you that long, I will go crazy. Perhaps I'm nuts already — sometimes I think so. Dan is spending the night with me. He will be going back to his outfit tomorrow morning. His outfit and my outfit use the same hospital, so I will just send him in on the ambulance in the morning and maybe he can get his ambulance over to his outfit.

I received both an Air-Mail and a V-mail from you today — was certainly glad to hear from you. However I was sorry to hear that my Honey had been so down in the dumps for two days that she considered it better not even to write to me. Don't worry about something like that, Honey — I don't expect you to be in tip-top humor every time you write to me. You know I love you whether you are on the top of the world or under that proverbial wooly worm that you always refer to.

It's raining again tonight, but I shouldn't gripe — we have had a few good days recently. Tomorrow it'll probably be better cause we are supposed to fly. It'll probably be colder though — do I remember that ride from El Paso to Del Rio — gosh yes! It's a wonder my wife and that poor little Kay hadn't frozen to death. That was pretty cold. I was colder that night that I hunted down that crew that bailed out in Nebraska though, I guess. And I was pretty cold one time in Marrakech, French Morocco. I don't remember being so cold since. Well everyone is now in bed and I'm being forced to write by candlelight. So I'm going to say goodnight to you and write my two daughters a letter.

All my love,
Joe.

Dear Kay and Carol,

How is Daddy's two big girls today? I hope you are being good and nice and not giving your sweet Mama a lot of trouble.

Hey, did Santa Claus bring you what you wanted? I heard that

*Santa Claus was going to bring you a doll buggy if you had been good.
Did you get it?*

*Daddy is going to send you some more gum soon. Too, he is going
to send you some nice little gowns. Now, Kay, you be good and don't
fight with Carol. Some day pretty soon Daddy will be home, and you
and Carol and Daddy will get down on the floor in the living room and
play and play and play. Maybe we'll play out in the sand pile too.*

I love you,
Daddy

December 22, 1944
10 p.m.

My dearest Honey,

This morning I took Dan back to his outfit. I wrote in my letter to you last
night that I went over and got him yesterday. We had some rather enjoyable
talks together. He's so dumb about some things, though. He still figures that
Myrtis was a nice girl and that she was awfully good to him. If he only knew.

Got back to camp and found out that my squadron had two different auto
wrecks in my absence. One boy turned a truck over and it rolled down a 30
feet embankment — he was unhurt. The other fellow just ran his jeep off into a
ditch — he had a dislocated shoulder and a separated joint — just goes to show.
Tonight I went up to Group to look over my two patients that I have up there
and while I was up there went to the show, which was fairly good for a change.
It was Gary Cooper in "Casanova Brown" — very entertaining.

The west front doesn't look very good, does it? I have been down in the
dumps ever since the Germans have been putting on this counterattack. I realize
that Hugh is somewhere around this sector — so as soon as you hear from him,
please mention it to me. I surely hope things aren't too rough — I know Mom
and Dad Black and you are worried, but there's not anything that worry can do
— just wait and everything will be O.K. I certainly hope the tide of the battle
changes. Don't worry too much for I'm sure the big-wigs must have known
it was going to come off, but just didn't have what it took to stop them at that
time.

My little radio quit playing a few days ago. Think a tube has burned out,
and I can't get any similar tubes over here. However, in a way it was a good
thing for since then Zinn has gotten out and rustled up an airplane radio, which
is very good and we can now get many more stations. Some enlisted man has
got mine now working on it — don't know whether he is going to be able to fix
it or not.

Zinn is giving me hell right now. He and Reese have both had the itch for
the past couple of months. Reese treated himself real diligently and finally got

rid of it. It, at that time, wasn't bothering Zinn very much, so he just let it go. Now he has got it real bad, so he's fussing at me about it. I tell him all I can or intend to do is tell them what to do — not do it for them. I think what they are really griped about is that so far I haven't even had a scratch even though I live right here with them. Guess it'll hit me soon — don't know — I don't think bugs like my taste. Hope it continues that way.

Tell Kay and Carol to be good girls, and Daddy will write them a letter again real soon. Also I'm sure by the time you get this our eldest daughter will be 3 years old — extend my best wishes to her. I love you with all my heart and will continue to do so for all my life.

Joe.

December 23, 1944
6:10 p.m.

Dearest Honey,

They are playing "I can't put my Arms around a Memory" over the radio — Rudy Vallee, the guy who's singing it, doesn't know how true that is but I do. I only wish I could. I just got through going over every picture of you and the children. I have them all "posted" up on my wardrobe under a piece of plexi-glass where they are easily accessible for me to gaze upon. Every time I want to study your face I can. Just now I had that peculiar sensation while looking at you that "you were mine — only mine" and that the offspring was the product of us.

I finally did today what I have been trying to force myself to do for a week. I went up and had some teeth filled. He has one or two more to fill, and I'm supposed to go up there tomorrow. To tell you the truth, I even dread the thought of it. Today was too uncomfortable. We filled three today, and by the time he was through I was worn out.

Also I have some more news. It is possible, if not probable, that Hoppock is going to be transferred to the 27th group. Boy, I certainly hope so. I don't know who is in line to become the new C.O., but I don't care — I'll just take potluck and should get the better end of the deal.

Today when Peterson finished working on my teeth he told me not to eat until tomorrow. I'm sitting around here with my stomach so empty it feels like it is glued to my back. I probably am not so terribly hungry, it's because I know I can't eat. That's the only reason I even think about it.

Well Honey, I am not much good at writing tonight so if you will pardon me I'm gonna quit early. Tell Kay that her Daddy still loves her and Carol (but mainly you).

All my love,
Joe.

Good morning — It is now 7 a.m. Christmas Eve. Bing at this moment is singing White Christmas over the radio. It isn't a white Christmas here, but it's certainly cold enough this morning to be white.

Got up early this morning. One of the boys was pretty sick and I had to get up. Fairly unusual. I don't have to get up often for medical reasons. Gotta do a few things and then go up and see Peterson again today — hope he is able to finish up my teeth.

I love you,
Joe.

December 24, 1944
10:30 p.m.

My dearest Honey,

Christmas Eve Gift!

Well at this moment it is about 4:30 p.m. where you are. You probably haven't gone to Mother's and Dad's yet. I know you will enjoy watching all the children open up their packages — I always did. I would just like to see the kids go wild and look at their eyes get as big as saucers when they open up their gifts. Gosh, can't they look surprised and bewildered. Bless their hearts, I hope they have a hilarious time. I wish that you would too, but I realize that your happiness will be toned down with a bit of melancholia. Perhaps next Christmas we will be together.

Tonight the boys are really raising a rumpus. They are shooting off their pistols like mad (all in the air, I hope) but you can still hear the bullets sing. They are also firing their Flare Pistols and the sky has been full of red, green, yellow and parachute flares. Well goodnight, my love. Have yourself a swell Christmas and a Happy New Year.

I love you,
Joe.

December 25, 1944
Christmas Day, 12:30 a.m.

My dearest Honey,

I have been reading and have just finished my book — consequently I am a little late in writing you today. It has been another wet day, not raining hard but just barely raining. This morning I went out for take off. It was so bad that I was surprised that we flew, but we did. Later I went up to Group to take Reese up to the dentist. He had a toothache and had to have a tooth pulled. Had a couple of drinks with Wagner and Peterson and later had Christmas dinner up there. It was a delicious dinner. I had no turkey, however it was there to be had — it so happened that we arrived late for dinner and all the turkey was gone. We didn't mind though for the meat that we had was delicious. I then went to the runway for landings and then to my casa. Ever since I have been reading.

Four minutes (it's now 12:34) and one year ago, I was awakened in South America and told to get on my clothes that I was going to fly over the Atlantic Ocean that morning. I'll admit that it was quite a sensation 'cause then the Atlantic Ocean seemed like it was awfully big. When we went to bed an hour earlier, we didn't know that we were going to fly that leg the next morning. Well anyhow we finally got out to the ship — got plenty of water, strapped on our pistols, and our knives and our "Mae Wests" — got in old 442 or "Supermoose," checked the gas and revved up the engines. Going down that black runway was a peculiar sensation. You couldn't see the runway — too dark. All you could see was the two lines of dim lights on either side of the runway. When we finally became air-borne, you knew, or hoped, that the next time your feet touched the ground, you would be in Africa. Too you wondered when would you ever set foot in the Western Hemisphere again. Then started that flight over the Atlantic. Supposedly the "blue Atlantic," but the first few hundred miles it was definitely black. Then about five hundred miles or so out, you could see "the Rocks" and you wonder just what the Hell rocks are doing jutting out of the water away out there — but they were nice for us to see 'cause they were on our course, and we certainly wanted to be sure our course was right. Then we hit the storm in the middle of the Atlantic that everybody flying over hits. We couldn't get over it, so we decided to go under it — so we flew through the front 700 feet above the water — and did it rain. I was surely glad that there were some good pilots aboard and it was in their hands and not mine. You couldn't see anything, so I layed down just behind the pilot's and co-pilot's seats and went to sleep. Later we sighted land and found our field. Our Navigator missed the "estimated time of arrival" about 1 minute, and we hit the field right on the nose — pretty good navigating.

Well, again I hope you and the children and our families have had a good time today. Write and tell me what I gave you for Christmas and what it looks like. So far I haven't heard from Sis, so I don't know. I hope whatever it is that you like it and it's useful. I notice in the paper today that where the main German thrust has been in the West Front is south of where Hugh is. I am much relieved 'cause until right now I thought it was right in his sector. Take care of yourself for I love you with all my heart.

Joe.

December 27, 1944

My dearest Honey,

I went to town today. It's the first time I've been to town in a long, long time. It is now so arranged where we can get our rations out here in the squadron, so I now have no reason to go into town unless for supplies, or to mail a package, or just to get away from here. Today I wanted to get some supplies (medical) and also to send the little nighties to Kay and Carol (which look more like dresses) and to send some more gum so Kay can stick up the furniture and her hair more completely. However I got into town too late to send the packages. They are all wrapped up though and ready to go, and I will either carry them in or send them in soon. Gotta keep my girls in gum — that's the least I can do.

Looks as though we are going to get a new C.O. (thank goodness). Looks as though Hoppock is going to Group as Deputy Group C.O. and we are going to get a Captain. This boy we are getting knows Mahon — he used to be in Mahon's flight back in the States. If he is half the guy Mahon was, he will make a good one and he will have my help in doing so. Anyhow this Captain is moving down here in the morning — as soon as he catches on I'm sure Hoppock will go up to Group.

I did you bad last night and didn't write. However, take it from me, I had an excellent excuse. I was mountain climbing under very unfavorable conditions. About 4 p.m. yesterday, a couple of Italians turned up and said an airplane had crashed over in the mountains. We set out for it in the ambulance. Found out the closest we could get to it was 2 miles and two mountains over. It was dark before we finally got going up the mountains — had no flashlights, no boots, no nothing. The wreck was on top of the second mountain. It was freezing up there. My feet were sopping and my pants legs frozen up to my knees. And believe me, mountain climbing is no fun in the dark even had I been in the best of shape. Wasn't able to accomplish much for I was unable to get any mules to haul out anything, but was able to find out that they weren't from my group anyhow. So this morning I turned the job over to the right group. But it was an experience and when I finally got home, I was too tired to write. The snow up there was quite deep but was frozen over in most places. I really had a time — only had a couple of enlisted men along with me.

By the looks of things I will get to be home by next Christmas. That is if we don't stop the German drive on the Western Front soon, they will have us pushed back that far. That isn't nice, is it? I hope very soon that we are able to stop that offensive and no doubt they will be. I've "writ" enough for one night. Tell Kay and Carol to be good little girls and to be good to their Mama, whom we all love so very very dearly. Goodnight.

All my love,
Joe.

December 28, 1944
Thursday, 8:30 p.m.

My dearest Honey,

I'm practically "newsless" today, but I will make an attempt to write a letter anyway. The best news of the day was a letter from my Honey, one from Mother, and one from one of the boys back in the States. I did hate to hear from Mother though that Aunt Lis was to come back up there to live with them. I pity both Mom and Dad — particularly Dad. However, I dread to hear from Mother now 'cause all I will ever hear about is how badly she feels and how Aunt Lis is driving her crazy. I don't see for my life why they don't get sensible about that affair and put the poor old lady in some sort of home before she wrecks all the families that she knows. Oh well, guess that's their business and not mine.

I see you are still cleaning up. In your letter you say if you can just get the kitchen fixed up you can call it a day until Spring. I think that's an excellent idea. I believe all you have been doing since I've been gone is cleaning the house. I admire you for your energy, but I wish you would take it easier. Take it from me — you can live in quite a bit of dirt and still be reasonably healthy.

In the letter from Mother today, she seemed quite proud of herself for finally being able to get near Carol. Hope Carol has decided to make a bigger circle of friends. The little rascal had better get more friendly cause I don't want to spend my first six months when I get back in the States trying to make up to my youngest daughter.

By the way, I finally got the girls nighties and gum off to you today. There were 29 packages of gum in the box, so that should hold you for a while (even at 3 sticks per chew as is Kay's allotment before breakfast). Yes, there's a few packages of Dentyne too. Would send more of your favorite brand, but Dentyne is always most difficult to get.

Can I take you into my confidence and do a little "bitching" without making you feel bad? There's no use to feel bad for you and I both know I'm extremely well off in comparison to most boys in this man's war. However, I'm so tired of this most obnoxious job that I hardly know what to do. I wasn't cut out to be a chaplain. I don't mind listening to people's troubles when I can do something about them, but I'm really tired of listening to things that I can't help in no way whatsoever. Boy, in future life if a good old psychoneurotic walks into my office, I'll probably kick him out 'cause I'm tired of the stuff. Sometimes this is a most distasteful job, and I'm fed up to the gills — tonight especially, for some reason. Perhaps it can't last forever, but I sometimes wonder. Well, now that I have that off my chest, perhaps I can sleep better tonight.

Well, Honey, I've run out of decent subjects to talk about. I'm a "thull dud" tonight anyway. You know how it is sometimes. The moon is out tonight in all its glory and it is a gorgeous night. It would be much too cold to ride around with the top down though — but, if we were together, we could make good use of the night anyhow — as if we couldn't any night or day. Much of the time is behind us though and maybe it won't be too long until my ambition is fulfilled again i.e. live with my wife and chillun and be happy.

All my love,
Joe.

December 29, 1944
Friday, 10:15 p.m.

My dearest Honey,

Another wet day. We flew a mission today, but I hardly know how. It has been bad and raining all day — the whole camp area is just a sea of mud.

This morning I went up to Group and got Wagner. He and I finally went to town. He had to buy himself a new blouse for some Italian had burned it up while cleaning it with gasoline. I wanted to go in and get some medical supplies. While I was there, I borrowed a pretty good book on medicine and intend to study it a little bit the next week or two. It's been so long since I've read any medicine that I probably won't know what some of the words mean — we will see. Came back to camp about 3 this afternoon in time for the planes to land, and it was a good thing that Wagner and I were here for we had plenty of business even though we weren't expecting much.

Then I went up to Group to see Colonel Jones who was sick and wanted me to come up and look him over. Had a couple of drinks with him and then had dinner with Wagner. Later I went to the show, but it was so putrid that I left before it was over. Nothing unusual — just the same old stuff.

Next night 10:50 pm.

My Honey — I'm a dull thud again tonight. The best thing I could do for the good of the country tonight would be to get in bed. Tonight the future new C.O. (he takes over the squadron Monday) and myself went to the picture show. We saw Deanna Durben in "Christmas time" or something like that. It was the second time I've seen it in the past month, and it was just a little bit "cornier" this time than it was the first time I saw it.

I haven't done anything today but stay inside. It's been pretty cold on the outside, and the wind has been whipping out of the northwest. It's so muddy that one can hardly walk. To cap that off, I had someone to steal my boots today — gotta get me another pair somewhere 'cause in this mud boots are more important than a pair of pants.

Tonight my feet are as cold as yours. They should be warmed up in a few minutes though for I just took off my shoes and put on those nice warm sheepskin bedroom slippers that my Honey sent me for Christmas. If I were with you tonight, it would be your job to get my feet warm instead of vice versa.

The news sounds a little better over the radio tonight for which I'm extremely glad. I hope there's some real reason for it and someone is not just going off half-cocked just because they think the public needs some good news (aren't I a skeptic!)

Gave myself a shot of Typhus Vaccine in my leg today, and tonight my leg is pretty sore and stiff. Don't guess I can take my own medicine. My boy, Rosie, wouldn't give it to me — at least you could tell that he didn't want to, so I gave it to myself. I'm going to bed now — but not before I tell you that I still and will always love you with all my heart.

Joe.

Will do better tomorrow.

December 31, 1944
New Years Eve, 11:25 p.m.

My dearest Honey,

I don't suppose there is any better way to start the New Year out than to be writing my Honey a letter. By 1945, I will still be writing this letter unless I make it awfully brief, which I have no intention of doing.

I have done nothing today except sit around the house and play Cassino. No kidding, we have spent the whole day doing nothing but that. This afternoon about 4 o'clock I went over and drew my $184 for my month's effort. Sent you $200 for I had about 20 or 30 bucks in my wallet. I'll bet you don't know how much money I've sent you since July 1 — $1375. That's not bad for my end of the business — particularly since all I've drawn in total pay is $1012. Looks like I've been lucky playing cards, etc. Well, you gotta do something to pass the time away. Since we are talking about money, I don't think I've ever got a letter from you telling me of the $125 I sent you November 31. Let me know when that comes in and also when you get the $200 I sent you today.

The boys are shooting flares all over the place now — they have twenty minutes to go before New Years. They will probably flood the place with flares at midnight. I wonder how many tents they will set on fire. If someone gets a tent burned down tonight, I really pity them. It has been sleeting all afternoon, and now it's really snowing. The ground is so wet that the sleet and snow aren't sticking to the ground too well, but if it keeps snowing like it is right now, we will have quite a bit of snow by morning nevertheless.

One year ago tonight, Blackie and I were spending the night in Oran. We were lonesome then, and I'm still lonesome. Things haven't changed for me one iota. I certainly hope and pray that 1945 holds more in store for me than 1944 did. However I should be thankful, and am, that I have been as fortunate as I have. I still have all the parts of my anatomy and all are in good usage — I still have a good wife and two good healthy children to go home to. Think of the guys who started over here thirteen months ago with me that can't say that. Yes, I am fortunate. Think of the poor guys up on the front on a night like this, and here I am in a nice warm casa about ready to turn into a nice warm bed. I haven't got a decent gripe coming. Nevertheless, not to be ungrateful for my well being for I'm not, I certainly do hope that the beginning of 1946 finds me with my wife and children and finds all of us healthy. The paper says tonight that the drive on the Western Front has been definitely checked. I certainly hope the paper knows what it's talking about.

It is now midnight. If you were present, we would start off the New Year with a kiss, but as it is, I can only wish you a kiss. May God see fit to end this war this year, end the misery and suffering and bring us together again healthy and happy — Happy New Year!

All my love,
Joe.

P.S. The boys are letting loose with their pistols now.

Dearest Babe,

1945

January 1, 1945
10:30 p.m.

My dearest Honey,

Well we started the New Year out right today. Hoppock left the squadron and went to Group. I can't say that many of us on the staff hated to see him go 'cause most of us didn't. The fellow that we have in his place so far seems to be the right sort of a guy. I'll reserve my opinion until later. The first couple of days I thought Hoppock was going to be O.K. — and I guess he was O.K. except in the little things that count so much when one is forming an opinion of someone.

Some more good news — today I sent you another $225. That's $200 yesterday, $225 today and $125 that I have not heard whether you got yet. I think it goes without saying that the money is yours to do with as you see fit. The $225 that I sent today was won in a big crap game up at the club last night. I'm afraid you are going to get the wrong impression. I'm not a professional gambler by any means. I haven't done any gambling except two bit stuff since the doctors meeting at Bari until last night. It just so happens that I won both times, so I might as well send it home to you. As far as the nickel, dime, two-bit stuff goes, we do that day in and day out just to pass the time away.

It snowed most of last night. When we got up this morning, we had three or four inches of snow, and it's been snowing off and on all day. It's a delightful condition in comparison to the mud, but of course it will thaw out in a day or so, and it will be muddier than ever. Something to look forward to. I've simply gotta get me another pair of boots. I left mine on the outside of group headquarters a couple of days ago when I went inside. Sure enough, as I should have known, when I came out to get them someone had made way with them. Money is worthless over here, it's the article or merchandise that is valued. Those boots wouldn't cost $5 in the States but they are worth $50 easily here in the black market.

All our water pipes to our plumbing system have frozen up. Certainly hope you haven't had similar troubles at home. As soon as the snow gets off the ground, we are going to have to bury our pipes better and wrap the exposed ones. So far we have just been too lazy. This is the first time they have ever really frozen. Most of the time it is cold here because of the wind, but now it is just plain cold. This is much better than when the wind is blowing.

Didn't get any mail today — they didn't have mail call on account of it being New Year's I guess. Nor did we have a "Stars and Stripes," today which I read as diligently as I ever read the Appeal or the Press. Hoping I have some mail tomorrow — a couple of juicy, fat Air-mails would do the trick nicely. Gotta go in tomorrow or the next day and get me a haircut too — my hair is almost hanging over my ears now.

There are so many reasons I need you that it is impossible to enumerate them — most all of them due to the fact that you are the essential cog in the wheels of my happiness.

> I love you,
> Joe.

> January 2, 1945
> 9 p.m.

My dearest Honey,

Received the Air Mail today that you wrote me December 14 — you were still wrapping up Christmas packages. Today it has snowed a little bit more but in general is warmer, therefore sloppier and "messier." At present, it is either raining or sleeting, I don't know which. Today, Zraick and I went in town. I got a haircut, which I have been needing for weeks and tried to buy me some shirts which I couldn't do. I have all of 2 shirts now (wool), one of which is too small in the neck. Thought I would break down and buy me a green wool shirt, but the PX doesn't have any. I'll try to get one the next time I'm in Bari or Naples. I simply gotta get me some clothes even though I hate to.

I have been up to Group tonight (again with Zraick) helping Colonel Jones drink up a quart of Scotch that he has acquired from somewhere. Don't know how we finally came out but I did my share. Gotta shave, take a bath and go to bed. Tell the chillun hello.

> I love you,
> Joe.

Dearest Babe,

January 3, 1945
Wednesday, 9:30 p.m.

My dearest Honey,

Didn't hear from you today but got an Air-mail from you yesterday and also one from Mahon. Mahon was really griping — said they were sending him through B-24 school and some second lieutenant was trying to teach him how to fly a B-24. Can you imagine such foolishness? He is now at Smyrna, Tennessee and says he might be getting down to Memphis sometime. I don't think his wife is with him now.

I haven't gotten that picture of you which you promised me in your letter of yesterday, however I'm eagerly awaiting it. You know a nice picture of you would certainly be an apt gift to your husband since he doesn't have one. You don't have to be in your black and gold outfit to show me how beautiful you are. I know you are beautiful anyhow. And, since I'm talking about clothes — "Clothes do not make a woman" any more than a man. Maybe Sis bought you some good looking clothes for Christmas. I suggested clothes but told her to use her own discretion. Hope she got you a real good looking, expensive dress — something you would be proud of. If whatever she got, you don't like — take it back and get something you do like. She certainly won't mind, and I'm sure that's what I want you to do. Did she get you some flowers too? Hope so.

Tomorrow I gotta work for a change. I have got to examine about 8 or 9 enlisted men that we are trying to get promoted to officers. Tomorrow's the last day — gotta do them then. I hate to do routine examinations so much that I always put them off just as long as I can.

Did you see where the Huns are starting another drive against the northern part of the 7th Army? Also notice where we were supposed to have destroyed 354 planes of theirs yesterday but had a loss of 29. I hope that's right at least. Gosh, will this damn war ever end? "I nant to go home."

How are our precious little cherubs tonight anyhow? Tell them that their Daddy still loves them. Most of all I love their Mother though. Goodnight.

All my love,
Joe.

January 4, 1945
11:30 p.m.

My dearest Honey,

Worked more than usual today. This morning I did six 64 exams. At one o'clock, I talked for about 1 hour to the officers and enlisted men of the squadron on venereal disease, women, life in general, etc. Sometimes I'm a pretty good chaplain. Today I must have made a good talk. I certainly held their attention. At least I got them thinking for a short time even though they have forgotten by tonight. Something has gotta be done — my V.D. rate is tremendous.

Tonight I've been up to the club drinking tomato juice (believe it or not) and playing poker. The new C.O. got promoted to a Major today, so he was "setting them up for the crowd." Kiss the kiddies for me. And kiss their mother many times.

> I love her,
> Joe.

> January 5, 1945
> Friday, 8:30 p.m.

My Darling,

Hey, I certainly intended to say something to let you know that I remembered January 3, 1942, and Kay's birthday. Somehow when I wrote to you that night I forgot. I'm sorry — no excuses except dumbness. Many happy returns to the eldest daughter.

I got in town today and finally was able to get me a pair of boots, thank goodness. No more wet feet until I let someone steal these from me. Boy, they are worth a million over here. I also bought two pairs of socks. You should see the socks. The sole of the foot of the sock is thick and the rest is not so thick, and they come up to your knees. The Army can really think up some bizarre patterns.

I was going to fly with our new C.O. in the morning, but I just got notified that at 10 a.m. I am to be on a Court Martial Board. I certainly do have some pleasant jobs. I also have another job for the next week — tonight Capt. Bell, the surgeon for the attached service squadron, came over to tell me he was going to Capri and to ask me to take care of his outfit while he was gone. I really don't mind for he has very little to do. Lately, I have been working a little more though. It has helped to pass the days a little faster but not much.

The war news sounds a little better on the Western Front tonight. Certainly hope there is something to it.

Can't tell you how much I enjoyed your letters today for they are priceless to me. I just needed them tonight. I could almost reach out and feel you and wanted to so very much. Knowing the value of letters, I was sorry to hear that you only got three from me in three weeks. I can assure you that I wrote better than that, and perhaps by now you have received them.

I'm talkative tonight and being talkative by candlelight is none too good on the writer's eyes. It's all probably to tell you again that I love you and miss you and want you so terribly much that it hurts.

> All my love,
> Joe.

> January 6, 1945

Dearest Babe,

Saturday, 10 p.m.

My Own Honey,

This morning I went up and sat on a Court Martial Board. After the first case, I got off it by "pulling a string" or so. I hate to sit on something and help to decide how much to fine a guy or how long he must serve in prison. It's a nasty job, and I don't feel qualified to be passing on some guys freedom. I'm supposed to be on that Board permanently, but you can count on it that I will get off it somehow pretty soon.

It's been raining all day and it rained last night too. As I drive the ambulance along, the mud "squishes" on its bottom. That's pretty deep for the ambulance is built pretty high. Tonight I was riding up to Group in a jeep and the mud splashed over the radiator. You have never seen such mud — up to your knees and over your boots in some places. We surely do need for the sun to come out for a few days for the ground has taken up all it can for the present.

Tonight I went up to Group again to go to a U.S.O. show. It was very good — about the best I have seen or perhaps it was just because it had been such a long time since I'd seen one. These shows certainly help to entertain you for the hour that they are on the stage, but after they are over, somehow they leave me depressed. Seeing men and especially women in nice American clothes and seeing good entertainment makes one realize more acutely what he has been missing. I don't know what it is exactly, but they do leave you depressed rather than exhilarated. I know a number of the boys feel similarly.

In the morning I have to be at a Group Surgeons meeting at some outfit near here, but I don't know where. Wagner is supposed to be at two meetings in the morning at 10, so he asked me to make one of them for him, but none of us know where the place is. All we know is that it's somewhere near here. I'll be able to find it O.K.

So Adrian showed up on December 22? Well I'm glad. I know Mother and Dad enjoyed him being around for Christmas. Guess after all the rumpus of the last visit this one was relatively quiet, and I certainly hope so. You mentioned that Babe's (Adrian's, of course) folks were going to move back to Texas — I didn't know they ever lived anywhere else. You see I haven't heard from Adrian since the Helena days — not even through Mother and Dad as I remember. He should at least write Mother and Dad more frequently, but he doesn't. Guess he's sorta like all the Kings and just doesn't write. I'll have to admit with the exception of writing you, I'm certainly no good at it myself.

So the Christmas tree was "gogeous" to Kay. Gosh, she must be a knock-out. How I would love to see her and listen to her try to express herself. I won't miss anything for I will see Carol going through the same stage probably. Watching them Christmas Eve night and Christmas morning must have really been a show. I haven't gotten your account yet, but I'm eagerly awaiting it.

I got the lock of Kay's hair today. She has much more of a reddish tint to her hair now than she did the last time I saw her — but isn't it lustrous? It looks so alive and glistening — am I just seeing things in it cause I know it came off Kay's head? — but to me it is real pretty hair. Thanks for sending it to me.

I must again say goodnight to the only girl I ever loved and the only one I ever care to love. Hope some day not too distant I can tell you goodnight in a more appropriate manner. Tell the kiddies Daddy said hi.

All my love,
Joe.

January 8, 1945
Monday, 9 p.m.

My dearest Honey,

The days happenings (My Day-Eleanor) have been inconsequential. The only thing happening was me pulling an inspection on the kitchen help and making a surgeons' meeting up at Group this morning. I did hear up there that we are getting a new Squadron Surgeon to take over Kremer's old squadron. Who he is, we don't know — all we know about him is that he's coming in a couple of days and that he has been a patient in a hospital for quite a while lately.

Was glad to know that you got a corsage of gardenias from me for Christmas — hope and think you got more, but as yet I haven't heard. Should be hearing any day now.

We had quite a party at the club last night. What the boys call a "skin show." Quite an appropriate name considering how much of the skin was shown by a couple of Italians questionables. (Not very questionable as far as I was concerned). I had to stay up there until the so-called "artists" left to see that no other Italian art besides dancing was forth-coming. Do you remember the Chinese strip-tease act at the Carnival that you persuaded me to remain at? Well that gal was fully dressed at the end of her performance in comparison to these two. In fact, I believe I could say that this one was slightly on the indecent and vulgar side without being called a liar. No, I'm afraid I didn't particularly enjoy the performance, and I don't think the performers will perform here again if I can help it.

I'm enclosing a small picture of me that was taken by Lt. Blaschke this summer while he was staying with me after he had been liberated from Romania. He must have had them developed in the States and sent me this back. As you see, I was working hard at that time trying to hold down the beach chair and look comfortable at the same time. It was taken inside our casa. Don't know why the inscription on the back unless I was cussing everybody and everything while he was here, which is quite probable.

On the back on the photograph: To Dr. King, Hi ya Doc—chin up?

Since I've started writing this letter, it has turned from a relatively still night to a howling one. The top sounds like it may leave the house any minute. I'll bet the temperature has dropped 15 degrees. Well, I gotta get busy and put up my laundry that the Italian brought back today. It is out on my "trunk" and soot from the stoves is falling on it. They have put in a Quartermaster G.I. cleaners in town, and my last cleaning was done there. It surely is nice to put on steam cleaned clothes and not gasoline cleaned, and steam pressed clothes that hold a press instead of hand ironed. Too, I have got to take a bath

Wonder where I'll be this time next year — hope it's not Italy. Really hope it is somewhere back in the States with my three Honey's right by my side. Guess if I have to, I can stand being away from you for another year, but I surely hope I don't have to. For I sincerely need the help, peace of mind, love, loyalty and companionship of my wife more than I could ever express.

I love you,
Joe.

January 10, 1945
Wednesday, 7 p.m.

My dearest Honey,

Honey it's been snowing here like I have never seen it snow before. Most of the so-called snowflakes look like balls of cotton. They are the biggest hunks of snow that I have ever seen. This stuff has been coming down almost constantly for the past 24 hours. I'm afraid we don't quite need the snow — as yet I haven't seen anything drying up around here from lack of moisture. The snow is so heavy that we are snow bound as far as the mail is concerned. Neither yesterday nor today could they get the mail across the mountains to us. In fact, today we didn't even get the "Stars and Stripes."

I've just finished writing your Mom and Dad and my parents a V-mail apiece. I'm afraid the letters seemed like carbon copies 'cause there certainly isn't anything to write about. The weather has been so bad lately that we have been doing very little flying — in fact, it's hard to get anyone out of their house or tent unless they just have to. Think of the poor guys up on the front lines on a night like this — how they live through it is a mystery, and no doubt some of them don't. I'm losing what little knowledge that I have acquired day by day, but gosh I'm fortunate.

Talk about knowledge — I had to do two circumcisions. Wagner and I did my "two little operations" today, and both of us felt completely lost. I never felt so much like a ditch digger in my life. I think Wagner and I are going to make a habit of trying to do something as often as possible for a while anyway — at least until we run out of warts, moles, cysts and such. There isn't much more that we can handle and stay out of trouble with the facilities available.

After I get through writing you, I think I will do some reading. I have about three little pocket books around here that should be worth reading. Better watch out — by the time I get home, I am going to be better educated in music and books but far subnormal in my own field. I have even gotten to where I listen to symphonies and know there are such guys as Sebelins, Delins, Elgar, etc. — rounding out my education some. When I get with Hugh again, I'll let him straighten me out on some of these guys.

Next day — Look, Honey, I didn't finish this letter last night and I'm going into town right now — so I think I will close this letter up so I can mail it. I will do better tonight.

All my love,
Joe.

January 13, 1945

My dearest Honey,

We have been snowed in from our A.P.O. for quite a while now, and today was the first mail we have received in four or five days. I was expecting a big stack of mail but only received a V-mail from you written December 28 and a postcard from Reader's Digest telling me that I was to receive the Digest for 1 year — that I appreciate very much too. Thanks — you couldn't have possibly thought of a thing more appropriate for reading material is, as I have said before, devoured not only by me but others. No one throws any reading material away until it has been read by all who might be interested. As yet I haven't heard of your Christmas, which I am very eager to hear about. That's when my Christmas will occur.

We have a visitor living here with us now. He is a boy, a mere kid, from the front that is down here on detached service to observe what the Air Corps does and how it does it. In return, we have one of our officers up there observing the same thing about the Infantry. He is a nice kid from Ohio and is a veteran of 22 months front line service in spite of the fact of his youthfulness. I dare say he isn't over 21 or 22 years old. We are trying to make him enjoy himself back where he can eat warm food, listen to a radio, and take a bath. I think he will be with us for about a week.

I have been reading all day. Read Douglas' "Magnificent Obsession," which I enjoyed very much. Royce sent it to me. In case you think about it while you are around him sometime, thank him for me 'cause I may forget the next time I write Sis. I did enjoy it though.

Zinn is standing over here a few feet from me naked rubbing his itch medicine on. We are still having an epidemic of the stuff, and it is damn hard to cure. These boys have had it for months now, and regardless of what we do, it doesn't seem to want to respond to treatment. As yet, the bugs still don't like me. If it were something less repulsive, I'm afraid my feelings would be hurt by their complete indifference.

Tomorrow I want to get a couple of nice, big, thick, juicy Air-mails from you. I believe that is exactly the medicine that the doctor needs. Of course that is only a substitute for the real thing but a definite help nonetheless. I need you, Honey, more than I can ever tell you, but I want you to know it anyhow.

All my love,
Joe.

January 14, 1945
Sunday, 7:45 p.m.

My dearest Honey,

No mail again tonight, doggonit. Guess it still can't get over the mountains for there weren't any "Stars and Stripes" either. If they can't get over those mountains pretty soon, I'm going nuts for I want to hear from you. It's still raining — we're having terrible weather. We are having more continual rain than we have ever had I believe and a corresponding amount of mud. Surely can't do much flying in this sort of weather.

I have managed to keep myself slightly busy today. This morning I was lazy and wouldn't get out of bed until about 9 a.m. (Don't you envy me). It was raining and dark and quite evident that we wouldn't be flying today, so I made the best of it. Finally got up and took the remainder of the morning to shave. You know how I can procrastinate when it comes to shaving time. I'll bet I can think up more reasons why I shouldn't shave in a shorter period of time than any other man of similar shaving experience. Finally about 11 o'clock I finished shaving, so I then wrote Sis and Royce a note. This afternoon I did some 64 examinations and then went up to the Group Hospital to see a couple of patients and shoot the bull with Wagner. The above has been "My Day."

Do you know that I have now been in Italy for 1 year and 2 days? Last January 12, I flew up from Algiers to a little town close to Taranto, Italy. That was a long, long time ago. Certainly seems like I have been in Italy longer than that. Hope it won't be that much longer before I get out of here, but it's certainly possible.

January 17, 1945

THE STARS AND STRIPES
MEDITERRANEAN

Vol. 1, No. 168, Thursday, January 18, 1945 — Printed In Italy — TWO LIRE

Red Army Frees Warsaw;
Vast Offensive Under Way

My Honey,

I started this letter the other day and then forgot about it. In the meantime I have written you a couple more. However, I will send the above along to prove that I was thinking about you whether I finished the letter or not.

I, too, read of the prediction of the buzz bombing of New York — also of Italy. However, I can't lay much store in either prophecy. It is my opinion that such statements were made to prod the so-called "home front" a bit and make them realize that this war is not over yet. In my way of thinking, I don't believe most people of the States have ever realized there is a war — with the exception of the relatively small inconveniences it has caused them. Don't get me wrong, Honey, I'm quite aware that you and your family know there is a war — I have no ulterior motive for saying such a statement to you other than to express my opinion. I do know that over 60,000 war workers quit their jobs while our armies were going across France because they wanted to get in something more permanent. So I believe much of the talk in the States concerning manpower legislation is to curb the persistent feeling of optimism — and if it will do that more power to them. Personally, I am more optimistic about the war tonight than I have been in quite some time. The radio has just announced the fall of Warsaw to the Russians. Now when the Russians take Crakow, they should be able to roll (I hope). To me, things look much brighter than ever before. Excuse me for the lecture on the "home front" and the personal briefing of the war situation — it may not be too interesting.

Sorry to hear that your Grandfather and Grandmother aren't doing so well. Apparently your Grandfather has a Decompensated Heart and certainly must go to bed. The bedrest will do him even more good than the digitalis, even though digitalis is vital within itself.

I'll bet you can't guess what kind of weather we are having. Nope, you're wrong — it is still raining — I can hardly see how the oceans keep from overflowing.

In a few minutes, I have got to go up to the Group Hospital to see my patients and to do a little lab work. I'm really getting to be a surgeon — yesterday I even opened a Rectal Abscess — can you imagine? While I'm up there tonight, I am going to the picture show. I think "Kismet" is on, which I have been trying to see for quite a while.

I'm going to be without transportation for a while. My ambulance is broken down, and it's going to take a few days to get it back into running order.

Looks as though I will be in Rome for my birthday. They are having some kind of medical meeting up there beginning January 29 and lasting until February 3. I'm going up there for the Group. I'm not too anxious to go up there for there is no heat nor any warm water up there in any of the hotels, and all one can do is sit around and freeze. Personally I'm not too anxious to go anywhere until next spring or summer.

Flash! The radio just gave out an unconfirmed report that Crakow has fallen. Those Russians must be really rolling. Their breakthrough, to me personally, is the best news since I've been in Italy — probably overshadowing D-Day. Well, Honey, I've talked too much about war tonight. All I've talked about you have probably already read in the newspaper. Take care of yourself and remember that I love you with all my heart.

I love you,
Joe.

January 19, 1945
Friday, 7:15 p.m.

My dearest Honey,

The Lieutenant who is down here from the 5th Army Front is leaving us tomorrow, heading back for the front. He has proven himself to be a regular sort, and we sorta hate to see him go back. Since he has been down here, we have really let him rest up — every morning, we have let him sleep until 9:30 or 10 a.m., and I think he has really enjoyed the rest.

Some of the boys just called me up. They want to come over and play some two-bit poker. I'm not exactly in the mood tonight, but I told them to come over anyhow — that will serve to pass away some of the time. In fact, I don't know what kind of a "mood" I'm in — today I have read a little of Aristotle's "Politics." Can you imagine me indulging in something of that sort? I'll have to admit, I didn't find it too interesting, however I did enjoy the rather lengthy introduction, which was written by Max Lerner.

I have a big inspection Monday, which I'm not ready for at all. The weather has been too bad for us to make any particular preparation or to keep things in very good order. Too, my records aren't in the shape that they should be in, but I'm not too bothered. Something I'd better get bothered about is the venereal disease rate in this outfit, but I can't do much about it. These guys know as much about the subject as I do, but they still partake so — I can't run around and hold their hands for them. We are still leading the Air Force. Well, I'll do the best I can and lecture until I'm blue in the face and that's all.

Hey, Honey, how about sending me a pair of leather house slippers — size 8 or 8 ½. These you sent are swell for winter time — but by the time you get this letter and send the other ones, and then the two months it takes for them to get here, it's going to be spring, and these aren't going to work too good in the summertime. Also you can send some canned goods anytime — sardines, anchovies (high-toned, I am) or anything — some more of your jellies (particularly jam) would also come in handy.

The last couple of times I have written you I have talked more about the war than anything else. I think that was probably on account of the Russian breakthrough. Seems the Russians have been slowed down somewhat, but they are still doing plenty good — certainly hope they continue. Just since I have been writing you it has started pouring down rain again — that's typically Italian. The stars were out bright when I sat down here. Only the good Lord will know how thankful I will be to get out of here and back to the States — no, not on account of the rain, mud, poverty, etc., but just because I want to be with you. However, I'll admit I certainly would enjoy a quart of good, cold, pasteurized, Grade-A milk. I think that's about the only foodstuff that I really miss.

In your last letter, you mentioned a new maid, Josie. Don't know when you got her nor how much she works, but I certainly hope it's every day and that she is a somewhat permanent installation in our household.

By the way, yesterday I played our new C.O. a couple of games of gin rummy. It certainly did bring back some very pleasant memories. Remember how we'd go on a trip and sit around and play gin-rummy just like we did at home? Just goes to prove how happy we were with just one another even in the so-called routine home life. That was the first game of gin rummy I had played since the last game we played together. What fun we used to have and what fun we are going to have when we are together again.

I love you,
Joe.

Dearest Babe,

January 21, 1945
Sunday, 11:30 p.m.

My dearest Honey,

Last night was an exciting one. We had either a wind, gale, cyclone or tornado. I don't know which, but I do know that the wind blew very strongly for about four hours. At about 2:30 a.m. this morning, part of our roof went with the wind and part of the wall went with it. We were afraid the whole house was going to cave in but it didn't. It did shake. When the top went, we all had the same reactions — at that moment all of us just pulled the covers up over our heads and stayed that way for a few minutes until we were sure that nothing was going to be falling down on us. The place today was ruined due to cement and mortar dust all over everything — boy, it was a mess. Along with part of the top of our house went the top of the officers mess completely, a 30 by 40 foot top — and part of the top of the kitchen. These tops landed on top of the enlisted men's mess and crushed it in, so it might as well have blown away. A couple of houses blew completely down, and a few tents blew down too.

We were fortunate, I suppose, that no one was injured. As luck would have it, my inspection comes off tomorrow — you can imagine what kind of shape my messes and kitchen are in. The officers and enlisted personnel are both using what is left of the enlisted men's mess — it will probably take two weeks to fix our mess back up again. Today all of us have been working putting a top back on our house, so tonight we are tired for a change. We haven't completed it, but we have gotten it to the point that it should keep most of the rain out.

Tomorrow every doctor in the outfit has to begin a series of lectures on V.D. We have got to lecture one hour a night for seven consecutive nights. That's going to be more difficult than you think for lecturing for seven hours on one subject without bringing in pathology, prognosis and treatment, etc., is going to be a job. We won't be able to repeat ourselves due to the fact that we will be talking to the same guys all the time. I dread it because it's gotta all be from memory, I don't have anything to read to refresh my memory.

I notice that the Russians are still progressing. Since they are already in German Silesia, they have done away with one of our tougher targets. We are all watching their progress with much interest.

Tonight we all went down and saw Abbott and Costello in one of their movies. I don't believe it was as entertaining as their pictures used to be. I think I must be getting to be an old sourpuss. However, if so, a few days with you will sweeten my disposition and personality up to its former level, I'm sure. I wonder how much emptiness and loneliness affects one's personality — probably quite a bit. Well Honey, I must go to bed. Hope I hear from you tomorrow.

All my love,
Joe.

January 22, 1945
Monday, 8:45 p.m.

My darling Honey,

I finally received one letter from you. Got the letter today that you wrote me January 1. It was a good letter too — one of those kinds that raise your morale, makes one feel near and makes one feel dear.

It's been snowing all day. Thank goodness we got the top on our house before it started. I wrote you last night that part of our roof blew off the night before last. Part of someone else's roof just passed the front door a few minutes ago, so someone besides us lost their roof too.

I just returned from lecturing to the boys concerning the Anatomy and Physiology of the male and female. The boys got a big bang out of it. I have lectured so many times to laymen about medical things that there's no telling how I would talk in front of a bunch of doctors about a medical subject. The boys really enjoyed it, however, cause they gave me a lot of applause when I finished and that's something that I've never heard before. I have never heard a bunch of G.I.'s applaud a lecture that they were forced to hear. I heard that tomorrow night they are going to hook up a loudspeaker system so they all can hear better. They have got to where they say that "Doc is going to give us a show tonight" rather than "Capt. King is going to lecture." I don't have to tell you that I derive some amount of pleasure out of that feeling myself. There is nothing more horrible than to have to lecture to a bunch of guys who are disinterested and bored and show it.

You know, Honey, I am inclined to agree with you. Apparently, considering Kay's antics, doings, and undoings, our eldest daughter is no fool. I read every word about them in amazement and wonder. I certainly would like to see and be with them. However, don't feel bad about it because the stage that I'm missing in Kay I will probably be able to observe in Carol, and of course, I had already observed Carol's stage in Kay. What I want to do is get home and realize that Carol is a child of ours too. I'm afraid at present (I'm ashamed of myself) I am partial to Kay and read with more eagerness what she does than I do Carol. I don't like to feel that way, and as soon as I get home and get to know Carol, I'm sure that feeling will pass. I just don't know Carol's personality as I do Kay's. We have been as fortunate, or blessed, in our offspring as we have in one another, and, as I can remember, all other things since we have been married. Hope He sees fit to continue our good fortune.

Well, Honey, I certainly was glad to hear from you and, as I said before, I enjoyed this letter even more than usual 'cause it had been a long time since I had heard from you.

All my love,
Joe.

Dearest Babe,

January 23, 1945
Tuesday, 10:30 p.m.

My Honey,

Got another letter from you today. It was written Christmas night and you told me of your and the children's Christmas; consequently, in the year of 1945 I will have two Christmases — one today when I heard from you — the other this December 25 (I hope). I also received a letter from your Mother and your Dad. I certainly enjoyed hearing about the kids' Christmas. Perhaps it wasn't as good as being there and seeing it in reality, but I saw a lot in my mind. Imagination, used properly, is a wonderful thing. Next Christmas will be even better, for Kay will still be young enough to get a real bang out of Christmas and Carol will be old enough. Your letter was very descriptive, and I had been looking forward to getting it for a long time.

I lectured to the boys again tonight. Tomorrow night I'm going to lecture on syphilis and finish up. I have written you in the last couple of letters that I was to be in Rome on the 29th to attend a medical conference. Well, that medical conference has been cancelled, however, I am leaving the day after tomorrow (the 25th) for a two-day medical conference in Naples. I will probably see the Memphis boys while I am over there.

I had the inspection today. I think the snow helped me out considerably. You know how clean a place looks when there is four to six inches of fresh snow on it. Well that's the way it was here this morning, and the area looked real nice. If the thick, deep mud that was under that snow had been showing it would have given a bad impression even though there is nothing one can do about it. As far as I know nothing of any importance was found wrong.

The war news sounds extremely good still. It was just announced that the Russians were on the Oder River on a 37 mile front. I'm afraid the Oder will slow them down some, and they will probably have to wait for supplies to catch up with them. However, it is difficult to ask any more out of an army than they have contributed in this drive. I hope I am wrong and they can just keep on going. Don't you know there is chaos in the German High Command? Too, the fighter bombers are apparently playing havoc with the withdrawal of the Ardennes salient — perhaps this war won't last forever after all. Don't be too optimistic. As I have said before, we have sold the Germans short many times.

I think of you so much and want to be with you so much that sometimes it seems that I can hardly stand to be away from you any longer. They are singing "I'll Get By" over the radio right now — and I guess that 'I'll Get By As Long As I Have You' whether I'm with you or not. It's the happiest thought I have to know that I have you.

I love you,
Joe.

January 26, 1945
12:20 p.m.

My dearest,

Yesterday, just before I pulled out of my field to come here to Naples, I received two Air-Mails and three V-mails from you. The Air-Mails were written December 26 and 27. It was mighty good to get all that mail from you, and I won't attempt to answer it here.

I am now in Naples, and will be here for two or three days attending a Medical Conference. Last night I stayed with Harvey Carter, but due to crowding them, tonight we expect to stay elsewhere. Didn't have much chance to talk with Harvey. He has only been back from the front a few days. Slept until 10:30 this morning and then came down here to the Red Cross to get a cup of coffee and to get a shave. This afternoon I expect to report for this conference. The main reason, as far as I am concerned, for being over here is just to get "away from it all" legally for a few days. Kiss the kiddies for me.

All my love,
Joe

January 27, 1945
Saturday morning, 10 a.m.

My dearest Honey,

I am here still at the medical meeting but at present am "cutting" a few lectures just because I'm not in the mood to listen to something I'm not particularly interested in. In a few minutes, there is going to be a lecture on cholecystitis (Gall Bladder Disease), which I intend to make and one on Pilonidals, which I also intend to make. This afternoon Johnny Hughes is giving a paper on Bronchiectasis (a lung condition), which I also intend to listen to. After that I'll probably take-off from here and go to town.

This is a very nice hospital. No doubt it was at one time one of Mussolini's best sanatoriums. It is the Vanderbilt unit that is running it. Dr. John Shaw, a Memphis urologist, is on the staff, and I was very happy to see and be with him. He was a good man. Remember, I used to send Mrs. Weaver, a big woman who had been operated on for cancer, to him while I was practicing.

Guess who sat down in front of me last night at the supper table? — Brock! Were we both surprised after we finally recognized one another. He didn't know me and I didn't know him — finally I noticed a similarity and asked him if he weren't Brock and sure enough it was. He didn't know I was in Italy, and I had heard that he got killed in a crash but he is alive and kicking. I was with him last night and part of the morning. This afternoon, he and I will probably take-off together and take in a picture show or something. He informs me that Shirley is doing well and still working for the telephone company. Also saw Bleecher and the Hughes twins yesterday. I'm still not too fond of either of the Hughes boys though. Well, Honey, I'm running out of paper. I'll do better when I get to a place that I can write. Tell the kids hello.

I love you,
Joe.

January 28, 1945
Sunday, 7 p.m.

My dearest Honey,

Here I am back at my home base. However I had a hellova trip today in order to get here. When we started from Naples, the weather was pretty. About half way across the mountains, it started raining and hailing. Of course in a jeep, we got pretty wet. Then about 20 miles from our base, it started sleeting and freezing. I don't have to say that by the time we got home we were almost frozen. It took me about 2 hours to get thawed out even though I had taken off my wet clothes.

I had a very good time at the meeting. I made quite a few good talks and better than that saw many people that I knew. I have already told you of seeing Brock, Harvey, the Hughes boys. Yesterday afternoon, I saw Bedford Otey and was with him quite a while. He asked all about you and also about Bess. I told him all I knew about everyone. Also saw Dr. Tulley, who used to be associated with Henry G. Hills clinic, and Dale Fox, who had a fellowship at Campbell's Clinic while I was at Gaston. Last night one of the old John Gaston Nurses looked me up, and we had quite a confab about practically everyone who was at Gaston. It was Miss Colley, who was a student and a graduate nurse there while I was an intern. All in all I had a very enjoyable time, and it was worth getting half frozen for.

Got in here about four-thirty this afternoon and ate three eggs and drank a couple of cups of hot coffee and told the boys that I didn't want to start doing any work before tomorrow if I could possibly help it — so far I haven't had to and I hope it keeps up. As soon as the 9:30 news comes on and I hear how the Russians fared in today's news, I intend to hit the sack.

When I returned tonight, I had one V-mail waiting from you, one from Burr Craycroft and a Christmas Card from Frank Jones and his wife and baby. Burr didn't have much to say except that he had seen Avent and that it didn't look as though he would ever get back home. He said his living conditions were still perfect. Apparently he is still living on that English estate — lucky guy!

I certainly am glad to hear that you (the Black Family) have been hearing from Hugh recently. I know Mom and Dad Black — and you, feel 1000 percent better. The relief of anxiety is certainly a blessing. I really am glad — still looks like Hugh should get rotated pretty soon.

You mentioned the house filled with the washing — the weather being so nasty that you had not been able to hang the things out. You mentioned the well-remembered unmentionables hanging around, which, to my recollection, always being in the house whether it was raining or not. It's been a long time since I've seen unmentionables — it certainly would be nice to see some — modeled of course by a good looking blonde woman — say about 5'5," Varga legs, etc. I love you, you know!

Honey — think I'll quit and get ready for the news and bed — I hope the Russians are still going strong. I want this damn war to get over and let me get back to you so bad that I can taste it. I am so tired of not living, and being unhappy all the time I hardly know what to do. We have a lot of "making-up" to do and we are going to do just that.

All my love,
Joe.

January 29, 1945
8:30 p.m.

My dearest Honey,

This morning I arose at 7 a.m. (rather early for me lately). After I had been up for a while, I got a call from the orderly room telling me that I had been "invited" to have dinner with the Colonel (the Group C.O.) along with the squadron C.O. I was filthy and had a two day beard and couldn't shave for lack of water. Finally, with the help of a couple of gasoline torches, I was able to get the water pipes unfrozen and was able to get cleaned up a bit. Looks as though the pipes are frozen up again already, but today we filled up a couple of five gallon cans so we are prepared for any eventuality. Tomorrow morning I will at least be able to wash my face.

I just got through drinking a beer that was really ice cold. I put it out in the snow for a couple of hours, and it was so cold that it was hard to drink without hurting my teeth. You know over here we always drink beer warm — it's ghastly stuff that way until you get accustomed to it, but once you are used to it, it's not bad at all. I guess man can develop a taste for most anything — except carrots.

If the Russians would keep up their good work, perhaps we could enjoy one another's company again before too many years elapse. Gosh, I don't see how those guys fight in the frigid temperatures. They may finally get Berlin, but the war with Germany isn't over then. Something might happen to the 15th A.A.F. 'cause Major General Twining, our C.O., is in Washington. Well, honey, I must say goodnight. Tell Kay and Carol that their Daddy sends his love.

I love you so much,
Joe.

Dearest Babe,

February 2, 1945
10:30 p.m.

My dearest,

Tomorrow I am going to Bari to carry four men down to the hospital. I suppose I will return here the following day. I'm going down in my ambulance so the drive shouldn't be so bad.

Got a V-mail from you today telling me that you had received a letter from Hugh which was written January 9. I certainly am glad to hear that! I have thought of him a lot. Too, you told me that your Grandfather was still pretty low — I certainly hope he has improved since then.

The weather has finally let up a bit around here. Yesterday it was even good enough for us to get in a punch at the Krauts. Today our plumbing thawed out, so we can wash again and brush our teeth.

This morning I had to lecture to a couple of new crews. I've said practically the same thing over so many times to new crews when they come to do combat that it is almost a memorized spiel now. I think that is what is wrong with all of us over here — we have just done the same damn thing so many times that everything is monotonous. Well, Honey, this apparently is one of my "bitch" letters. Since you are my chaplain, I guess it's O.K., but I think they are probably coming too frequently. If you were an Army chaplain, you could give me a "T.S." card. One of these days I will tell you what that means but since it's a wee-bit vulgar, I won't write it — but for morale it has been worth a lot in this man's Army. Even though I gripe much, it is certainly not because of you or ours. I love you all (you especially) more perhaps than you may ever know.

All my love,
Joe.

February 5, 1945
Monday, 8:30 p.m.

My dearest Honey,

Got back from Bari about 5 p.m. yesterday, and I was dead tired when I got here. Driving a military ambulance is exactly like driving any other kind of truck — it's a job and a tiresome one. I surely would like to get in behind a wheel of a nice grey, sleek, Chrysler convertible coupe and be able to drive with one finger along a nice smooth U.S. concrete highway. You probably say to yourself that I wouldn't enjoy it long for I would soon run out of gas.

I'm so terribly sorry to hear that Josie hasn't been turning up for work. I felt good that you now were so lucky to have someone working for you. I hope that she either turns up again soon or someone to replace her does. My poor Honey has more trouble with the kids. Kay getting paper clips caught in her teeth and my Honey having to get them out. I can just see you — probably more scared than Kay, but trying to act as if there wasn't a thing in the world to worry about. When I get home I promise to take such details off your shoulders.

I did have a boy today who had an unusual hot appendix. It took me a little while to make up my mind as to what it was. There was no doubt by the time I sent him into the hospital this afternoon. I haven't seen many appendices since I've been over here. I think this was probably the fourth one I've had.

The news hasn't been announced as yet tonight. That is the Russian communiques which hold more interest at present. Looks as though they are being slowed down some now.

All my love,
Joe.

February 6, 1945
8:15 p.m.

Dearest Honey,

Haven't been doing a thing all day except sitting around and playing Cassino and Solitaire, of all things. Went over to the dispensary a couple of times and up to the Group dispensary to see my patients there — that has been the total of my efforts today. I should have gone into town to get my cleaning, but my tail is too sore to ride over those rough roads. Either I will have to send for my cleaning or wait a few days to go for it.

Yes, Honey, I have read about the so-called National Service Act. However, at present I only thought it was a proposed Act and not passed yet. I figured it would have a hard time getting through Congress even with the President's backing. Like you, I think that should have been enacted December 8, 1941 instead of the present time. Just between you and me, I wouldn't be surprised if I weren't "bitter" against the guys that have been dodging the draft, the 4-F'ers, and the so-called war-workers. I never thought I would be that narrow-minded either — but somehow or another it "p's" you off to hear all those guys back there trying to run things, telling how the war should be run, etc., when all they are really doing for the so-called war effort is buying a bond occasionally with some of their overstuffed paychecks. It is my opinion that it is unnecessary to have the young male war-workers. Such fellows as my Dad and other elderly guys could hold down their jobs if they would really pass an effective National Service Act. As I have said before, I'm probably "bitter" and biased, and I had just as soon no one knew my views — especially my family — they may not understand. The thing that burns me up is that so many millions of people in the States haven't the slightest of an inkling of what the war is, and certainly don't want to. When we gain a mile here or there, people forget the number of lives that were ruined or lost in that process. I'll hush — I'm bitter.

No, I don't think that Japanese are "sucking in" the Americans in the Far

East campaign. I, personally, think MacArthur is much too smart for that. And in a warfare like that they would have to have a navy of formidable proportions to trap us. As far as I can tell, they don't have that. MacArthur is fighting a brilliant war — it's a personal affair with him. However even with the Philippines, (they announced storming Manila today) we are still almost 3,000 miles from Japan proper — that part of the war is going well, but far from finished.

As far as this European War is concerned, whenever they announce that hostilities have ended with the Germans, I still won't believe it. How it continues is difficult to see. There is only one answer — discipline of the masses — either voluntarily or forced. Perhaps both factors are present there, but fear is a mighty weapon. Now there is the Joe King Commentary tonight. I'll probably give you my views again soon which may conflict with these materially.

Again, it's goodnight, my Love. I love you with all my heart. Wish I was with you. Tell the babies hello for Daddy.

I love you,
Joe.

Mission #186: 7 February 1945
Target: Korneuberg Oil Refinery, Austria.
Results: 1st force bombed primary, hitting to south and east of target, cutting rail lines. 2nd force bombed Bratislavia, hitting quays and docks adjacent to oil refinery. Since lead ship is missing, can't explain why Bratislavia was bombed instead of primary. 5 A/C of 725 lost over Vienna. 1 A/C 726 lost. (Maj. Reichenbach) 1 A/C of 724 crashed near base. Spreha of 727 bailed out all crew, left plane headed out to sea.

February 7, 1945
Wednesday, 10:15 p.m.

My dearest Honey,

Today has been a terrible day for us — one of our worst — so my feelings and humor aren't quite up to par tonight perhaps.

The news was just broadcasted — saying that "The Big Three" were meeting someplace. I had a "scoop" on that — I knew it was going to happen about a week ago. The news didn't say much about the East Front, which has been for the past month our main source of "entertainment." We, being "arm-chair strategists," decide which way the Russians should move and when. Of course you realize how much we know about it, but it does help to pass away the time and is interesting in that it is fast moving. I'll certainly be happy if they can break over the Oder River in the vicinity nearest Berlin. We are anxiously awaiting the developments. The capture of Berlin may not win the war — undoubtedly, it won't. But, it should be a mighty shock to civilian and military morale which is certainly important — particularly in Germany.

Zinn and I have just been discussing and "cussing" economics, world-wide problems, politics, etc. You know, he is everything but a fool — regardless of his lack of tact and sometimes individualism. He has a rather deep insight to many things, and I enjoy discussing them with him when in certain moods. The peculiar thing to find out in our discussion tonight was that about three days ago he wrote to his wife for the first time that he was becoming bitter and intolerant to certain things. I wrote you the same thing only last night, and, until tonight, such things had not been discussed by us.

Anyhow after so much philosophizing, we were hungry, so we cooked a couple of eggs and made some toast and coffee (all ingredients being conveniently smuggled from the mess hall). He is now laying over on his cot sound asleep — me? I'm writing probably a very disconnected letter.

I was glad to hear today that you took the remainder of that money that I sent for your Christmas and bought you a coat. I daresay the coat was probably a necessity even though you don't put it that way. Well, if I'm gonna write Kay and Carol and take a bath and listen to the 11:45 news, I'd better get at it. Goodnight, my Honey.

I love you,
Joe.

Dear Kay and Carol,

I got a letter from Mama today. She told me what smart girls you are. She also told me of how much both of you are growing. Mama never forgets you — she tells Daddy about you every time she writes.

Today Mama told me what a big girl you had gotten to be, Kay. She told me that you could even dress yourself now, and put on your shoes all by yourself, and bathe yourself. You must be getting to be a real big girl, and Daddy is sure proud of you. Mama also told Daddy of how much you help her and how you read out of your Goose Book to Carol. That's the way Daddy wants you to be — he wants you to help Mama all you can. He wants you and Carol to play and play and have a lot of fun, and he wants you to do what Mama tells you to do without fussin'.

Mama told me that Carol was getting to be a big girl too — how she now ate her food with a spoon and didn't get it all over her — and how she walked with just holding on to one of Mama's fingers.

Now, Kay — you and Carol keep on being real good girls and do what Mama tells you, and help Mama. And pretty soon Daddy will get to come home — and then we will get in the car and ride and ride. And we can get on the floor of the living room and play together — I'll bet Mama won't mind if we mess up the house at all if we will pick up the toys when we are through playing.

I love you,
Daddy

February 10, 1945
Saturday, 8:45 p.m.

My dearest Honey,

Today has been just as dull, drab, and monotonous as usual. The only thing unusual was that I had to go up to Group at one o'clock to have a picture taken for another A.G.O. identification card. We are so damn tired of this monotony that it's just about driving us all nuts. I'm becoming such an old grouch that I'm sure it would break my face just to barely smile.

Zraick and Lather are going up to Rome in a few days on a five day pass. I could go along, but I think I'm gonna hang around for a while and be in line for a pass sometime in March or April. Through the grapevine I hear that a rest camp hotel for Air Force personnel is going to open up on the French Riviera about the first of March. I think I will try to make that either in March or April. I'll see if the Riviera is all it's shot up to be, and if it is, one of these days we might make it together — that is if you can ever get me out of the southeast U.S. With the way I feel now, it would be a job to pull me out of that area. But that feeling will probably pass after I've been in Memphis for a while.

Today was ration day. We are only getting five packages of cigarettes per week now. We used to get seven. I guess five packages of cigarettes per week in the States sounds like millionaire's rations. We just finished drawing cards to see who went up to the Officers Club to get our weekly rations of beer. Zinn lost and he's moaning about it, but he's going nevertheless. You should see us — we are so bored with one another that we are like a bunch of cats. However, each of us lets the other get away with murder just because we realize what's up. Occasionally when one guy flies off the handle, the other two will accuse him of "just menstruating" and remark that such is to be expected. Since Reese has been on edge particularly for the past 10 days, we decided today, in front of Reese, that something must be pathologically wrong and that I probably should examine him to attempt to determine the malfunction. Such are our conversations — you can see that I am certainly becoming more and more "genteel" and refined. Lord only knows what you are going to do to me when I get home.

Just a couple of days ago, some of my boys were forced to bail out of their ship when returning from a mission. I was looking up and not noticing that a Red Cross girl was standing within two feet of me. I had told the tower where to tell them to bail out, so they would land somewhere close to the field. Whether the tower told them or not, I don't know, but the boys bailed out to the leeward side of the field rather than the windward, so the wind was just carrying them farther and farther from the field. I yelled at the tower a very obscene exclamation. As soon as I noticed that the gal (and I mean gal) was standing there, I apologized but the damage was already done. It didn't bother me too much 'cause the girl had just about as much business there as I would have had in her doughnut truck. My language is atrocious though, as is everybody's. I will try to improve it before I get home so I won't make you an outcast — too for our daughters' sake. Thank goodness another day is drawing near an end. Tell the kiddies that Daddy loves them both and impress upon their Mother that I love her best of all. Goodnight.

All my love,
Joe.

February 12, 1945
Lincoln's Birthday

My dearest Honey,

Today, it is beautiful. The weather compares favorably with any nice early spring day in the States — the sun is out, nice and warm. The breeze is just mildly blowing the smoke of the gasoline stoves away. The mud — yes, we still have it, and it is still most plentiful. However, if it were to decide to remain like this for a few days 'running' (as the British say), much of it would dry up. It's been so damn long since we have had a day like this that I can hardly help spoutin' off about it. Everyone greets you this morning with "It's a pretty day, ain't it Doc?" Somehow or another that helps the situation, and is much more uplifting to morale than the usual "More damn mud!"

The breeze is nicer to us this morning too. Rather than the usual cutting wind that bites at your face, neck and ears — it's just a breeze. Would be a nice day for a family washing, or the maid to hang out the baby's panties and diapers. Now if it would just stay this way for a few days and let some of this mud dry up.

All the little Italian boys are running around the camp airing out blankets and bedding of the tent in which they work. Those blankets need it too 'cause they probably haven't had any sun on them since last fall. About 100 feet from me, the whole contents of a tent are getting aired out — that is what's left of it. It caught fire last night and burned down along with most of the boys' belongings. The unusual (I started to say funny) thing about it was that the occupant of that tent was a brand new replacement crew that had only been here a couple of days. When they arrived, they were raising the devil because they had to sleep in a tent with a dirt floor. We will now get to see how they like the dirt floor with the sky over their head as their roof. Some boys expect to live in resort hotels and fight a war. However, it usually doesn't take long for them to settle down to the "facts of life." I just sorta wish those kind had been with us last winter instead of this one.

I have a flock of sparrows that live on my roof and have since the house was built. Somehow they know that it's a pretty day too for they are hopping around and chirping more today than they have done in a long time. They roost between the corrugations of the tin roof somehow. Sometimes they get in the house by crawling through the layers of overlapping tin — then we have to catch them and get them out before they kill themselves flying into the windows. They surely make a lot of noise on a tin roof, and this morning they are really trying themselves.

I guess being in Italy for 14 months has been good for me. I was taking everything too much for granted all my previous life. You, a home, children, all the luxuries that I had become so accustomed to without a thought of what it would be like without them — heat, milk, cokes, sanitary facilities, nice paved streets and sidewalks, beautiful stores and parks, shoes, etc. I know that after I'm back there that at first I will appreciate those things intensely — but even after the "new" wears off, I believe I will still have a more profound respect for my average daily way of living. I can almost understand now why all people outside of the U.S. say that all Americans are rich. We probably are not in terms of that money that Europeans are thinking of but in things much more important

— we just don't realize it.

I'd better get out of here into that nice sunshine outside and do a little work for a change — sunshine is good for the soul somehow, isn't it?

I love you,
Joe.

February 13, 1945
10:30 a.m.

My dearest Honey,

Didn't write you last night — so I'll try to get a note off to you this morning at least. Last night I went into town to hear D'Artege and his all-girl band. They were playing at the Red Cross theater. They were very good and have some very talented musicians in their troupe. For instance their first violinist is none other than Memphis' own Peck's Bad Girl, Hope Brewster. Or at least I think her first name was Hope. When they announced that her name was Hope Brewster from Memphis, Tenn., it certainly brought a lot of memories to me of the many screwy things that she and the rest of her family have done that you have told me about. I spoke to her after the show for about two minutes. I told her I was the husband of Olive Black — see how convenient it is to be married to a "celebrity." They had only been overseas for one month, and I was hoping that she had been in Memphis and had just happened to see you recently but found out that she had been in New York for the past two years. That's about the extent of my conversation with her. Of course I started to walk up to her and break forth with "I know you — you're that screwy dame that rowed that skiff from Memphis to Helena — I was practicing in Helena at the time," but I was a nice boy instead. I wonder if all the members of that band are as nuts as she used to be. Did you know she was in this all-girl band?

The Russians announced a new big offensive north of Breslan last night. I'm sure glad to see the Russians get upper Silesia — now if we could only get the Ruhr and Saar Basins. That particular Russian wedge ought to go to town — maybe. Even in the presence of all the so-called good news, we are all inclined to be skeptical but intensely interested.

Honey, it's lunch time. With the frequent interruptions, I have written this in one and a half hours. I want to get this off in today's mail, so I'll take this by the mailroom now. Kiss the kids for me and take care of yourself above all.

All my love,
Joe.

Dearest Babe,

February 14, 1945
Valentine's Day

My Honey,

I won't write and request that you be my Valentine — I think that would be superfluous. However, I may ask that you remain my Valentine forever.

I was terribly sorry to hear of your Grandfather's death. I was rather surprised because I was somehow under the impression, due to a letter from your Mother a few days ago, that he was some better. There is nothing consoling that I can ever say about a death. Nothing one says helps much. However, if it's any consolation at all, when a person dies of a "senile heart," they don't actually suffer as much as it might appear that they do. The blood passing through the brain slowly and irregularly probably depresses thought processes and pain considerably. I'll write your Mother, but kindly convey my sympathy to her. I am glad that you had a chance to go to Cape Girardeau this summer, so your Grandparents could see Kay and Carol.

Honey, what happens to your Grandmother now? Will she come to Memphis to live with your Mother and Dad or remain in Cape? I imagine she will have to come to Memphis since I don't believe there is anyone in Cape Girardeau to take care of her. Let me know what the final outcome is. I wish I were there to help you in any way possible. All I could do is hold your hand and let you know, or reassure you, that whatever happens or whatever the trouble is that I am on your side. You know that already but reassurance at the right time means a lot.

I did something a little different today, and the day slid past much faster than one has in a long time. At noon, I got in the B-25 (twin engined medium bomber, just in case you don't know) with Lather, and we flew over, across the Adriatic to an island off the coast of Yugoslavia. Landed there and only stayed for an hour or so before we came on back. I enjoyed it very much. I don't care to just go up and fly for the sake of flying anymore, but I still like to go places fast and see how other people live and what their country looks like. That's the first time I'd been to Yugoslavia — the 10th foreign country I've seen. I have no particular desire to be in another either, unless it is necessary to traverse it in order to get to the U.S.

Well, again Honey, don't let everything get you down. Some of these days all this chaos will be over and perhaps it won't be too long. Then we can be together again and things then will be as they should be once more.

I love you,
Joe.

February 16, 1945
Friday, 7:30 p.m.

Dearest Honey,

Thought perhaps I would go to the show tonight — that would pass away 1½ hours, but the 300 or 400 yards to the place we have the show is mud that is knee deep, and now I don't feel in the humor to get out and fight that stuff when I'm so comfortable in here. Now you can just see how terribly lazy I am. Last night I passed away a couple of hours playing gin rummy with three other guys. We played partners — after two hours playing we finally ended up breaking even in number of games won and lost.

I'm without transportation again. My old ambulance is in the "truck hospital." Don't know how long it will be there, but it's hell to be without transportation. The whole squadron is damn near out of transportation. This mud just ruins trucks and jeeps. It's impossible to take care of them in stuff like this.

Hey, what do you think about the attack by the Navy Air Task Force on Tokyo? That sounds pretty good to me — but news since the Russian advance. Of course that's only a drop in the bucket, but it's a drop nevertheless. Perhaps if they step business up enough in that theater, Hugh and I may have a chance to get back to the States for a while before going to the Pacific. That's probably asking too much — and I'll be damned if I'm going to allow myself to become disappointed. I'm afraid I couldn't stand it.

It missed raining again today — miracles will never cease to happen — however the sun didn't come out all day long. Guess tomorrow it'll probably start raining again. For four days straight we have been able to fly missions. We were hardly able to do much better than that the whole month of January. Hope the sun will shine again tomorrow. The other day when it did, I felt more like a human being than I'd felt in a long time. Somehow it just helps out.

I love you immensely,
Joe.

February 19, 1945
Monday, 9:30 p.m.

My Little Honey,

Today I got three letters from you. One of them was so pitiful that you filled me up with my emotions. When you said how much you needed me — how unsure and insecure you were without me there about some things. The funny thing about it is that, not three days ago, I wrote you a letter either saying practically the same thing or expressing the same thing.

It was certainly a fortunate day for both of us when I made a date "ambiguously" with you that night back in '40 — when you didn't know whether you had a date with Dan or me. Have you ever thought about if that hadn't been done we might not be man and wife right now? It's almost shocking to think that some little incident like that might be responsible for one's whole happiness. In fact, it scares the hell outta me to think about it. We had such a whirlwind romance that it's a wonder you accepted my proposal — or even it's more of a wonder that your folks didn't raise some holy hell. Suppose Kay or Carol wanted to marry some guy that neither you nor I even knew and she had only been going with him just a couple of weeks or so. I don't know whether we would just sit by and watch or not (just between you and me, I hope that doesn't happen). There were so many places that there was a possibility of a slip-up back in those times. I have wondered many times what would have been the outcome if either of us, me particularly, had let that ounce of will-power slip a couple of times. Guess I can sit back and say that even though any "marriage expert" would have advised against it, we have got that most successful marriage. It is swell that we are so interdependent upon one another for our happiness. And that is what it boils down to. You were quite able to be independent prior to '40 as I was — both of us are quite able to tend to the usual problems without any outside help — but we are so dependent on one another for the most valuable thing man can possess, happiness.

Of course, I didn't mind you paying off your grandmother's grocery bill. If you could have helped her out further financially, I wish you would have done that too. Honey, always do with the money at your disposal as you see fit.

Glad that Kremers called — and don't worry about me doing just "pretty well." I'm quite healthy. Don't know how much I weigh but assure you that I probably weigh as much as I ever did if not more. When Kremers said "pretty well" he was probably talking about my mental attitude — which, I admit, gets depressed frequently. I am quite healthy.

I am sorry to hear that Carol is allergic. I'm rather surprised. Honey, don't let it all get you down. Everything will all be hunky-dory when we are together, and we can hold out until then with the courage and patience that we have.

> I love you so very dearly,
> Joe.

My Honey,

Here I am down at ▮▮▮▮ tonight. Guess I will go back to my outfit tomorrow. If I don't tomorrow I will the following day — at present I don't feel too anxious to get back. However if I sleep cold tonight in the hotel, I will make every effort to return tomorrow.

Got down here about noon today and have completed most of my business already. Also took in a movie this afternoon which was entertaining — it was Bob Hope in "The Princess and the Pirate." I am now in the Red Cross, which is the sorriest Red Cross unit in Italy (even though they do serve ice cream daily).

The fellows out at the ▓▓▓▓ General Hospital have invited me to come down on Detached Service for a while if I can make it. If things continue to be as boring as they have been lately up at the field, I might try — they seem to be a nice bunch. Will tell you goodnight 'til tomorrow.

I love you,
Joe.

February 21, 1945
10 p.m.

My Honey,

Tonight I am weary. Today has been one of those days that a Flight Surgeon earns his money and no one envies him his job — just one of those days. I always feel low after a day like this — so consequently not in a very conducive frame of mind for writing enjoyable letters. However, I'll write anyhow just so you will know that, as usual, I am thinking of you as I always do — many, many times a day — and as the day draws nearer to its end, I think of you most of all. When, during normal times, I would derive the pleasure of being in your company — the incentive that had been keeping me going at a pace all day long I still need someone to "talk it all over with" and get it all off my mind, just as I needed it in private practice. You have been worth your weight in gold to me many times. And I appreciate you very much for having the virtue of being a good "listener." It is a virtue which few people have — especially women, I believe — and which I should have developed more when I was young. Perhaps "listener" is not the right term — it is probably patience. You could come to my aid tonight like it would be impossible for anyone else to — and I love you for that characteristic too.

Received a letter from Thea today. It was the first time I have heard from him since I last saw him in Riverside. He wrote me a long letter, 11 pages, which was very interesting and very newsy. I think he must have sat down and tried to think about what I would want to know about in Helena, and wrote his letter accordingly. Thea is quite a fellow and a much more all-around guy than Dan, and I like and admire him very much. He says he is going to drop in to see you the next time he is in Memphis. He also said that the evening he spent with you in Helena was the most pleasant evening he has spent since he has been in Helena since returning from the Army, which I accept as quite a compliment.

Honey, you tickle me with all those "gadgets" to remove wrinkles etc. I may be wrong, for I'm not a beauty expert, but I'd bet my bottom dollar that all that stuff is "hookum." And I mean "hook 'em" instead of "hokum" — I believe though it's both. Don't get me wrong though, for, on my account, I wouldn't have you quit using them for the world — you're cute. But, if you have developed any new wrinkles, which I doubt that you have, I shall love you even more for them for I shall know that they appeared there from trying to raise two children as they should be raised or because of worry over me and Hugh. If for me, don't worry for I'm as safe as Kay or Carol. I can't positively speak for Hugh but I believe he is reasonably safe too. If you have "definite indications" of getting older — as you call them — so have I, and I don't like the idea of getting older by myself. I don't have the zip, pep, energy that I once had — and somehow a soft easy chair with a foot stool sounds more appealing to me than ever before (or the sofa).

According to all the boys, I am getting old. You can readily understand why I'm considered an 'old man' when 99% of these guys around me are less than 24 or 25 — most only around 20. But damned if you don't begin to feel rather old when in their midst. Being the Doc, of course, contributes to that. If they weren't kidding me about my "age" they'd be razzing me about something else. It takes even a sharper tongue and a much quicker wit to keep ahead of them.

Now I'm going to call it a day and go to bed. I've had insomnia lately (not conscience) and haven't been able to go to sleep until after midnight.

I love you so very much,
Joe.

February 23, 1945
7 p.m.

My dearest Honey,

Last night I didn't do you so good. I sat down and started to try to write but nothing worthwhile would come forth. Received a letter from you today — also I got a Valentine from you and one from Kay and Carol. Enjoyed all of them very much.

It's still colder than a well-digger's _____. You know how cold that is. It's not raining though so that's certainly a help. The ground has just about dried out to the extent that one can walk around without his boots on.

Honey, do you remember that pair of shoes that I bought at Phil A. Halle's when I bought the rest of my uniform and went into the Army. Well I made up my mind today to finally give them to some Italian. They have surely lasted me a long time. I've had them half-soled once over here but now a big hole is in the sole. The uppers are still good, and it hurts my scotch soul to give them up but I guess I will. I've worn out all of those old undershirts and shorts too. No, I don't need anymore, I still have gobs that I would like to wear out.

Some of the news is not too good. As yet the radio hasn't said much about how we are coming out on Iwo Jima, but I think they are preparing us for rather heavy losses. That must be heavy warfare — 35,000 Marines and 10,000 Japanese on an island no more than 8 square miles — it's hard to visualize. Well I hope it gets over soon as I hope the same for the whole hateful business. Wish the tempo of the war over here would pick up some and this part would draw to an end. Mother wrote me that the war, as quoted by the radio, had a good chance to be over in three weeks. Such rubbish! Of course you know our opinion of such talk.

Well I gotta get up from here and take a bath — either that or the boys are going to soon liquidate me. Reese isn't here at present though — he has gone up to the Front for a week to be with his brother. Well, Honey, I'm gonna kiss you goodnight early tonight and go take my bath. Tell the kids I love them. Hope that it won't be too long before I'm with you again.

All my love,
Joe.

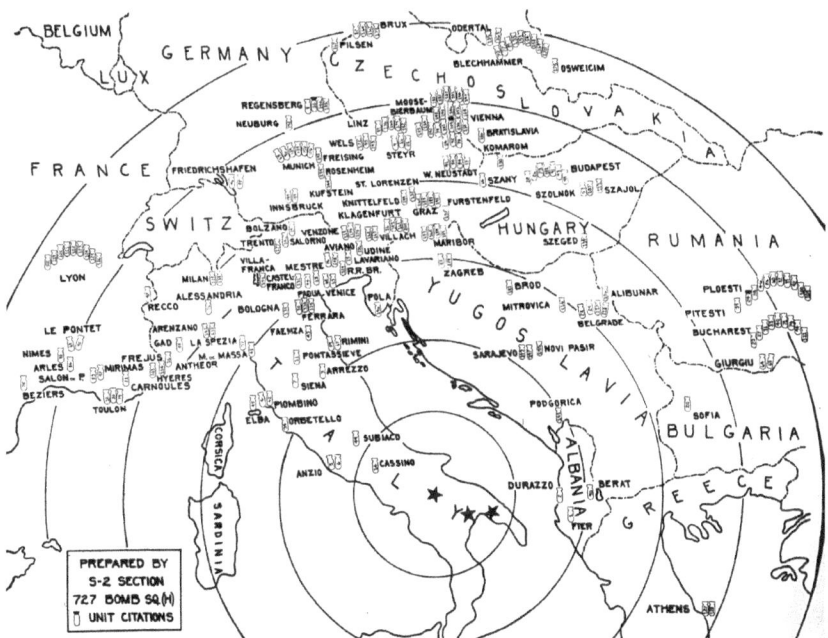

On February 24, 1945, the 451st flew its 200th Mission. Map of the first 200 bombing missions.

February 24, 1945

My dearest Honey,

Today we flew our 200th mission over enemy territory — I hope we never celebrate another 100. This morning was nice and sunny, so Zinn and I got us a pick and shovel and got in my ambulance and went out to get us some trees. We went to the highway which is about 6 or 7 miles from here, and started digging up shrubs that old Musso had planted along the road. We dug up 10 nice "skinny cedars" and brought them back. Planted five of them this afternoon and tomorrow we will plant the other five — hope they live, but I'm doubtful. So far we haven't been able to get any grass seed, but we have some wheat as I guess we will plant wheat for the "sod" of our front yard — it's gonna look pretty good. This hill is as bleak as it can be so a tree will look good (too it gives us something to sink our teeth in). I heard that some guy over in one of the other squadrons has some flower seeds. If so, I'm gonna try to get some and we will plant those too. No, I'm not crazy or a lover of nature — it's just something that is interesting to do and will give us something to look forward to — also will make this place look like somewhere someone lives (foolin' ourselves, see).

Didn't get any letters today, but I did get the box you sent me with the ping-pong set, fruit cake, sardines and crackers. Thank you so much. We have already eaten the cake. It was delicious — I didn't know my honey was such an excellent fruit cake baker. We haven't eaten the sardines yet, but tonight I imagine one can will go — when we can get us a good ping-pong table made, we will put that to use too. Thanks again for everything — it's very nice and thoughtful of you.

Noticed in the Commercial Appeal that was in the package that you sent me that they were advertising for a steward at the Broadway Beach Hotel at Biloxi. Remember that place — it was that swanky bunch of Hotel cottages just down from Colonial Tourist Courts — we used to eat breakfast there occasionally. Brings back very pleasant memories. The Gulf always brings out pleasant memories to me. We just gotta spend some more time down there. I love you with all my heart.

Joe.

From Joe in Naples to Babe in Memphis

February 28, 1945
2:30 p.m.

My dearest Honey,

Two days ago I left our base and came over here after a jeep. After getting over here, I find out that the jeep isn't here at all, but back at a town nearer our base. Now I'm having another jeep worked on in order to get back to the base — quite a mixup. It will be a couple of more days before I can get outta here.

I have spent both nights over here with Harvey C. I have seen the Hughes boys and John Wilson. In fact yesterday I had lunch with them. Phil Bleecher has been sent home with something wrong with his back — something pretty serious apparently. He is back there by now I guess. Anyhow I wouldn't say anything about it for I don't know how much his wife knows.

This morning I slept late. Got up about 11 o'clock — long after Harvey and his cell mates had gone to work. Went down and walked along the water-front and watched the little Italian kids stand by the shore — take off all their clothes and wash them in the sea. Put them back on wet and start off again on their daily routine of scavenging, stealing, and begging — boys and girls alike. It is a warm sun-shiny day, and I dare say the first time those kids' clothes had been washed since it was last warm enough to do the same thing.

This town seems much more alive — two-wheel wagons, ox-carts, Italian trucks, street cars, taxis, and army vehicles are rumbling and buzzing around more than ever before. It is interesting and I dare say educational just to stand by and watch. I can always stand around and say to myself either that "I wish you were here to see this" or "I'm damn glad that you don't have to see this." This morning it was the former — I believe you would have enjoyed watching these people with me and strolling along the waterfront in the nice, warm sunshine watching the street urchins and everything in general.

Well, my honey, I gotta get out here and get me a ride somehow out to the medical center area. Perhaps tomorrow that jeep will get out and I will have some transportation. I'll write you a good letter when I get home if not before then. Tell Kay and Carol hello for Daddy.

> I love you,
> Joe.

Dearest Babe,

Spring 1945

In the spring of 1945, the Allies advanced across Europe, and Joe frequently wrote about the war news. He started doing the physicals and paperwork in preparation for the unit to be moved at the end of the Italian campaign.

* * *

From Joe in Naples to Babe in Memphis

March 1, 1945
8:30 p.m.

Dearest Honey,

Wrote you an Air Mail yesterday from the Red Cross. Didn't know the date for sure then but it was the 28th. I'm still here in this town and still staying with Harvey C. I know he and his pals are getting tired of me, but I can't get out of town. Hoping that this jeep will be repaired by tomorrow. I told them when I left camp that I would be gone for two days and I've already been here four. When I get back, I might be Private Joe W. King instead of Captain. I'm doing nothing but just wasting time, so I'll be glad to get back to camp — much better to do nothing in one's own environment. Kiss the kids for their Daddy and remember that I love you with all my heart.

Joe.

Castelluccia di Sauri

March 2, 1945

My dearest Honey,

Finally managed to get back to camp this afternoon. And, even though I had an enjoyable trip to Naples, my casa looks pretty good — my cot particularly. I'm going to try to remember to enclose a piece out of the Arkansas Medical Society Journal. I think you might be interested. I'm sure we could get help for one year and possibly more — according to the last paragraph. If so, with what little I would probably make, plus what little we have — perhaps we could make a go of it without you and kids having to go in rags. As yet I haven't done any writing concerning it. Apparently I won't be able to do anything about it soon, and maybe we can talk it over face to face before it is necessary to worry about it. This is an interesting article nevertheless.

The weather here is still good. As you have possibly read in the news, we have flown 18 straight days — that is a record for us in Italy I can assure you. The sun didn't shine today, but it didn't rain — we can surely stand it! The mud is just about all gone. That answers one of your questions you asked me. No Honey, I don't need any boots, or socks, or undershirts, or shirts or anything.

Don't worry, Honey, when you hear lectures on diseases such as Tuberculosis, you feel sure you must have it. That's certainly the normal reaction — especially on T.B. I can hardly imagine Carr talking to a woman's meeting on a medical subject though. I never think of Carr without thinking of some guy who goes completely nuts in an operating room. I've assisted him a few times while at John Gaston, and I'd rather assist anybody except likely W. Simpson the Nose and Throat Man.

Yes I know what you mean when you say that perhaps we all have "Dr. Jekyl and Mr. Hyde" in us. I have them both in me too and I frequently surprise myself with my short temper and lack of patience. I know it will right itself when we are together living the normal life that we are made to live.

Hey, they just announced over the radio that, in some places, the Ohio River was above the flood stage. Certainly hope that the Mississippi doesn't go rampant. Glad Memphis is up on a bluff. Now I'd better stop and take a bath — gonna go to bed early. I love you so much — I wish I could put it across but I can't — just know that I do. I miss you just as much even still and want to be with you more every day.

All my love,
Joe.

387

Dearest Babe,

March 5, 1945
Monday, 7 p.m.

My dearest,

My back is about to break. Practically all day I have either been digging with a pick, shoveling dirt, or shoveling and hauling gravel. We have really been working on our front yard. If we aren't too sore tomorrow, we will work on it some more. About one or two more days should finish this yard up. It should look pretty nice when we get through.

To answer a couple of questions for you — Giovanni is not with us any more. His Daddy got nasty, so we ran him off and made Giovanni go too. I gave him quite a few clothes before I sent him away. He did not lose his eye, although his vision was impaired considerably. His Dad thought the U.S. Government should pay him something — when some other Italian threw the stuff in his eyes — just because he was working for me, I guess. Anyhow he made me so provoked that I made him leave. We have another boy about the same age who is fair but not as good as Giovanni.

A service squadron is a squadron which is attached to a larger tactical outfit to help the tactical outfit keep their airplanes in shape to fly. They take care of the major repairs of airplanes which take too long for us to fix. They are not tactical — even though sometimes they might find themselves in the "thick" of things — such is the outfit that Bill is a Medical Officer for.

You mentioned the black earrings that I bought for you in Rome. I was thinking that I had sent them to you, but I don't guess I have. But they must be lost for I haven't seen them in a long time. I'll get you some more the next time I'm up there. I haven't gotten you anything in a long time — don't think I have forgotten you, but there just isn't anything worth picking up.

Glad you got the colored pictures from Cairo. We took some here (not colored) a couple of weeks ago, and we have been trying to get them developed. When we finally succeed, I will send you some prints to prove to you that I am still quite whole and hearty.

Yeah, I'm sorta like you. If I ever get a chance to come home for a month, I'm gonna take it I think. After one has been overseas for 18 months, he can put in for a 1 month furlough at home. Of course it takes three or four months for that to go through. As soon as that month is up, you return to your same station. If you don't accept this, when you have two years overseas you can put in for rotation, how long it takes for that to come through varies but usually it takes about 6 months. Then you are transferred to the States on a permanent transfer — which means that you will be in the States at least 4 or 5 months before you are sent overseas. Of course on the latter plan, you may not ever get sent overseas again — but you probably would be. So that's the whole story. Anyway, as much as I want to see you, I would probably jump at a chance to be with

388

you for four or five days (or less) and then just take my chances on what might happen after then. I really want to see you bad — just about as much as I have ever wanted anything. Enough is enough, and I've already had too much of this separation.

Tell Kay and Carol that Daddy says hello and to be good little girls and not to give their Mama too much trouble. Take care of yourself.

I love you,
Joe.

March 6, 1945
11 p.m.

My dearest Honey,

Just a shortie tonight cause I just got in and the other boys are asleep, and I want to finish writing as soon as possible and get the lights off.

Worked today. Got up about 7 (unusual for me when we don't fly) and started working over at the dispensary. Did two 64 exams then went up to Group to take care of a couple of sick boys of mine — one has pneumonia and the other a possible appendix. Got back about lunch and this afternoon I have been working on our front yard again. Hauled a big load of rock to help fill in our walks and did a lot of digging and laying brick. About one more load of fine gravel should do the job — then I have got to level it off — plant two more trees and plant my wheat and then I'll be through — silly, ain't I — but it'll make the place look better, it gives me something to do. By the present soreness of my shoulders and back, I need to exercise. Running out of paper, Honey — just enough room for me to tell you I love you, but there would never be enough room to tell you how much.

Joe.

March 7, 1945

My Honey,

I've been working on the yard again today — planted two more trees and brought another load of gravel. I thought I'd finish in a day or two, but now Zinn has gotten interested in it (after it's shaping up and he can see how nice it will look), and he wants to make it much larger and extend it all around the house. This suits me for it will give us something to be interested in, keep us busy, and provide some much needed exercise. It looks real good already but will look better after our wheat comes up. Since we have been working on it, I haven't had any insomnia — I hit the sack and go to sleep.

In your letter today, you told me that Carol took seven steps — about time, the little rascal. I wonder what has made her so afraid of walking. She must have a sensitive bottom and doesn't like to sit on it too hard. I would certainly like to see both of them — second only to their mother.

Can you imagine being in one location in the Army for 11 months? Eleven months to the day we have been right here. If they don't move us pretty soon, we should apply for citizenship where we could demand our rights (There's one thing that Uncle Sam will never have to worry about).

Sounds like things are moving on the West Front. The 1st and 9th are on the Rhine and the 3rd is coming up fast. A blow to the heart is the capture of the Ruhr, but a tough job for the Allied Armies — may it come soon. Honey, I'm about "writ" out tonight. Tell the kids that Daddy still loves them and remember that I shall always love you.

Joe.

March 8, 1945
8:30 p.m.

My Honey,

I'll try to write a letter — don't know how much luck I'll have for there's too damn many people in here talking, jibbering, asking dumb questions and doing exactly the things that get on one's nerves when he's tired of being around a bunch of peppy youngsters and wish to hell they would go home. They have been over helping Zinn censor mail. That's a job that I talked myself out of when I first got overseas, and I'm certainly glad I did, it is most boresome. There is no doubt about it, I am definitely getting old.

Haven't done much today. I finished leveling off our yard and finally planted the wheat. Now I am eagerly awaiting for it to come up so I can see how it's gonna look.

I thought we were going to have an early spring. To prove what an excellent weather man I am, we had about 2 inches of snow last night, and the surrounding mountains are covered. It's funny — for yesterday it was sunny and warm and today it was also. Consequently, the snow melted quickly, and now it's just muddy again.

I'm certainly sorry to hear that the film that you took didn't turn out. I, too, have been looking forward to getting them. In this next roll, the one that you have now, be sure and get some of yourself. Those snapshots that we took here haven't been developed yet. In fact the sergeant who is developing them is getting peeved when we call him about them. We have called him so often, and he promises to have them done "in a couple of days," but when the couple of days are over the answer is always the same.

News Flash — Just this instant announced that the 1st Army just crossed the Rhine south of Cologne. I believe this really means something. Now Patton (our most able general in my opinion) will cross too. Then if Zukov gets across the Oder (and he will), Germany will be cut in two in short order. Of course these are only my hopes — I know no more about what is going on than you do. But crossing the Rhine, through the Germans so-called "prepared defenses," certainly must be good news.

I've been studying Italian on my own today. Trying to learn the conjugation of some verbs. It's no use. I can't get my mind on it, and I find it too much trouble. Who the hell wants to be able to talk Italian anyway. Well Honey I must finish this off. Zinn just sat down here and is eating peanuts, and if I have to sit here and listen to him "suck up" into his mouth ten more peanuts, I'm gonna scream — maybe I'm the guy menstruating tonight.

All my love,
Joe.

March 9, 1945
8:30 p.m.

My dearest Honey,

I haven't done anything constructive today at all — just sat around and saw a couple of "goldbricks" and neurotics. Had one pretty sick kid. Went out to take-off this morning and landing tonight. This has comprised "My Day," and I couldn't add anything else to it with the exception of just plain boredom without lying. Tonight it has turned exceptionally cold. This afternoon it hailed and rained. I'm not going to predict spring again until I'm certain it's here.

I'm sitting here trying to write and I can't make the pen go where I want it to. I'm shaking, I'm so cold. Two of our stoves are going (we have three in the casa), but one of them is out of gas, and it's cold here by the table. I'm sitting up here with a fur lined flying jacket on, but I am still chilly and my feet are freezing. The wind is blowing very hard and cold outside — I can never tell about the weather in Italy.

Today was a little better for I've been busy practically all day. Early this morning I had to do a 64 exam, then a kid came in that had gotten his finger caught in a machine and I had to amputate part of it. That's the "secret" to my humor — when I'm busy I can make out O.K. — it's just when there isn't a damn thing to do and I'm bored stiff. Tomorrow shouldn't be so bad for I have a little operation lined up in the morning (gonna take a cyst off a fellows face) and I got a couple of other things to attend to. Monday I should go to Bari but don't know whether I will or not — I dread the ride down there, and I don't like to fly down there 'cause then I don't have any transportation while there.

The news is still good. Tonight they report unrest and rioting in the German population. We will never be able to whip the German Army until the German people are ready to be whipped. Even in despotism the will of the people is probably the strongest force there is only harder to express.

Dearest Babe,

Certainly glad that you have recently heard from Hugh. I'd like to see him almost as much as you would. There's hardly any doubt, in my mind, that Hugh will come home when this mess in Europe is over. Yes, send him movie magazines or anything else he might want that might help to pass a few minutes. It might be nice to save the Sunday comics and send them to him if he enjoyed the comics while at home. Then, of course, any good long book that it takes time to read would probably be welcomed by him. Those guys have had it tough, have made the best of it, and done a damn good job. I'm rather proud of him and am eager to see him again. He and I are gonna have something in common to talk about the next time I see him.

Before I go to bed I want to tell you again that I love you so very much. Tell the kids hello for Daddy.

Joe.

March 11, 1945
10 p.m.

My Honey,

Today hasn't been too bad. I was busy all morning, and this afternoon I had enough to do to keep me somewhat busy. This morning I did a neat little operation on a fellows face when I removed a cyst. I felt right proud of myself. It was the sort of operation that in civilian life I wouldn't have sneezed at and would have been bored by having to do it. However this morning when I finished, I got the same self satisfaction out of it that I would have had in '41 and '42 by removing a brain tumor successfully or a real good abdominal job. Life is funny sometimes — if one is able to look on the humorous side. For instance at the moment, I'm sitting up here drinking that one coke we get per week and feeling like I'm Mr. Rockefeller. Almost envy myself every swallow — and once upon a time, I used to drink 6 and 7 per day — yeah, life is funny. There! I just took the last swallow of it.

Tomorrow I am going down to Bari. I'll probably be down there for two days. I will try to get at least one letter to you while I'm there. I certainly hope the weather will be good. Today it was ideal.

The war news is good tonight. We still have our Rhine crossing, and they announced that the whole western bank of the Rhine is in our hands. Things look better now than they have ever looked before, I guess. After the Allies have had time to consolidate their gains, they can start a drive that will be the final one — perhaps! I know you are wishing for it as much as I. Today in the "Stars and Stripes," I noticed that General Somervell says that many of the troops will come back through the States on the way to the Pacific. I hope that it turns out that Hugh and I are in that bunch and feel sure that Hugh will be.

I'm running down. I think I will hush and go take a bath and get in the sack early. Too I gotta get my blouse and pinks out and see if they are in shape to wear — cause I want to wear them tomorrow. Hope you are feeling better by now. Tell the kiddies hello for Daddy.

All my love,
Joe.

The Ides of March
8:15 p.m.

My dearest Honey,

For three days now I have not written you one line. I have been in Bari. I could not obtain anything to write on. To make up for it in part, I will write you a long one tonight. Went down three days ago to send home this kid of mine who had been wounded. He was unable to fly anymore. I was successful and I suppose he is on his way by now. While I was there, I had a nice visit with some of my friends. Was with Bassett a good portion of the time. He was one of the boys at March Field. I also sat in on a couple of Disposition Board Meetings and just "got around" in general. Sorta enjoyed the first two days down there.

One of these days, you must get a chance to sleep on an Italian bed. They are all always short — I guess about 5 ½ ft. long. One's feet always hang over the end. They all have another common characteristic too — they sag badly in the middle and sleep much like a hammock. It's impossible to sleep on your back, tummy, or side, so you have to get in one of those unusual positions where you're partly on your back and partly on your side. I started to put my mattress on the floor and sleep. However my roommates were three lieutenants that I had never seen before, and I knew they would think I was crazy if I pulled a trick like that, so I just fought it out the first night — the second night I was so tired, due to no sleep the first night, that I could have slept on a pile of gravel. Italian beds and hotels are certainly uncomfortable.

The drive down now is pretty though. The Italians have many orchards, vineyards and farms. All the orchards are in full bloom now, and they are real pretty. Something pretty in Italy is so unusual. The drive back wasn't pretty though. There was a heavy fog the whole way, and I didn't think we would ever make it but we did — my eyes were really tired and stinging though. And last night, I slept like a baby.

Got some additions for the house while I was in Bari. There were some glass lamp shades in the room on the bed table which we were able to confiscate for our light globes. After washing them up, they look really nice and help to concentrate the light somewhat. We are trying to fix the house up so nice that we will hate to leave it. We know that if we get it in that shape, we will move — since the Army never lets anyone be where one wants to be.

Today Zraick was transferred up to Group. Looks as though he might get to be the next Group Intelligence Officer. That leaves me as the oldest officer in the 727th. I'm the only "original" now. The rest were assigned to this squadron after Tucson. Well, Honey, all I can tell you most about now is how much I love you.

All my love,
Joe

March 16, 1945
7:15 p.m.

My dearest Honey,

At this moment my belly feels like yours probably did in December '41 and August '43. I've unbuckled my belt and unbuttoned my trousers but I still am uncomfortable — something's gotta happen soon! All I ate for supper must have turned into — gas.

Today has been nice again. However it rained a little late this afternoon. This morning I got out and pitched a ball a little while without a shirt. The sun doesn't come down hot enough yet for a sun-tan though, but it surely felt good.

Today I haven't had much ambition. Take off was late, so I couldn't get started doing anything until late in the morning. Then I washed out one of my caps that was a total wreck. It was so greasy it was slick and you could hardly see the cloth. I washed it in airplane (100 octane) gasoline, and you should see it now. Honestly, it looks brand new. After it turned out so good, I thought I would wash out another one — which I did — and it looks good as new too. So see — you can afford to spend that unmentionable amount for the little black straw job 'cause now I won't have to have another hat until I hit the States again — that is, if they don't keep me away too much longer.

I'd surely like to see you in the blue denim slacks. I can't imagine you in blue denim anymore than I can imagine myself in blue denim overalls. Sounds very practical however. Do slacks still look as good on you as they used to? You used to be one of the few women that I have ever seen that wore slacks gracefully — I remember how nice those white ones looked on you.

Every day it does look like this European war is going to end. They have been broadcasting that the people are revolting, which, if true and in large enough numbers, is certainly a good omen. The Germans are certainly a well-disciplined people. "Fear" is a remarkable weapon. Churchill today predicts the end of it by the end of the summer if not sooner. Remember he also predicted its end by December 1944. Apparantly the so called "big shots" don't know very much more about that than the average non-political person. However, the German people must revolt before the end of this war — otherwise, it will probably take forever.

My Honey, glad to know that all of you are finally throwing off your colds. Since Carol is walking does Kay continue to want to crawl? Honey, not a very good letter — will do better mañana. Meanwhile take care of yourself.

I love you,
Joe.

March 18, 1945
Sunday, 3:45 p.m.

My Honey,

The weather was too bad for flying so there was nothing to do. Since all of us and most of the fellows next door overslept, we decided to have breakfast here — if I could get some eggs, cream and such from the mess hall. I'm always the goat when it comes to beating the mess hall out of food for I can go over there anytime and get just about anything from the enlisted men and they aren't so free with the stuff with everyone — so I went over and got a dozen eggs and two cans of milk and we had scrambled eggs, toast and coffee finally. We cooked the eggs with onions — did you ever eat any that way? It's good — I learned how to fix eggs that way in Africa — one of these Sunday mornings when I'm home, I'll fix us some.

Well, to go on with "My Day" — as soon as we finished breakfast I began getting bored and the longer the day has passed the more bored I have gotten. So I just sat and moped around this morning and let Zinn get on my nerves (He has a manic-depressive type of personality and this morning he was in his manic cycle and I wasn't in the mood for it). So finally out of desperation, I went up to Group to see my patients — piddled around up there for a few minutes and returned here for lunch. Of course I wasn't hungry, having eaten 3 eggs for breakfast, but I had a cup of tea (I still don't care for it) and a bowl of thin soup. Then I came back to the casa to listen to the news and resume my boredom. (Sounds like I'm enjoying it, doesn't it?) Then I heard that we were going to show some short subjects down at the enlisted men's club at 2:30, so I went down there so I could more easily pass the time. Don't know what has gone wrong with me but I never go to a show anymore — possibly because they are all so decidedly poor. Well after the show, I came back to resume being bored but decided if I talked to you for a few minutes, I would pass away some time profitably. I'm also having a drink at the same time. I feel better already, but it's because I'm talking to you — this is my first drink, and it is a case of "too little, too late" so it can't possibly be the reason.

Another thing which has accentuated the "Sundayness" of today has no doubt been the weather. It is quite cold and has been drizzling rain most of the day. Drizzling is unusual for Italy — like Morton's Salt — "when it rains it pours" here.

Well, my Honey, it is now time for chow. I could thank you for helping me pass a most unbearably boresome afternoon, but you might not appreciate it put in that manner — and, I don't mean it in that manner either, 'cause you know I love you.

Joe.

March 20, 1945
7:45 p.m.

My dearest Honey,

I have been trying to get a letter off to you for the past couple of hours, but there is a drunk sitting on the other side of the table who is doing everything in his power to distract me. I dare say he is doing a very efficient job. I've listened to him rant and rave and fume for so long that I'm shaking all over and so damn nervous I could knock hell outta him and get a lot of sheer joy out of it. He is one of those nasty drunks, which always comes out of a nasty person when his inhibitions are released. It has gotten to a point that it is very difficult for me to exhibit patience anymore — when I do, it isn't patience at all — just sheer will-power. Sometimes I wish I could get out of the environment of egotistical brats, and I believe some people the age of 19 to 22 are worse than that age in kids from 8 or 9 to 15. Sometimes they can really irk you.

Tonight quite a few of them have finished their missions and are all getting drunk. In a way, I can't blame them much. After you have been shot at and shot at and still alive and finished, I suppose you have got a good drunken spree coming, but after you have watched this process for months and months, one can't feel the same hilarity about it as the fellow that has just finished. To me, I have seen the same thing many times — to the guy who has just finished, it's a one and only for him. Again, I guess I am just old and am getting older by the day, and due to an upset in my normal life with you, am not adapting myself as well as I should be able to — still, it frequently gets on my nerves.

I'll bet you one thing though — as soon as I'm back in the States for a couple of months living my (or our) normal life, I will be as stable as I have ever been — my patience will lengthen to the point that I can get up at 2 a.m. out of a warm bed and tell anybody that "the best thing to wash a bathtub out with is soap and water" or "that I have no idea when the boy scouts will be in" and still use a partially civil tongue.

I have rambled tonight, but I have enjoyed rambling to you — I always enjoy expressing my thoughts to you and no one else — wonder why? Maybe because I know you will understand and I love you.

Joe.

March 21, 1945
10 p.m.

Dearest Honey,

Here I am alone just thinking of you. Everyone else has long since been in bed — I just got in. Earlier tonight I had to go up to the Group to see a pneumonia patient of mine who is pretty sick. While I was up there, we started playing cribbage and ping-pong, so we played until right now. Wagner and I were partners in both games. I had a very enjoyable evening, which passed much faster than it would have otherwise. Talking about pneumonia, I'm having a regular epidemic — I've had five cases in the past two weeks. It isn't the old Labor pneumonia, which we used to have in the States but a relatively new kind that's caused by a virus rather than the pneumonia bacillus. As a rule it is much less serious than the old variety, but it doesn't respond to any of the sulfa compounds or to penicillin either. The old kind of pneumonia responds readily to both.

Today I finally went into town and got me a haircut and some cleaning. It was the first time I had been into town in a pretty good while. In fact — so long that I rather enjoyed going in. I certainly needed the haircut too — and the cleaning.

Last night I acquired some tomato, radish and lettuce seeds. I think tomorrow I will fill up some boxes and plant the tomatoes — raise them up to a little size and then transplant them outside. I'm gonna plant the radishes and lettuce as soon as I get the energy to go out and do the digging — course I don't care whether they bear fruit or not — I just want to see them come up for the hell of it. My wheat is coming up now. Don't know what kind of a "stand" I'm gonna have 'cause the sparrows eat it by the time that it peeks through the ground. I'll have to wait and see.

Haven't heard the news as yet today but intend to listen to the 11:45 BBC news. Hear that Patton is running wild in the Saar Basin and surely hope he is able to close his trap with Patch's 7th Army without letting those 80,000 Germans escape across the Rhine. The "Stars and Stripes" has been hinting at peace proposal rumors for the past couple of days, but I'm not gonna put too much store in that as yet. It does look as though something has just gotta happen soon though. When the eastern and western Armies start a drive simultaneously (which they claim they are going to do in "Stars and Stripes"), perhaps something will happen.

Well Honey, I must go to bed. Tell the kiddies hello for Daddy.

I love you,
Joe.

Dearest Babe,

March 22, 1945
Thursday, 11 p.m.

My Honey,

Today I received a nice long letter from you and also one of the nicest boxes you ever saw. Thanks for both. The anchovy paste on the Melba toast is delicious, and believe it or not, the Melba toast wasn't even broken up when it got here. The little Melba toast sandwiches are delicious too. I haven't opened the sardines, but they are always good. Hey, don't use up your ration points sending me stuff, Honey, 'cause I don't need stuff that bad. You are probably having more trouble getting something to eat than I am. The reading material always comes in handy too. When you wrote the "doctor's wife" into the title, I don't know whether you were aware of the complete title or not. Now it reads, "Just what the doctor's wife ordered — Fun in Bed." You wouldn't be slyly propositioning me, would you? If so, I accept — now what the hell are we going to do about it.

I also received letters from Kremers and Jimmy Hunt today. I enjoyed them both — especially Kremers'. He wrote me a letter trying to put down on paper his reaction of being with his wife and kids again. Kremers thinks a lot of his wife too, and he knew I would appreciate hearing from him in that manner. At the end he told me that "he hoped he hadn't made me homesick." Kremers is a good guy, and he and I were nearer one another than any of the rest of the "docs." I enjoyed knowing him even though we didn't spend too much time together. He is now at Palm Springs — presumably in a hospital. He didn't state whether or not they had found out anything further on him. I doubt if they did though for I dare say when he got to be with his family again, he was well on the road to recovery — and I don't mean that "nasty" either. I know that whatever might be wrong with me, seeing and being with you would be the best possible treatment. Jimmy Hunt didn't have much to say except that he was still griping.

Today has passed much faster than usual. This morning I was kept busy with all my sick boys up at Group. This afternoon I kept myself busy planting my seeds. I planted two rows of radishes out by the side of the house — the rows being about 12 feet long. I also planted three large boxes of tomato seeds. Put them inside the house by the window. When they come up and get some size, I will transplant them outside. I guess tomorrow I will plant the lettuce somewhere just for the hell of it.

Honey, it's pretty late, so I'd better hit that sack over there — again all by myself. It's a hellova note, isn't it? Wonder if I will have to get used to sleeping with you again like the first time. I don't know, but I'll bet not. If so, I'm ready and willing to start again — and what a joy it will be.

I love you,
Joe.

March 24, 1945
7:30 p.m.

My Honey,

I received two Air-Mails and a V-Mail from you today. Mail has been coming through pretty good now for quite a while — certainly hope it continues.

Can't help but mention the news of the day — Montgomery's and Patton's crossings of the Rhine. Remember my opinion of these two Generals. Of course I haven't the slightest idea of what it's all about, but the crossing of the Rhine in Force is music to my ears. This war has dragged and dragged until I'm damn near "dragged out." I wish for everybody concerned, and my own selfish reasons no doubt, a speedy victory. There must be an end to it soon. I am eager to see the developments.

You know, I just thought about it today when I received your letter telling me that you had gotten me some bedroom slippers. I'll bet you had to use your shoe ration coupon for them and, if so, I feel very badly about it. I could get along very easily without them. If you did have to use your coupon, I really am sorry. I guess it's too late to do anything about it now except express my feelings.

Have worked pretty hard today. Spent the morning lecturing to some brand new crews — trying to get them oriented, so they can profit from our experience instead of having to learn by it as we did. However, it's rather hard to put it across sometimes to a bunch of "know-it-all" kids in a way that will stick. Sometimes I think that they think a fellow lecturing to them is standing up there so he can relate his experiences rather than try to help them out. It doesn't take long for them to change their minds.

This afternoon I went up to look after all of my sick boys — and I have quite a few of them now. Not very interesting though for they all have the same disease — every one of them. A couple of them are pretty sick boys though. Also had to take off Mason's (my right-arm sergeant's) thumbnail this afternoon after he smashed his finger between a ton trailer and the ambulance when we were hauling some gravel.

Got gobs of other work to do now — those damn fool records — enough to keep me busy for a while anyway. So I suppose I will have to give up my landscaping and gardening. By the way the wheat is coming up good in the bed on one side of my walk, and I can't even find a sprig on the other side — same kind of seed, same type ground and planted similarly at the same time. I don't understand it, but so far it's true. Oh well, in Italy anything can happen and usually does.

Tomorrow I have to go in town to "school" for the whole day. Also have to carry one officer from the squadron — it's about malaria control, so I suppose (in fact I know) that it's about time that I'm going to have to start worrying about malaria again. Someone backed into my ambulance today and crushed the radiator while it was parked up at Group. Now I gotta get out and beat up a jeep from somewhere to use tomorrow.

Glad to hear of Carol's accomplishments with her pedal extremities — sorry to hear of yours and Kay's colds — and, Honey, if I were home I wouldn't want you to be hoarse, so you couldn't talk. I'd devour every word you said.

All my love,
Joe.

March 25, 1945
10:30 p.m.

My dearest Honey,

We got in a few fountain pens in our rations this week, and we raffled them off. I won one and this is it — a nice Lifetime Shaffer but for me, it writes too damn fine. I'll try to trade it off for a pen that writes stubbier. This is by far the best pen, but it doesn't hold enough ink.

Just heard Jeanette MacDonald sing our song over the radio. I wondered if you heard it and know if you did, that you were thinking of me. You know "One kiss, one love, etc." It was very good, and I sat back and just listened.

Today is another Sunday and, thank goodness, it's another one that is almost gone. I went into town to hear a lecture on malaria today — was supposed to stay there all day but got too bored, so I didn't go back after lunch. Went down to Wing Headquarters for lunch and shot the breeze with Colonel Nelson, the wing surgeon. He's a pretty nice guy. Finally came back here about three o'clock and have done exactly nothing except go out when the ships landed.

Today has been a lovely day. This morning it was nice and warm, and the sun was bearing down hard enough to get a sunburn. Had I been able to, I would have gotten out in it. I haven't forgotten that it is necessary for me to get a good dark suntan so that Carol will take up with me, if perchance something unbelievable happens and I get to come home this year. In the afternoon it cooled off considerably, and tonight is quite chilly even though the wind isn't blowing and the stars are out brightly. I have two of our stoves going, and I'm still a little chilly. Goodnight, my Honey, cause I'm going to bed.

All my love,
Joe.

March 27, 1945
9 p.m.

My dearest Honey,

Just returned from a U.S.O. show up at Group. It was a pretty fair show, and I enjoyed it very much. There were three Chinese Girls on it who called themselves the "Loo Sisters," who were very good. They could be in the same class as the Andrews' or the King Sisters or any good girl trio. A couple of other acts were good too.

Today I changed my routine a little bit and sent a bunch of my pneumonias into the hospital. I sent 7 of my sickest in, and henceforth, I think I will send them all in. I still have about 7 up at the little Group Dispensary, but they should be up soon. It takes a lot of bother off me, and they are sometimes pretty sick. You certainly miss trained nurses when you are taking care of real sick people. I don't care how well you train your boys, a nurse, as a rule, can just do it better and more willingly. Of course that is easily understood — that is their work and their "elected" job. These boys have been pushed into it — even so, some of the boys are pretty good.

It hasn't been as pretty today as it has been for the past couple of weeks or so. It only rained a tiny bit this morning, but it has been cloudy all day long. We didn't fly today either. We needed the rest however for we have been flying rather steadily.

The news is just coming over now and, lady, it is good. Patton is running wild apparently in the true Patton fashion. Hodges' First Army is not doing so bad either. Just announced unofficially that Patton's tanks may be nearing Nuremberg. Hope it's so — cause that's a good 100 miles east of the Rhine. With the dwindling size of Germany, a hundred miles is a long way. Yes, the news sounds better than it has in a good while. Looks like the war is going to move fast for a while now. That's good for it gives us something to look forward to, and one feels like we are getting somewhere. The next couple of weeks should bring intense news — wonder when the Russians and Americans will meet. Enough of the war, but the news was so good tonight that I couldn't help mentioning it.

In the "Stars and Stripes" lately, there have been lots of articles about how the newspapers and magazines back home are saying how to "handle" all the soldiers when we get home. Makes us sound like a bunch of psychopaths. Let everyone know when I get home, I'm gonna be me and won't need to be handled. They can ask me anything without feeling as though I'm being offended, and if there was something that I didn't want to talk about (I can think of nothing at the present) I'm sure I could evade talking about it. I believe the papers are playing up that angle too much.

Hey, don't work too hard in the yard. I wish I were there to help you — I would be glad to be your yard boy. Tell the kids hello.

I love you,
Joe.

March 28, 1945

My dearest,

Guess what I got today — you missed, for I got a quart of Scotch. How'm I doin.' It's good Scotch too. I know, for I had a couple of drinks out of it this afternoon. We didn't fly today — two days in succession — unusual for us lately. The weather has been closed in, but it hasn't been raining.

Hey, I needed you last night badly. I had two nightmares — and I dreamed the same cockeyed thing twice. Both times I woke myself up groaning some weird sort of sound. It was awful. If you had been around, you could have hollered "Hey, Joe, roll over and go to sleep" twice and gotten even. I certainly don't want to have such dreams as that again. I think I was sleeping too hot and that afternoon I had a nap, so I just wasn't sleeping very soundly. If I thought I was going to have a similar experience tonight, damned if I'd go to bed.

This afternoon I went into town and saw a very good show that I enjoyed a lot. Good shows are such a rarity. It was Irene Dunne and Charles Boyer in "Together Again." I think you wrote me a few months ago that you had gone to see it. I like Irene Dunne, as a rule I don't care too much for Charles Boyer, but I did like him in this picture. Another good thing about the picture was the complete lack of an Army atmosphere. I don't even remember seeing a single uniform — not even a doorman. It was also light, well-casted and good acting.

Just got called over to the dispensary to look at some kid that someone had really hit in the mouth. Had his lips cut in a few places. He was "well-oiled" of course. It's funny, but it is rare that the boys ever get to scrapping with one another. I don't think I have seen more than three or four boys due to fights since we have been overseas. Looks like we would have more with a whole bunch of boys living together so closely — particularly when all of them are more or less dissatisfied and irritable.

Over the radio tonight, they announced that there's a blackout on the news concerning the whereabouts of the thrusts of all the Armies on the Western Front. That usually denotes good news is coming. The morale of everyone is up a hundred percent I think — including myself. In fact, I think we are doing too much wishful thinking again and letting our expectancy override our common sense. But it's very difficult to keep thinking level about the situation when everyone else is being so optimistic. I'd rather consider the thing a little more conservatively and be happily surprised than to be too optimistic and wind up disappointed. But, no doubt, the situation does look good for the Allies at present. Well, Honey, I'm gonna listen to the news and then go to bed. Wouldn't it be swell if this thing did finally get over with soon and I got to be with you for a while sometime this year?

I love you,
Joe.

March 30, 1945
Friday, 6:30 p.m.

My dearest Honey,

Another day is in the process of passing away — can't say that I am sorry. It hasn't been as bad as some of them have though. Some things have happened to make it pass faster.

To begin with, about 9 a.m. this morning, Zinn was working on one of our stoves and set the house on fire. It's a good thing it is made of stone and concrete primarily for if there had been much wood in it, tonight I would have been sleeping with the sky as my roof. Zinn had gasoline all over the place, and somehow it caught fire — black smoke was pouring out of all windows and doors, but we finally extinguished it. Zinn burned one of his hands pretty badly in the process — but we got it out. However soot was over everything, and the walls were smoked black. Doesn't matter much, the house is still quite liveable. This afternoon, right after lunch, I took off all my clothes and got out on a cot in the sunshine. Don't think it was hot enough to tan, but it felt good nevertheless. I stayed out for about 1½ hours.

You see in the "Stars and Stripes" today about half of the paper was devoted to the "end of the war," "the falling to pieces of the German Armies," "how London would celebrate VE Day," "Secretary Stimpson's declaration that the enemy was whipped," "the restlessness, revolt, and panic in Germany." The war must be nearly over — however, I have seen evidence to the contrary. I will admit though that the "Stars and Stripes," being written by G.I.'s, is not such an optimistic newspaper as a rule. I certainly hope that the people aren't going off half-cocked. Just a few minutes ago, it was announced that Montgomery's Armies were threatening Hannover. Hannover is midway between the Rhine and Berlin and is a hellova advance. Of course all armies are and have been smothered from news. So, when we find out really where they are, maybe we will be pleasantly surprised. Are you getting tired of hearing two week old news and my opinions concerning the war? I realize you hear the news six or seven hours earlier than I do by the clock, but we spend so much time studying the maps, looking and speculating, and arm-chair "generaling" that it is difficult for me to write about anything else. Perhaps it will end soon, and I can write about something else.

I didn't get any mail today. Apparently it's falling off for everyone. I imagine all the air and boat space is needed for last minute supplies on the Western Front, so I can't gripe. Your new light grey suit sounds pretty. You bought it rather late in the season though, didn't you? Why? — 'cause it was cheaper, no doubt. Think, if you had bought it earlier, you could have been wearing it all this time — glad you got it though. The gold job must be a honey — you have mentioned it so frequently.

Well, Honey, it's time for me to go to the show if I'm going. Don't know whether to go to the show or sit here and listen to the 9 o'clock news. It's a cinch I can't do both. Goodnight, Honey. Tell Kay and Carol goodnight for me.

All my love,
Joe.

April 1, 1945
Easter and April Fool Day
Sunday, 9 p.m.

My dearest Honey,

Well today I put on my "brand-new" old, worn-out, run-over, scuffed shoes; my "brand-new" mismatched, dirty, socks; my "brand-new" shiny, dirty, small, green pants that I bought at Phil A. Halles a few years ago; and a dirty O.D. shirt that is too small and went on my Easter Day Parade. My "parade" consisted of a couple of trips to the runway and the same old routine stuff. No doubt my Honey and my two little Honey's "strutted" enough for the whole King Family, and I know you all looked just as sweet as you are (almost) in the spring duds. Wish I could have seen you.

We did celebrate somewhat though. We had an excellent supper tonight of biscuits, chicken, potatoes, and peas. It was very good, and we all enjoyed it. The chicken was too tough to eat, but the meal was nevertheless good.

I'm listening to the round-up of the news now. The going on the Western Front has been slowed down considerably — possibly due to the lack of their own supplies. Too the news black-out is still on. Wonder how Eisenhower expects me to keep up my morale with the news black out.

Honey, today I sent you $175. Guess you thought that I had forgotten that it took money to live, but I hadn't. Henceforth I will try to do better than I've done by you in the past couple of months. Haven't heard from you in three days now. However, I got a very nice letter from your Mother today. She always gives a good deal of her space to you and the kids. Consequently, her letters are always especially enjoyed and appreciated.

Tonight we get gypped out of an hour's sleep. At 2 a.m. we are supposed to set our watches up an hour and go on daylight, daylight saving time. No, I'm not stuttering. We operate on daylight saving time all the time, but in the summer we do 1 hour better. Gotta get in bed early tonight, so I will be able to get up tomorrow.

Honey, I'm gonna do you bad tonight and "chicken-out" on you somewhat by cutting this short. I have a hellova headache and have had it all day. Aspirin is just dulling it and that's about all. Tomorrow I will do better.

All my love,
Joe.

April 2, 1945
4 p.m.

My Honey,

Here it is 4 o'clock, and it seems as though the sun is only just overhead. When you first go on this new time it certainly makes lunch and supper come at a peculiar time. This morning when it seemed like it was mid-morning it was lunch time. In 30 minutes it's dinner time (supper) and it seems mid afternoon.

I have been quite busy today. Had a little more than usual to do this morning due to the fact that it's around the first of the month. This afternoon Major General Twining, himself, came up and presented the Group with its 3rd Presidential Citation. Of course we had to hold a review for the occasion. The boys can march pretty good too even though they don't do it except on such an occasion. Three presidential citations are pretty good for an outfit. As far as I know, no other outfit of any kind has more than that. I only know of one other that has that many, and that is an outfit that has been over here for about 3 years. They were in England, North Africa, and Italy. The only thing sad about it is that on the occasions that each of those citations represent, our losses were pretty high. So, the guys who really won them didn't get them.

Honey you mentioned "sharing" me when I returned. Don't worry, if I can make it possible, you will meet me somewhere other than in Memphis whenever that time comes. I don't like to make a spectacle of myself either. And I, too, want your company alone for a few days. The folks probably don't realize it, but if you all met me somewhere and I just rushed up and kissed you about 20 times and didn't even see anyone else and started walking off with you, everybody would feel mighty bad about it in the end — and such as that might happen when I see you again. As you stated though, this is certainly crossing many many bridges before we get to them — but it is certainly pleasant to even think about being with you again. It gives me a feeling of well being which I haven't experienced in a long time but is vaguely familiar — it's an exciting feeling too. That is what it does to me, Honey, when I even think of you — you can imagine what it will mean to me to be with you. Never, prior to you, did I think that the feeling of oneness could be so alive, be felt so keenly, be so vital as it is with you. Must be love.

All my love,
Joe.

April 4, 1945
3 p.m.

My dearest,

This morning I received two great big fat Air Mails from you. They were certainly welcome too. In one, you were telling me of your trip to town with

405

Miss Kay King. Sounds as if she is quite a young lady — she wanting gloves, pocketbooks, shoes (like her Mama), etc. I was glad to know that you also went to Gerbers and had a picture taken of Kay. While you were there, why didn't you pose too? You are the most difficult gal to get a picture out of that I have ever seen. Do you realize that I haven't had but one batch of pictures of my three women since last Easter? Even so, I have a perfect picture of all of you in my mind.

Honey, I received the box of jellies that you sent me. They just came in today and they look delicious. Thank you very much. I am sorry to hear that Royce is to again report for his induction physical exam. I'm afraid that this time he will be inducted. I had hoped so much for Sis' sake that he wouldn't be called. It will probably be the Infantry or Navy for him. I wonder how in the devil Sis will be able to make out on that allowance. Many other people are, so I suppose she can. She is usually pretty good at handling the situation. She is a good gal though (my favorite Sister), and I hate to see her have to go through with it. Yes, I know that you are going through the same thing, but there are small differences that means a lot. One, you know I'm reasonably safe. We make a decent living for us and the kids even though I'm in the Army.

If I were with you, I could probably make myself much clearer on the above subject — you know already about my philosophizing — I usually get to talking in circles. And now again I draw near the end of another letter to my Honey. Wonder how many more I will write before I can tell you face to face that I love you.

Joe.

April 6, 1945
9:45 p.m.

My dearest Honey,

Another day is almost gone. One year ago today I moved up to this place — remember I rode my motorcycle up. Little did I think that I would still be on this same plot of ground looking at the same hills for over a year, but I am. One year ago yesterday was a memorable day. I'll never forget April 5th even if I live a long time — I'll tell you all about it at some later date if you care to hear it — it was the day of the first Ploesti raid. Certainly hope I'm not in Italy next April and don't think I will be, thank goodness.

I just got back from doing some malaria smears on my mess sergeant. Sure enough he has got malaria, so I sent him on into the hospital. He is a key man, and I'm gonna miss him while he is out more than I would most any officer. I'm afraid we are gonna have quite a few more too. Hope it won't become as rampant as this damn virus pneumonia has been. Between the latter and the itch, I have had a headache.

I will enclose another picture I have here of me. As you can see Reese was trying to put me in the trash barrel. It looks as if it were posed, but it wasn't — the guy with the camera just hollered, we looked around and he snapped. The building that you see on the right is my dispensary — that is from between the middle two windows to the end

that you can see. The shack on the left is housing a generator and engine that we have built which furnishes us with lights. I think you can also note some good ole' Italian winter mud too. That's just a mild sample of what it really is in the winter time.

Well the pace of the advances of the Western Front has slowed down considerably. From what is given out over the radio and what one can read, it isn't due to stiffening resistance though. I certainly hope that is correct, and I hope that they can advance fast enough to where the enemy won't be able to form a stable line of defense. No doubt taking care of so many liberated people is quite a job and a drain on already diminishing supplies. Maybe it won't be too long though.

Well, goodnight again, my dear. I love you with all my heart.

Joe.

April 7, 1945

My dearest Honey,

Here I am sitting in this house by myself. This is unusual, but occasionally I enjoy it. Reese left for Rome this morning, and Zinn has gone into town. Even the little Italian boy, Martino, has stepped out for somewhere. It is only 2 o'clock but my routine daily work is completed, and there is nothing in particular that I've got to do except meet the planes this afternoon when they return. More boredom, no doubt. It's a pretty day, a little cool and a fair breeze — cool enough to wear a jacket even with the woolen clothes.

Hey, do you remember that I told you that my tomatoes weren't coming up? Well, I lied. I just had not given them sufficient time. In the past couple of days, I have had about 30 little tomato plants to sprout up out of my boxes. Gonna have more too. They sprout out so fast that it is both interesting and amazing. I go over every couple of hours and take a peek at them, and each time I'm surprised to see new ones. I'll have some tomato settings yet.

Guess what I dreamed just before I awoke this morning? I dreamed the war was over. Boy the people were making so much noise shouting, shooting

pistols, yelling that one couldn't hear himself think. Apparently that will never happen in this war though 'cause it looks as though there won't be any formal surrender. Looks as though we will just have to fight until there is no one else to fight and then, I suppose, the war with Germany will be over — for a time.

I'm enclosing another picture. I'm not very modest, am I? I think I have had a picture in almost every letter I've sent you in the past week. I just want you to see that I'm still altogether and how beautiful I am.

Honey, do you remember Manoogian? In case you don't, he's the ex-football player that used to bring his wife to the dances at Fairmont (Doris), and would stand up leaning on the wall reading the sports page of a newspaper while his wife danced with all the boys. Now do you remember? He and Doris were exact opposites of Arnold and his wife. Anyway, Manoogian is leaving for the States in a couple of days. He is going to some school somewhere and is supposed to return here in 2 to 4 months. He told me he was going to call you up if he passed through Memphis.

There's another guy who will call you up too. It's a nice young guy named Ferguson who lives in Jackson, Tennessee. He just finished and is on his way home now. If you see this little devil, you will see the kind of young squirts I have to live and put up with. He has an excellent personality, nice smile, cocky, egotistical, efficient, likable, young but is full of so much devil that one wonders where it all comes from. He's a good kid though and has done his job well, and I'm sure he will at least call you and probably will come by to see you. He's also a pretty sharp gin-rummy player.

Hey, you girls had better cut down on your chewing gum and make that last batch last you a long time for I can't get it like I used too. They have cut down on our ration. We only get 5 packages of cigarettes per week now too, but I find 5 packages plenty. Well, goodnight again, My Love. One of these days I'm gonna tell you that instead of write it. That will be the day of my lifetime.

I love you,

Joe.

April 8, 1945
Sunday 9 p.m.

My Honey,

The officers and the enlisted men just finished a quiz program, opposing one another, down at the E.M. club. It wasn't quite fair for the Enlisted Men knew of it for the past week and have been studying the "Stars and Stripes." However they only picked the officers 5 minutes before it started. An enlisted man and myself were the final contestants, but he beat me in the home stretch. I was sorta glad cause the prize was a 2-day pass and, of course, I can take a pass anytime and they can't. So it's just as well, but that didn't keep me from trying. I finally missed out on "who recently replaced James Byrnes as War Manpower Commissioner — the answer was Vinson, of course, but I had missed it in the news. It all was fun, and I think the boys enjoyed it — anything to pass away a little spare time.

Wish you were around right now. I would try to persuade you to do some buttons for me. All of my shorts are minus one, two, or three. I wear 'em as long as they have 1 left, but when that last button is gone, I throw them back. Now, either I've gotta get busy with some buttons or go "drawerless." The latter would never do, so I suppose I will get busy with the buttons. A couple of nights ago, I let out the neck of a shirt by sitting the button hole over some and found out that making a good, tidy button hole is no snap. You probably wonder why I moved the hole instead of the button — well, it was because it would have ruined the shape of the collar — I know, for I tried that too. I also gotta repair a buttonhole on this one that I have on for one button won't stay buttoned. All of this is a very minor reason why I would like to be around you to be sure — but I have some major reasons too many to enumerate. They can easily be summed up in — I need you. One thing you must always know is that.

I love you,
Joe.

April 11, 1945
11 p.m.

My dearest Honey,

I have been writing, or trying to write this letter for the past five hours, and every time I get ready to write, someone drops in and brings a sandwich or a drink, or just drops in for a chat. This house sometimes is like the Broadway of America. Just then Major Lather came in and brought an enlisted man home-town friend of his to spend the night with us. I just got him to bed, so now I'll start over again.

Things look pretty good again on the Western Front. The radio just announced that the 9th Army reached Magdeburg today. Magdeburg is only 70 or 80 miles west of Berlin. That makes only 100 miles between that thrust and the Russian Front. That's pretty good news! Perhaps it won't be too long, my Honey. Of course when it's over in these parts, it certainly doesn't mean that we will soon be together, but it is a step in the right direction at least.

Guess what — my lettuce is already coming up. And I gotta get busy and transplant my tomatoes 'cause they are getting pretty big now. I've enjoyed puttering around with these plants. It has given me something to do. I hope I never see them bear fruit, but I probably will.

It was tough that Easter was so wet there when we had such a nice one here. It would have been nice if the twerps could have gotten out a little bit in their new outfits. The weather here lately has been beautiful. Not hot but certainly not cold — must have a fire every night but very comfortable in the daytime with all the windows open. Ideal weather even though it is in Italy. It would be nice weather to get out and play a few rounds of golf or to go on auto trips — I would like to see myself swing at a golf ball — I'll bet I would be as graceful and as rhythmic as a 200 pound woman pregnant with twins — probably would miss the ball completely.

Honey, I'm glad you are again able to take your music lessons. I'll bet if you are able to keep at it long enough, that sooner or later you are going to be able to play. There's one consolation — it probably won't be very long before Kay will start. Then surely you and your daughter can learn together. I will stay home and take care of Carol and get better acquainted with her and keep the

lightning from bothering Rachel, while you and Kay take your music lessons. (Some soprano just sang "One Kiss" over the radio!)

Here is another picture of Zinn, Reese, and me. Again I look like I think I'm the cock of the walk with that pose. If I ever get another snapshot, I'll see to it that I don't assume that characteristic pose. Well, Honey, again I bid you goodnight by pen. Don't worry, it won't always be like this. Now do me a favor and put a big smile on your face, and I'll bet that you will feel better — if you do, let me know and someday I will try the same treatment.

All my love,
Joe

April 12, 1945

My dearest Honey,

Haven't done much today — just "piddled" a little. This morning I transplanted 20 of my tomato plants. Got a whole lot more that I could transplant, but there's no rush — I guess I have plenty of time. Then about 1:30 this afternoon I took off all but my hat, shorts, shoes and socks and went out in the front yard and just sat for about an hour in the sun. The sun wasn't too bright, but I got a little sunburn nevertheless. I can't tell it by looking for I had quite a tan all through the winter, but I can tell that it's stinging a little bit tonight. Feels good.

Zinn is leaving for the Riviera the first thing in the morning. I had a chance to go but gave my place to Zinn. I will try to go later when someone I know is going up there and when it's warmer. Before the summer is over I would like to go over to Capri for a couple of days too. Reese is still in Rome. He was supposed to be back two days ago, but so far I haven't seen him.

Well, again it has been quite a while since I started this letter — about 4 or 5 hours, in fact. Since then the group dentist has dropped in with a bottle of cognac (Serti's — which is the best) to celebrate his promotion to captain. Very obligingly, I have helped him drink it. He's a good kid. He is an Italian and has only been with us a couple of months. Peterson, the dentist that came overseas with us and was with us at Wendover and Fairmont, has been transferred to the Air Force.

When I tell you this, you will probably faint. Don't — because it isn't that important — but in the past week I have really cut down on my smoking. For the past week, I've only been smoking 4 or 5 cigarettes per day. I had gotten to the point where I was smoking much too much. Believe it or not, after smoking for 17 years, to cut down to 4 or 5 cigarettes per day is no job at all. It's so easy to do that I'm considering trying to cut it out all together. I never dreamed that it would be so easy to cut down.

Next morning —

Honey, what do you think of Pres. Roosevelt's death? It's quite shocking, and the U.S. is going to miss him. I'm afraid we have no one who can even partially fill his vacancy — especially in post-war world relations. He will be remembered as our greatest president 25 years from now — overshadowing even Honest Abe. He died at an inopportune time for the U.S., but a very opportune time to die a hero's death.

The 1st, 3rd, and 9th Armies apparently roll on. Just announced over the radio that they are 38 miles from Dresden and 50 miles from Berlin. With the blackout on the front, they must be even closer. Our part in this war is nearing its end, I'm sure. The more I think of it, the more I'm convinced that we will head directly for the Pacific too — we will just have to wait and see — pray and hope. Well, Honey, I want this letter to get off in the morning mail, so I better hurry and get it over to the mailroom.

I love you so much,
Joe

President Roosevelt Dies

New York   *London Edition*   **Paris**   **Hemorrhage**

THE STARS AND STRIPES

Daily Newspaper of U.S. Armed Forces   in the European Theater of Operations   **Fatal in Georgia**
Vol. 5 No. 137—1d.   FRIDAY, APRIL 13, 1945

Harry Truman 32nd President

April 14, 1945
9:30 p.m.

My dearest Honey,

Honey, they have been counting up the number of "points" we have for the demobilization plan. I don't know how I stand but, don't worry, I'm quite sure I'm not going to be demobilized. They have also asked us again whether we wanted to stay in the Regular Army, the Reserves or get completely out. Again I told them that I'd take tweeds or stripes, wing-tipped shoes with perforated toes, and a slouch felt hat in preference of O.D., forever and ever, if I had my choice — so that's that whether I'm right or wrong.

In the morning we are supposed to hold some kind of memorial service at Group in honor of President Roosevelt. They are doing it everywhere. Don't know whether I will go or not — it all depends upon when they have it in relation to when the planes take-off and land. The war news looks especially good. Perhaps by the time you receive this letter the Russian and American Forces might have already met, and organized German resistance ceased. I wonder when we (my outfit) will cease operations. Then we probably will start sweating out the Pacific. If they send me to the Pacific without a peek of the States first, they are going to have an awfully sorry flight surgeon on their hands — guess I can take it if the next guy can though.

Well, Honey, I'll say adieu until tomorrow. Perhaps tomorrow I can write an interesting letter. Take care of my best love for me and tell the two young loves that Daddy said hello and for them to be good girls.

I love you,
Joe.

April 15, 1945
9:30 p.m.

My Honey —

Just a shortie tonight and I'm doing pretty good to get that out. At present I feel terrible — hurt all over and have a headache — probably about to catch a cold. I'm gonna take a nice hot bath in a few minutes and hit the sack. Think the bath will help.

Got another V-mail from you today (2 yesterday), also I received a very nice package of both salted and fresh nuts. Haven't opened the fresh ones yet, but if they are as tasty as the salted ones, then they are delicious. Thank you so much. You just never forget to do some little thing now and then for me to let me know that you are trying to help me along. In the next package, I could use some more hair-oil (not the expensive kind — just Vaseline or Rose Hair oil). I have done a little more work this afternoon than usual. Perhaps that's what's wrong with me. I'll tell you about it tomorrow in a letter.

All my love,
Joe.

April 21, 1945
2:30 p.m.

Hello My Honey,

Yesterday about 6 p.m. Zinn and I got in his jeep and drove over to a little isolated town over here in the mountains where there is a photographer that he knew about. We arrived there about 7 and "sat" for some portraits. We both had three poses apiece and then the photographer insisted on one together, which we did to please him. They do not have "proofs" over here, so Friday we go back to get the finished products I guess. How many of each pose he is going to make, I don't know. Nor do I know what they will cost — probably very little for I'll take them some soap and toothpaste when we go for them.

We didn't leave the place until about 9:30, and I enjoyed being there very much because it was just different. They had two kids which I enjoyed playing with a whole lot — both little girls — one 2 years old and one only 4 months. I think I enjoyed the little one the most 'cause neither she nor I could talk Italian, and she seemed to enjoy American "goos" and nonsense as much as Italian — I couldn't fool the older one so easily, but she and I got along O.K. anyhow. Playing with the kids sorta emphasized my "homesickness," but nevertheless I enjoyed it. They opened up a couple of bottles of pre-war, cheap, Italian champagne, which we drank and fed us a couple of fried eggs apiece and some fried potatoes. One hates to eat any of their food. There is so little of it, but if you don't their feelings are severely hurt. These were simple folk but undoubtedly good — ignorant and misled perhaps but good people at the core.

I think I paid for my "outing" yesterday though. Today I don't feel so hot — yesterday I felt grand, so I guess I overdid it. I'm either a damn sissy or a damn neurotic, I guess you will find out which soon enough after I am home — personally I wouldn't be surprised if I wasn't both.

Tell Kay and Carol that I think their new Easter dresses and their bonnets and especially those new white Easter shoes are mighty pretty. I love you all and you especially.

Joe.

Joe and Eli Zinn, 1945.

April 23, 1945
11:30 p.m.

My dearest Honey,

I've got some news for you for a change. At 7:15 in the morning, I am leaving here for France (the Riviera) for a five day rest. I can use the rest — not that I've been working too hard for that is certainly impossible over here, but five days away from here will be nice. I want to go primarily so I can see what it is like up there. Then, at least, I can tell you about it.

I shouldn't go — I have much work to do getting things up-to-date. Of course my enlisted men can carry on, but I naturally feel that they can't get as much done as if I were there. But I may not get another chance, so I'm going now. I didn't know I was going until just a little while ago. I knew that I was going to be on orders to go up there April 29, but one of the fellows that was supposed to go up tomorrow isn't going to be able to go, so I'm taking his place. Again, I wish to hell you were going to be with me — perhaps someday we may spend a few days there together. I will let you know tomorrow night whether we want to or not. So far I haven't seen but three places that I would like you to see — Cairo, Rome and Naples, and Naples isn't much. From what I hear though the Riviera is another place that is worthy of your presence.

I am feeling O.K. now — absolutely normal for a change. That damn pneumonia is tough to shake off, but I think it's finally left me and I feel good again. I've managed to keep pretty busy lately — paper war, you know. Still have a lot of things to do within the next few weeks — physical exams on all personnel, immunizations up-to-date, teeth, glasses, etc — just like it was back in the States before overseas. At least it will keep me busy and I welcome it. Well, Honey, I must go to bed for I've got to get up before 6 and drive 17 miles over the roughest road you can imagine to catch the plane by 8:15 a.m. Tomorrow night I will write to you and tell you my impression of the southern coast of France. Take care of yourself and always remember that above all —

> I love you,
> Joe.

From Joe in the French Riviera to Babe in Memphis

> April 24, 1945
> 10 p.m.

My dearest Honey,

Tonight I am in a very modern, up-to-date, French hotel on the French Riviera. I haven't been around much yet, but I will try to describe what I have seen so far. To begin with, when you hit the southern coast of France, it is certainly altogether different from Italy. It just looks better kept at the start. We landed in a make-shift airfield outside of this city. I can't tell you which city of the French Riviera it is, but I can send you a postcard of it. As I said the airfield is a make-shift surrounded by mountains on three sides and quite a job to get into — thank goodness I was in a C-47 instead of a B-24, which wouldn't have a chinaman's chance. They then brought us to the hotel, which is run by the Army as a rest camp hotel. Men from everywhere are here — from England, France, Germany and Italy.

We have exquisite rooms — quite on a par with any very nice hotel room in the States — two to the room. The hotel, as I said before is very modern — it has a nice lobby, bar, dining room, barber shop, ping-pong tables, elevators, hot and cold running water and all the luxuries. Our room has one double bed and one 3/4 bed (I have the double one). Nice plush carpet, a lovely bath with two wash basins, bathtub and shower (not connected), closets, easy chairs, and mirrors galore, large French doors which open to a balcony which lets in beaucoup sunshine. It is very nice.

Just across the street is the sea and a beach. The beach is nice but narrow. I would say that is about the same width as Biloxi just in front of the Buena Vista Hotel — nice sand though. This place doesn't halfway compare with Miami Beach. I would say there are about 10 nice deluxe resort hotels here on the bay and that is all. The town is well-kempt though, which certainly gives one a good impression. Too after looking at the Italian towns for 17 months, it is nice to look at the French ones for a while. The place hasn't been marred by the ravages of war, which is something I can't say for any place that I've seen before.

This afternoon I put on my trunks and went out and layed on the beach for a while. The sun was nice and warm, but the water was much too cold for me to get into. I have never seen such abbreviated swimming attire in my life as the French women wear — never. My swimming trunks would furnish enough material to make two full suits for the French dames. They wear bras which cover the nipples barely — that is all. Their panties consist of a bare G-string that would make Sally Roud blush, I'm sure. I took a couple of pictures just for your consumption if they turn out. I wonder how long it will be before we adopt their swimming garb. Too, 90% of the French women wear high pompadours — hair is piled up 6 to 8 inches — at first they look real queer. They are fairly well dressed on the whole though and show no effects from malnutrition as the Italians do. Something tells me that the Germans kept them up well while they were vacationing here.

Prices here are so sky-high they are unbelievable. This afternoon I paid 110 Francs — or $2.20 for a shave and a massage. In Italy that costs me 15¢. Everything else is correspondingly high. Some women's clothes are displayed in show windows, but they are out of my reach. As it is, I'm sure I won't be sending you money this month on account of this trip. Things are even higher than Cairo. A guy bought a tennis ball this afternoon for $12.00 — a used one at that.

We get these bottles of beer while we are here and all the cokes that you can stand up to the counter and drink. The only restriction to the cokes is that you have to drink them there on the spot. I've got to develop a coke thirst, so I won't feel like I'm being gypped. Well, Honey, tomorrow I will write you more. Right now I will say goodnight and that I love you so very much.

Joe.

April 25, 1945
11:30 p.m.

My Honey,

It is impossible for me to convey to you how much I miss you and wish that you were here with me. If you were here, we could have a wonderful time — it is such a nice enjoyable place.

Breakfast is served here from 8 until 10:30 a.m. Of course for 1½ years, I've been eating before 8 or not eating at all, so I've been in a habit of waking up at about 7. This morning I awoke at 7 and couldn't go back to sleep, so consequently I was about the first in the dining room at 8 a.m.

Soon after breakfast, we went and rented us a bicycle, so we could ride around and look at the town. European models of course — with no coaster brake. To put on brakes, you squeeze a lever on the handle bar. Well here I was riding down the boulevard looking at the sights — turned around to look where I was going and saw myself heading for the curb. Naturally I put on brakes — but with my feet instead of the lever on the handle bar. So I hit the curb — went over the handlebars — skinned my elbows and head and bruised my tail sumpin' awful. I can hardly sit down, it is so sore. I was really chagrined at myself for I have always considered myself an expert cyclist. Perhaps I'm not as expert as I thought. From the pain I encounter when I sit down, I'm quite aware that I'm not an expert cyclist.

Then this afternoon right after lunch, we rented a little boat that had two pedals like a bicycle to make it go and went all over the bay. They are quite the thing, and I have never seen anything like them in the States. You and I could have an enjoyable time pedaling around in one together. They are called "pedalos."

About five we came up to our room, had a couple of drinks — took a shower went down to the bar and had a couple of scotch and sodas (1.00 per scotch and soda) and then went to dinner. After dinner we went to a stage show — an all French affair making a gallant attempt to speak English for their 100% American audience — which was very good. The name of it was "Toot Sweet." They had a wide variety of acts and it lasted 1½ hours. I enjoyed it very much.

After the show we came back to the hotel — I had a couple of cokes (unrationed) and then up to the room to write my Honey. If it weren't for her absence, I could be having a very good time and I'm sure she could too. Perhaps some day we can come here. It isn't so terrible in comparison with any resort coastal town in the States — but it is just so damn much in comparison to what one is used to seeing in Europe and Italy particularly.

Well if I can wiggle and navigate tomorrow, I think I'll get a little sun. If I can't, I'll just lay around in this big ole double-bed all by myself and think how heavenly it would be if you were there by my side.

I love you,
Joe.

April 27, 1945
Friday, 9:15 a.m.

My dearest Honey,

Didn't write you last night, so I will get off a note to you this morning. I walked around considerably yesterday, so last night when I returned to the room I just fell in bed. Yesterday wasn't as good as it had been since we arrived here. The sun didn't shine all day long. By the way I saw Rogers yesterday. You remember Rogers — the doctor who lived up at the same tourist courts as the Privers at San Antonio and also lived on Balboa. I have only spoken to him a couple of times. He, too, is in Italy and has been, but I didn't know it and have never run into him before.

Yesterday I lounged around, played cards, ping-pong, drank scotch, walked around the town, etc. Last night I went to the one and only show and saw Joan Fontain in "Frenchman's Creek," which was just another poor show. No good at all — not even worth sitting through.

This morning I have been smart so far. I got up at 7 a.m., shaved, and was ready to eat breakfast before 8, which is the earliest one can eat here — so I took time off to read the paper before breakfast. Then I went and ate, leaving Bernstein in bed (that's who I'm up here with), came back up to write you and now Bernstien is down eating. As soon as he returns, we are going out to walk around some of the town that we haven't been to yet.

There he is right now, so Honey, I'm going to draw this to an abrupt halt. I will write you more later today. I miss you so very much.

All my love,
Joe.

April 27, 1945
10 p.m.

My dearest,

An "extra" just came out of the "Stars and Stripes" announcing the link-up of the American and Russian Armies on the Elbe near Leipzig. That certainly is good news, and notwithstanding a near-miracle for the Germans, should be the beginning of the end. Each such step brings me just that much nearer to being with you — sometimes I hardly see how I can wait — but I can.

You are going to like what I've been doing today, I'm sure — even though I've been spending money that would have been coming to you on the first, I think you will still like it. Firstly, I bought you a pair of earrings, naturally. They are a little gaudy and a little large, but not too large and will certainly be O.K. for evening wear. Then I went about 20 miles this afternoon out to a couple of perfumaries and bought perfume. However, not only did I buy some for myself, but also for a couple of more guys back at the base to send home. I'm going to send it all to you and it's for all of you women-folk — our two mothers, Sis and Katy. I bought all different kinds, so all of you wouldn't smell alike. Too, I was aware that I didn't know which type was correct for blondes, brunettes and redheads, etc. So you five women can decide for yourselves. You pick out what you want and then split and give each of the rest of them a bottle. I think there's enough to go around with a couple of bottles for you. I got Channel #5, Channel #28, Supreme (at the suggestion of a Red Cross girl), Shalimar, Shocking, Gardenia — I think that's all. Too, I got you 6 little dice that contain different kinds of perfume paste that volatilizes on contact with one's body that is supposed to be the doins'. I will send it all to you soon. For shipping purposes, the perfumes are all in little aluminum sealed bottles. After the bottle is opened, it must be put in a glass container. Wash the glass container with alcohol if you are going to wash it first — not water. Just a few helpful hints. The perfumeries are very nice — however after one has smelled the first couple of "flavors," his nose ceases to work for him correctly, and from then on all smell similar. By the way, there's supposed to be two ounces to each bottle. Again, don't open it until you are ready to use it.

It has been a bad day here as far as the weather is concerned. I wanted to get some sun, so naturally it has been cloudy all day and the sun hasn't peeked through yet. There is a high, strong wind and the placid bay has turned into rough waves that splash upon the beach with fury. I hope tomorrow the sun blazes out for the next day we will be pulling out for Italy. There is one redeeming feature about going back there though — there should be some mail waiting for me when I get there. Tell Kay and Carol that Daddy said he loves them — and for you, I love you with all my heart.

Joe.

April 28, 1945

My dearest Honey,

Today has been a beautiful day. The sun has been bearing down all day long. This morning about 11, I got out on the beach and I stayed out there until three except for a short time-out for lunch, and enjoyed it thoroughly. Right now I don't know whether I was wise or not for there is one little place on my right leg, or hip, or whatever that part of the anatomy is that a pair of trunks sometimes covers and sometimes doesn't, that I'm afraid got blistered. I think I also got burned pretty badly right behind the knees where I'm ticklish. Otherwise I'm still in good shape. I have enough suntan left from last summer for the sun not to bother me a whole lot — especially from the waist up.

We are supposed to pull out for Italy tomorrow. However the weather between here and Italy is not flyable today — don't know whether we will be able to make it tomorrow or not. It doesn't make much difference to me one way or another. This is a nice place, and I sorta hate to leave it. But I know there is lots of work to be done back at camp, and there should be some letters from my Honey.

Glen Miller's Band is here today. They played out on the terrace right after lunch, and they are playing here at the hotel tonight. If you were here, we might take in a dance.

You know, honey, I must have needed this rest. I have relaxed more right here than I ever have previously. I've taken some pictures of the place — especially the beach, shore line and hotels for you. Don't know whether they will turn out or not for it is that old film that I bought at Cairo. Well, Honey, it's time to go down to eat so I'll quit.

All my love,
Joe.

Joe and Bernstein in Cannes, French Riviera, 1945.

From Joe in Castelluccia to Babe in Memphis

April 29, 1945
Sunday, 9 p.m.

My dearest Honey,

Well, I am now back at my "home" in Italy. After France, Italy stinks just as badly as ever. I mean this literally too. The radio just announced that Mussolini was executed by Italian Patriots in Milan. If he was responsible for how Italy is now, he probably had such a fate coming to him. In fact, according to man's standards, he no doubt had it coming to him anyhow.

The ride back from France was especially rough today. I think we rode on the top of the plane more than our seats. I believe it was the roughest ride in an airplane I have ever had and was glad when we finally landed. I'm also extremely tired. As soon as I finish writing you, I will hit the sack. Tomorrow night I will answer your letters. I love you.

Joe.

May 2, 1945
4:30 p.m.

My dearest Honey,

Didn't get to write you yesterday. Night before last I found out that I had to go to Bari, so I pulled out of here about 5 yesterday morning. Got caught in the rain down there so wasn't able to come back until today. In fact, I don't know whether I would have come back if I hadn't got caught in the rain for that trip is just too hard to make in a jeep both ways in one day. When I returned to base, there was a letter from your dad. He sent me a piece about Ernie Pyle's death.

Honey, I just got through sending you $150 — it's yours for May 27 and May 31 (two big holidays of mine this month). Buy you whatever you want for your birthday and anniversary. If I were there, I would buy you a nice luggage set to replace those two pieces that I bought you in 1940 for your birthday — only now I would buy you nicer pieces. You could use a couple of good leather pieces of luggage, for as I remember, the ones you have have just about seen their best days. In fact, they have had so much wear and tear that I think they deserve to be "retired" from active service and relegated to the attic. Too, perhaps you might start needing luggage again some day. If you buy some furniture for the house with it or a washing machine or such — when I get home you will pay for "frauding" me and that's a promise.

Honey, I also got off a box to you a couple of days ago. It contains your perfume (and the other ladies), and your earrings, and a few packages of gum for you and the kids. I can't get gum now like I used to be able to, so you and the chillun' better spread it out a little bit.

They report that Hitler is dead over the radio — wonder if it's true. If so, it looks like he and Musso went rather close together. Wonder what effect it will have in the ultimate outcome — personally, I doubt that it affects things one way or another. Well Honey, I'll talk to you again tomorrow.

I love you,
Joe.

May 3, 1945
8 p.m.

My Honey,

The news of the day, of course, is the formal surrender of the German armies in Italy. I'm sure glad for those boys of the 5th Army — think about it — many of those boys landed at Casablanca Nov 7, 1942. They have been fighting under the worst possible conditions too and certainly have a break coming. I don't know whether any of them will be coming home now, but is a cinch that if they had had to fight up through the Brenner Pass, it would have been plenty rugged. It might mean that it will be longer before I get to come home for the

422

more there are to ship out of Italy, the less chance we would have for shipping space. I don't know but damned if it doesn't look like this mess ought to be over pretty soon. When it is, I certainly dread to see the exhibition and celebration that these boys are going to put on. You can count on it that it will be a good one. Hope no one gets hurt.

Today I haven't done a cockeyed thing. Got a lot of stuff down from being stored on the rafters and went through it. Gonna turn in a lot of my junk and equipment that I won't be needing anymore. Thought I might send a box of stuff home but perhaps, I'd better not. My clothes are down to a minimum though. I did buy a pair of G.I. low-cut shoes a couple of days ago, but I had to have them or start going barefoot.

By the way we had radishes out of my "garden" today. Nice ones too. My lettuce is just about ready to eat now. Of course it is planted in beds where it would never head up. My tomatoes aren't doing too well for I had to put them outside and at night it's too cold for them.

We are starting malaria discipline again — having to wear boots after 5 p.m. — sleep under mosquito bars, use mosquito repellant, etc. This year we are not going to take Atabrine — why I don't know, but I'm glad.

Honey, I'm going to quit early on you tonight. Now I think I'll take a much needed bath. I can't keep my enthusiasm down over the outlook of things in general and over the possibility of perhaps being able to see you within this year. Hope I'm not disappointed.

I love you,
Joe.

May 4, 1945
Friday, 10:30 p.m.

My dearest Honey,

The radio just announced that Northern Germany, including Denmark and Holland, has surrendered unconditionally effective at 10 a.m. tomorrow — that's good, but I wonder how many men will be uselessly killed or maimed up there before 10 in the morning. War is hell anyway one looks at it, isn't it? Even in victory. Now it looks as though Czechoslovakia and Southern Germany is all that is left with some of Austria — very little in comparison of what it was not too long ago. Strategically, I guess, the European war is over — however I know it isn't to those boys up there still struggling. It shouldn't be too long before it is all over in these parts. Then (maybe — I hope) within a few months, I will be back with my Honey for a little while before going somewhere else. The waiting seems to be even more difficult.

I'm enclosing two 35 mm snapshots that Lather took of me the other day while I was digging up the front yard for the walk with a crash ax. Reese came and slipped up behind me with his pistol and the C.O. took a picture. Of course they were trying to intimate that in order to get me to work someone had to stand over me with a .45. That isn't quite the truth — however it may be pretty close to it. You can see a couple of my trees that I had set out. In the other, I was resting after swinging the axe for a while — consequently the look of anguish. Is my hairline receding more on the sides or has it always been that far back? Concerning that point, the picture almost frightens me.

Honey, enough is enough, I must get to bed, for the next five days I'm going to be working harder than I have since I've been overseas.

I love you,
Joe.

May 6, 1945

My dearest Honey,

I'm dead tired tonight. Yesterday we examined 600 men — almost every man in my squadron, and all of us are dead tired. Today we will examine Quinn's squadron and so forth until all the men have had a physical examination. We started at 10:30 this morning and finished at 9 tonight. We have three more days of continuous physical examinations to do before we can take a breath. I'm tired already and hardly see how I can take three more days of the monotony of physicals. I've got to go over to my dispensary now and do some stuff that must be done tonight and map out for the boys what they must do tomorrow. One never gets caught up with stuff of this sort — it goes on and on. How they expect for us to get it done in such a rush or why is beyond me anyhow.

Got four letters from you today, Honey — certainly was glad to get them for I have not been getting mail too good from you lately. I will answer these when I feel more up to it. In one was a couple of pictures of the kids — Carol does look a lot like Kay, doesn't she? She certainly has gotten to be a big girl too. Found out today that doctor's (Flight Surgeons particularly) were the most critical item in Uncle Sam's Army, doesn't sound too good, does it?

Joe.

May 7, 1945
7:30 p.m.

My dearest Honey,

We are celebrating, Honey. The Germans just announced that General Doenitz has just surrendered all his troops and has broadcast for the Germans to surrender and quit fighting. That is not officially the end of the war as far as the United Nations is concerned, but it's worth celebrating anyhow. When we announce that the war is over, we will just celebrate again but then on a bigger scale. I've even managed to get a quart of scotch from Cairo for the occasion.

We finished examining another squadron today. Now we have one more squadron and Group Headquarter personnel to examine, plus all of the stragglers. I think we will finish Wednesday afternoon sometime. I certainly will be glad for it's terribly monotonous. I've been examining for hernias and hemorrhoids and all allied diseases for two days and I've said "Turn your head and cough — now cough again — turn around, bend over, and spread your cheeks apart," so many thousand times that I dare say I will probably say it in my sleep tonight.

You should have seen me last night. I went over to the kitchen and got me some sausage, sage and pepper. We get good pork sausage over here, but it is always tasteless, so last night I mixed up some to my own taste and cooked and ate it. It was pretty fair stuff too. I went out in my garden and got some lettuce leaves and some radishes and we really had a feast about 10 p.m.

So my Honey has been laying out my clothes for me, huh? Sorta optimistic, aren't you? Personally, I'm afraid to be too optimistic. I sorta think I'll come to the U.S. before I go elsewhere, however, that is strictly my own personal opinion and may not coincide with the "brass hats" at all. So, Honey, please don't get worked up for a big disappointment. Well guess this finishes another day. I'll certainly be glad when the next two pass.

I love you,
Joe.

Dearest Babe,

VICTORY DAY SPECIAL

THE STARS AND STRIPES
MEDITERRANEAN

Vol. 2, No. 154, Tuesday, May 8, 1945   ITALY EDITION • • TWO LIRE

IT'S OVER
OVER HERE

Victory in Europe is ours. After more than five and a half years of the bitterest fighting this continent has ever known, the armed might of Germany, the Wehrmacht and the Nazi party has been defeated -- finally and utterly.

Today will be treated officially as VE-Day, it was officially announced last night. There will be broadcasts from the Chiefs of State of the Big Three this afternoon at 3 PM, according to Reuter's. King George VI is expected to broadcast to the British and Commonwealth peoples at 5 PM.

As the entire world waited anxiously all day yesterday for the VE-Day proclamation, there were reports, unconfirmed officially by SHAEF, that the Germans had signed an unconditional surrender agreement at 2:41 AM yesterday.

While SHAFF declined to confirm the reports of the signing of unconditional surrender, AP carried a report from Rheims, France, where General Dwight D. Eisenhower's headquarters is located,

(Continued on Page 8)

426

May 8, 1945
"V-E Day"
9 p.m.

My dearest Honey,

Well, Honey, it's here — officially. Finally there is victory in Europe and I hope peace for many, many years to come. If the politicians have what the American boy has, there will be too. I wish I trusted the politicians like I do the boys — Peace in Europe — at least the defeat of Germany is here — and even though I have hated every moment of the time I've been over here, I feel satisfied that I have done a small share. Regardless how we suffer financially and regardless of how much I have detested being separated from you, and regardless of how much of a burden I have put on you, I think that you and I both can feel that we have honestly done our bit.

I can imagine the celebration and hullabaloo in the States today. We have heard about some of it over the radio. King George is making an address now, but I can't hear him very well. You would be surprised at the boys today though. Yes I have seen a couple drunk — and that's all. On the whole they have just been lounging around and doing nothing — no shooting, no rowdiness. There are a number of reasons for this — firstly, the surrender is an anticlimax for us. We, like the rest of the world, have been expecting it almost daily. Too, the war has been over for us for about 10 days because there has been nothing that we could bomb without endangering our troops. The boys don't feel like the mess in Europe is over for they, of course, are still in Europe. They know that only the war in Europe is over and the war is not over at all — to the average guy, I don't suppose it makes much difference whether he is shot at by one or the other in the air. All of them are like me too — hoping to go home for a while but scared to wish for it too hard for they don't want to be disappointed.

Today, being V-E Day, we were supposed to have a holiday. However it was not a holiday for me for we have been examining men all day long again. Tomorrow we should finish up, and I certainly hope so for all of us are tired. I have to get into town so I can get some laundry and cleaning done and also get a haircut. Goodness knows, I need the latter. In fact, I need all three rather badly.

Well, Honey, it's goodnight again by pen. I hope it won't be too long before we can say it to one another like it should be said.

I love you,
Joe.

May 9, 1945

My dearest Honey,

Well I finally finished up on those physical examinations today. Have a few stragglers to do, but on the whole we are finished up and am I glad. Got plenty of more work to do though. It's nice being good and busy for a change — it certainly makes the time pass a lot faster. For the past five days, the sun has been beaming down good and hot, and I have been so busy I haven't been able to get out and get in it — surely would enjoy a little of it too. Tomorrow I will try to get an hour or so of it — don't know whether I will have time or not.

Honey, I just packed a box full of stuff of mine that I'm sending home. In it are such things as winter underwear (which I have never had on yet), sweat suits, coats, overcoat, excess towels, and such. You can open it up and put the junk away for me when it gets there. Oh, there's another thing — don't send me those things I've written and asked for — no room for it.

I've been getting up my laundry tonight too. I haven't sent any out since before I went to France, and I certainly have plenty of dirty clothes. Tomorrow we go into cottons, and I'm just putting away my woolen clothes dirty — very few of them are wearable anyhow — but I still hate to throw clothes away.

Just sent two of my boys into the hospital with malaria. If it isn't one thing to worry about, it's another. In the winter it's colds, ears, and pneumonia — in the summer it's the G.I.'s and malaria. Of course, it's V.D 12 months all year.

Your letter telling me about Carol yelling "Dadee" when you get after her amuses me. Kids surely do catch on quick. Imagine that little thing hollering for me when she has practically never seen me. You and Kay must do a lot of talking.

Honey don't worry about me being "demoted" to a civilian. Doctors are still very much needed in the Army — too there are plenty of M.D.'s in the Army that are over 50 years old, and they certainly should get out before the youngsters — plenty of them are overseas too. So I certainly don't expect to be let out of the Army for quite a spell. As for Anhalts "point system" in determining where you go or what you do — I think it's just a lot of "stuff." In fact, I know it is. It may mean something for enlisted men and some officers but not doctors. The determining factor is Lady Fortune alone — until one has 30 to 36 months overseas. I think those are "latrine rumors" too or better known as "latrinograms."

I wonder how long I will have to be with you before I fully realize that I'm not dreaming.

I love you,
Joe.

May 11, 1945
Friday, 11 p.m.

My dearest Honey,

Got two very welcomed Air Mails from you today. Letters from you are still my sustaining force — don't know what I would do without them. Still working pretty hard. Still I haven't had time to go into Foggia and get a haircut, which I need very badly. It certainly seems funny to be able to say "Foggia" instead of saying "town." Yes, of course I'm located near Foggia — 10 miles (air miles) south to be exact — now I can say it.

I've got a big inspection coming off tomorrow. The guy that's inspecting is a "so and so" too — I know him very well. After one has been over here as long as I have, he ceases to give much of a damn about an inspection. I do the best I can and if the inspector doesn't like it, I can't get too excited over it — but as yet they have never found anything really wrong.

Last night they announced over the radio about the "point system" of being discharged from the Army. I'm wondering how many points Hugh has. Of course it doesn't apply to officers at all — I wish it did for, if it did, I would be O.K. (if I weren't a M.D.). I have a total of 117 points — the two kids = 24, 8 battle stars = 40, Soldiers medal = 5, service and overseas service = 48. Perhaps I could give Hugh some — he's possibly well enough off on his own though.

They have announced over the radio and written in the "Stars and Stripes" that the rations back in the States were going to be somewhat relaxed. Can you tell any difference yet? How much gas do you think you will get? I hope you get a lot for when I get home (whenever that time might come) I want to ride in a good car rather than a truck. Of course all I can do is sit and hope that day will soon come or at least come before I go elsewhere.

Glad that Ferguson called you — he's a good kid and he will drop out to see you if he gets a chance — one of the most stable boys in the air that we have ever had. Well, Honey, I've got a can of tomatoes here that I've "acquired" from the mess hall — I think I will eat about half of them and then go to bed.

All my love,
Joe.

May 14, 1945

My Honey,

I intended writing you yesterday on behalf of Kay and Carol for Mother's Day, but I was very busy all day. Last night there were a bunch of fellows in here gossiping until the wee hours of the morning. I intended writing Mother too but didn't get to do that either. Anyhow again — whether I get to write or not — the kids and I both love you so terribly, terribly much. If there is a woman on earth that can feel that she has been a success of being a wife and mother, and making a home what it is supposed to be — certainly you can. The kids and I are glad that you decided to make marriage your career rather than be the so-called "career" woman.

I've got a hard tough 16 hours of work coming up tomorrow — start at 8 a.m. and will finish about midnight. After tomorrow I don't see what I'm going to do except the usual stuff. I guess then I will just spend my time lounging out in the sun and taking it relatively easy. I just returned from a dinner that was held up at Group tonight in honor of our present Group C.O. who is leaving us. I guess he is coming home. The new C.O. looks like he might be a regular sort, but I will reserve my opinion to a later date.

Well, Honey, perhaps I had better go to bed and get in good shape for that hellacious day that I've got staring at me tomorrow. This time tomorrow night I'm gonna be tired. Tell the little ones hello for their Daddy.

All my love,
Joe.

May 17, 1945
11 p.m.

My dearest Honey,

How's my Honey doin' today? Today things slowed down considerably for me, so I have been taking it easy. Of course that only makes the day pass more slowly. This morning I did practically nothing. This afternoon I got out in a pair of trunks and threw baseball for a while and then tossed a few goals (or at a few goals) with the basketball. And Honey that just about did me in. I'm in the worst shape physically that I've ever been in before — that is, I just haven't got any wind and I'm soft — either that or at 30, "I just ain't what I usta was."

Tonight I went to a show out here in the area. It was something about the Barbary Coast with John Payne and Ann Dvorak — at least 10 years old, but fortunately I had not previously seen it. It was pretty good too — at least one didn't want to get up in the middle of it and leave.

Honey I got the package from you today with the slides, undershirts, shorts, shrimp and turkey in it — also the peach pickles. A very useful box, it was too. The peach pickles (all of 'em) were gone within a couple of hours after I opened the box. I ate all of them but one — they were delicious. I have already been wearing the slides. Then tonight I opened the smoked turkey and now almost all of it is gone. Been sitting here eating smoked turkey sandwiches and drinking cokes — how's that? We got two cokes for this week's ration, and I drank them both a few minutes ago with turkey sandwiches. So out of your whole package, I have the can of shrimp and the underwear left. Thanks a lot for all of it — the pickles and the turkey were both delicious. It surely arrived at the right time 'cause today our mess has certainly been lousy.

Next morning —

Again I didn't finish my letter to you. I'll finish it now so it will get off in the same mail though. This morning I'm sore as a boil all over. I could hardly get up out of the sack. Guess throwing a basketball is too strenuous for me.

I'm gonna turn in a lot of equipment today that I don't need anymore — one item being 60 quarts of whiskey. We used to give the boys 2 oz. after they had completed a mission if they wanted it. I dread the ride into Foggia.

Lately it's been very difficult for me to go to sleep at night. Don't think I'm a silly kid 'cause I'm at least old enough to be out of the kid class — but every night I lay down and start thinking of you and say to myself that it shouldn't be too many months before I can lay down "double" — and cuss and discuss the day's doin's with a good listener and have that good sense of well-being again. It is certainly worth looking forward to. Now, My Honey, if I want this to get in today's mail, I'd better get it over to the mailroom pronto.

All my love,
Joe.

May 19, 1945
Saturday, 8:30 p.m.

My dearest Honey,

I haven't been writing you lately as much as I should have been, but there have been many good, valid reasons. I have been pretty busy, and now there are more guys living in the house here and there's too much of a hub-bub. The latter to you must sound like a pretty sorry excuse for I realize it is difficult to have a hub-bub comparable to one made by two youngsters. However the advantage of the youngsters is that by 10 o'clock at night, you can get them in bed.

I had thought in my previous letters that my work was just about finished, however, as I predicted as soon as I manage to get something finished, the "higher-ups" decide that more forms of some sort need to be done — it has just been a rat-race. Yesterday alone I made out over 1100 forms — can you imagine? Fortunately about all I have to do is see to it that they are made out correctly and sign my name. Mason can sign my name as well as I can, so he does a lot of that. I don't see how they can ask for much more data though.

Not only has all the above "stuff" got to be done, but since the cessation of hostilities over here, we have more to do from the medical point of view than we did previously — believe it or not. I think it's because the boys just haven't got enough to do to keep them busy, and I have a few who have been hitting the bottle a little too persistently.

Over the radio they are announcing that Japan has sent out "peace feelers." I wish they wouldn't announce such rubbish. It causes too much of a feeling that it is all over. What a mess this world has gotten itself into?

To return, though, to most pleasant thoughts — Honey, when I get home, how are we going to get away to ourselves for a while — who will take care of Kay and Carol for us? All of these things, of course, will be entirely left in your hands 'cause I certainly have no idea of the lay of the land. Gosh, I certainly am way behind on everything. Truthfully, I'm a little bit "scared" too — really. Cause my language is atrocious, and I've probably acquired many other obnoxious traits and habits that I will have to break that I don't even realize I have. However, don't worry 'cause I'll get "back in my traces" faster than anyone you ever saw, but I probably will make some severe blunders in the process. If you have some "hoity-toy" friends, you had better not have them come around for a couple of months after I get home.

Well, Honey, I've sat up here and done all of this wishful thinking on paper — I've had some very pleasant visions of you. Now I gotta quit and go over to the dispensary for a while. Just want you to know that I love you.

I love you,
Joe.

May 21, 1945
3 p.m.

My dearest Honey,

All the boys have gone to town, so I will try to collect my wits about me and write you a letter. Today has been just another day of sitting around and waiting for nothing to happen — and I mean just that. Got up at 7 as always and went and had that damn french toast — always French toast or sorry hot-cakes. I don't want to see either for the next year or so. Then went up to Group Headquarters to see what was going on — "nothing" is going on up there as it is everywhere else around here. Came on home about 11 a.m. took off all my clothes and got out under the nice sun until 12:30 — came in and dressed for lunch — had lunch and then came back and got out in the sun until 2:30. Came in the house to take a nice cool shower, and sure enough all the water has been run out of our tank. So now I'm filling up the tank, so I can take a much needed shower. After staying out in the sun and sweating a few hours, I smell like something I wouldn't want to stay around.

I've been reading a book for the past couple of days — "A Tree Grows in Brooklyn" — it ain't so hot. In fact I would throw it down for something better, but there isn't anything around here. All of us have already sent our books and such stuff home — perhaps we acted prematurely. At least it keeps me from being bored to tears.

Got a real big inspection coming off tomorrow morning. I don't know, but I think I am quite ready for it. The whole squadron should be for that matter. As before it isn't bothering me very much. Honey the days are dragging like they have never dragged before. I'll bet the next couple of months seem like years. I know the time is dragging just as slowly for you.

Well take it easy, my Honey, and don't be trying to do too much — please just for me. Hope I get some mail from you tomorrow.

I love you,
Joe.

* * *

The 451st was one of the most decorated units of the 15th Air Force. They were the only unit to receive three Distinguished Unit Citations and move all 62 planes overseas without losing any en route. They had the highest overall bomb score in the 49th Bomb Wing and in the entire 15th Air Force, and they were the only group to have a recorded perfect mission.

They flew troop support missions, "milk-runs," and bombing missions. They encountered fighters, flak, mechanical difficulties, and crashes on take off and landing. Just boarding a B-24 was dangerous. The 451st lost a total of 425 airmen from October 1943 until they were disbanded in 1945. Many were buried overseas.

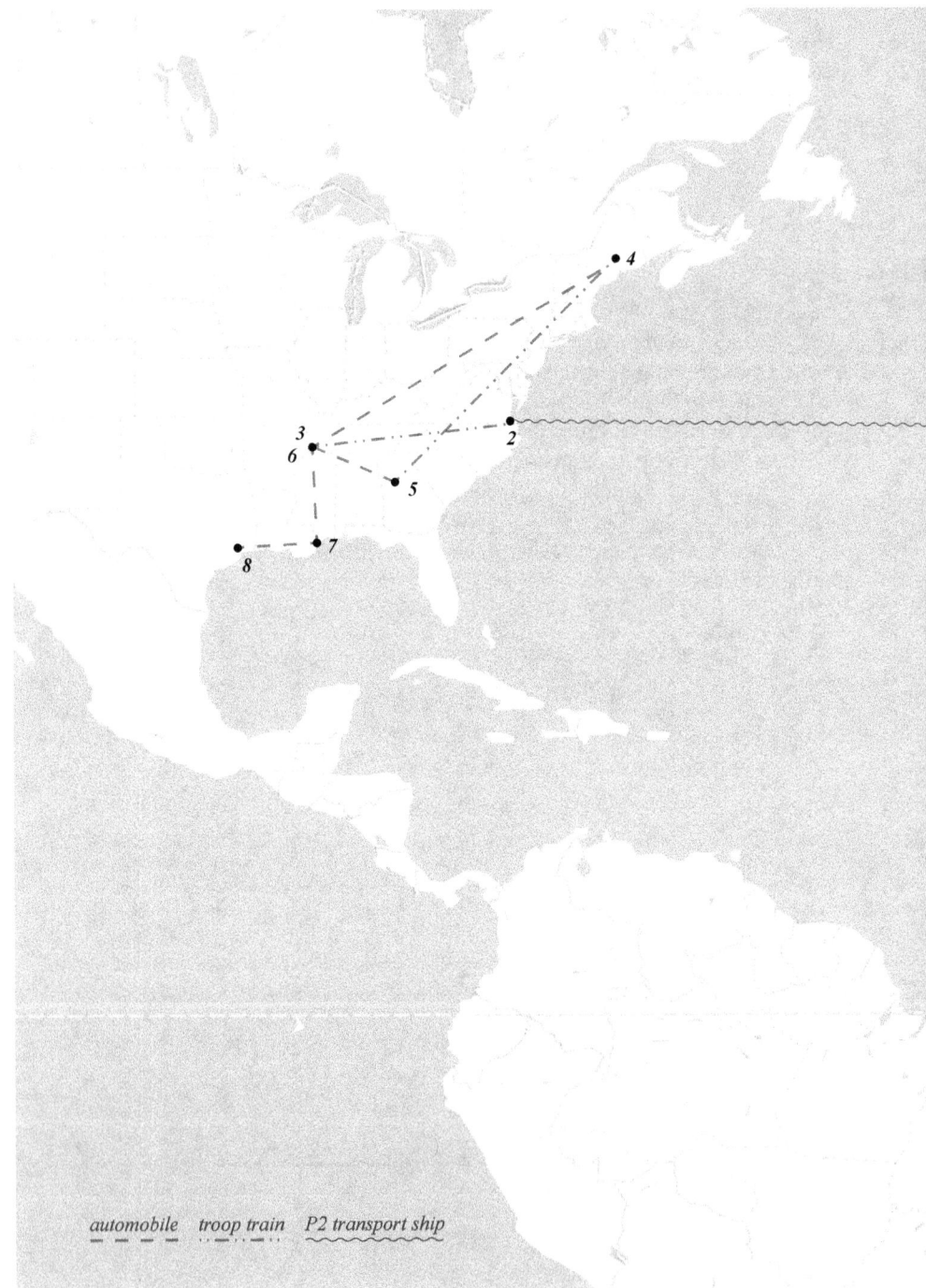

automobile   *troop train*   *P2 transport ship*

Part IV

1

Returning Home

On May 9, 1945, the 451st was alerted to prepare the airplanes for overseas movement. One week later, they were told they were going home. The airplanes left Italy beginning on May 20; however, Joe chose to sail home. Years later, he told his son-in-law, Jim Farris, that the airplane he was originally assigned for his return crashed on take off from Dakar, Africa. This was the only 451st B-24 that crashed on the trip back to the States. It went down on June 4, and all passengers were killed. Joe and the remaining ground crews cleared the base of tents and equipment and traveled to Naples on May 26, 1945.

Dearest Babe,

May 27, 1945
My Honey's Birthday

My dearest Honey,

Congratulations on your Birthday, my dear — and many Happy more! I received a letter from you today — the first I have gotten from you in a few days. I was happy to get it too — especially so since it contained pictures in it of you, the children, your Mother and Grandmother. I have not been able to write you for the past few days, so consequently the long lapse in between letters — nothing to worry about, I assure you I'm in perfect health, spirits, etc.

We have been terribly busy and are still — the time certainly passes much more swiftly that way too, for which I am thankful. Honey, I suggest that you don't write me anymore until I have further instructions. Don't know how many more times I'll be able to write you either.

Now if I were you, I'd be ready to make a quick trip to St. Louis on a moment's notice within a month or so. Just in case everything works out like we hope it will. If I get to the States, I will call you from wherever I hit first and you can meet me in St. Louis — if you can, of course. They just yelled that this joint is closing up. See you later.

All my love,
Joe.

May 28, 1945
5:15 p.m.

My dearest Honey,

I left the camp this afternoon. We are now living in a great big ex-university. Most of the boys are sleeping on two blankets on a board bed. However I'm much more fortunate. I have my pneumatic mattress along and my Arctic sleeping bag, which I'm using as a mattress — consequently it isn't bad at all. There are 8 of us in a small room — sleeping double decker style. That's O.K. but one of the fellows snores even louder than Mrs. Wade. This morning I finally managed to get to sleep around 3 a.m. This had better not last too long.

I bought you a little present today that I think you will like — that is if they fit. I'll not send this though. It will be much more fun just to hand them to you. I'm going by to see Harvey Carter in a few minutes. Am going to try to get him to go out to dinner with me. Don't know whether I will get to see him any more. Don't suppose I'll be able to see the Hughes' boys either.

Hey, do you reckon I'll get sea sick? I've never made a bigger voyage than down at Miami fishing or crossing the Chesapeake Bay — guess I will find out.

Wouldn't it be swell if in about 3 weeks I could hear your voice over the telephone — I can't believe it and refuse to do so until I do. Hey, do you think it might be possible to drive the car to St. Louis? I doubt if you can obtain gas that easily though, and perhaps it wouldn't be best any how. All of that is up to you. If it is inconvenient to leave the children and come up there at all, it's OK. It isn't but a few hours from Memphis by air or rail — I think if possible though you and I both would rather see and be with one another a while before Memphis. Of course we will have to do whatever is necessary.

Well, Honey, some of the boys are waiting for me so I'd better go before I let my anticipation run away from me. Please tell Mother and Dad that I haven't had time to write and why.

I love you,
Joe.

May 29, 1945
1:15 p.m.

My dearest Honey,

Just found out that practically all of the restrictions on censorship have been dropped, so consequently, I will try to write you a letter that makes sense.

Firstly, I am here in Naples waiting to board some ship for the States. How long I will have to wait I don't know, but I hope not long. This is our third day here — I came over here from Foggia the 26th. Don't know what kind of a boat we will finally board — don't know whether it will be big or small, a luxury liner or a tug. Nor do I know what its route or speed home — so consequently I don't know when I will be hitting the States. I think we will dock at Newport News, Virginia, but I'm not even sure of that and care less — all I want to do is get on some kind of a boat and head to the west.

As I have told you previously, I will call you if possible as soon as I dock. I will then be responsible for getting myself and a detachment of men to Jefferson Barracks at St. Louis. There I will get my leave orders and my orders telling me where to report when my leave is up. I expect to have a 21 day leave. After the leave, I think we will go to Manchester, New Hampshire. It appears we are going to become a part of the Air Transport Command for a while — of course, many things can happen between now and then, and the Army may change its mind many times.

We are indeed fortunate Honey. As far as I know and have been able to find out, we are the only Heavy Bombardment outfit coming back to the States. No other outfit is so fortunate. It is all very unusual — don't for one minute think that this is what is happening to everyone. Why we were selected, I don't know, but I do know that I'm mighty happy about it. We are all mighty anxious to get on the boat and get to the States 'cause all of us are afraid that this is too good to be true and that there is a catch in it somewhere or the "big wigs" will change their minds somewhere along the line.

If all goes as it should, you and I should be able to say goodnight to one another without picking up a pen around June 15. However if we aren't by then, DO NOT WORRY because as I have said before, I haven't the slightest idea of what kind of boat we will get. Lord only knows I hope that some change isn't made in our status of coming home because I couldn't even forgive myself for building you up like this and then having it all torn down. Count on it that I will be home only when you hear my voice over the telephone.

As I have told you before, if possible I would like to be able to meet you in St. Louis before coming down to Memphis — so be able to pull out for St. Louis on a moment's notice. I'm only repeating myself so that I will be reasonably sure that you will get one of these letters before you get a call from me, so you will know what is up.

Truthfully so far all of this seems like a dream to me. I can't — even yet — realize that there is even a vague possibility for us to be together again in a matter of a few weeks — and I won't really believe it until I am with you.

Too, Honey, if you can come to St. Louis, bring along some travelers checks 'cause I don't know how I will be financially by the time I get there. If I have any summer uniforms around, bring them too cause by then I'll be pretty short, I'm sure. Isn't it wonderful to be talking and thinking like this? Planning on being with one another again — I'm telling you all of this now cause I know when I call you up I won't have sense enough to say anything except maybe "Hello, Honey, I'm here." Well I've surely let myself run wild this time. There maybe won't be too many more letters from this end.

I love you,
Joe.

May 30, 1945

My dearest Honey,

Well, I'm still writing to you from Italy, so I'm still here — wish to hell we would hurry and get under way.

Rogers is sitting across the aisle from me right now — writing his wife, I guess. He is not so fortunate as I. He is going to Trinidad and not via the States either. Found out that our trip to the States may not be direct. Of course this might be just another rumor and probably is — just don't want you to worry if I don't get in about the time you think I should be. Wish we would get started — all of us are getting mighty restless. There is nothing to do now but wait. Just like the Army — you always rush like the devil to go someplace and wait. Tell Kay and Carol Daddy will see them pretty soon.

All my love,
Joe.

May 31, 1945
9:45 a.m.

My dearest,

I don't know when I was sweating it out the most — five years ago or right now. However, I'd guess it was five years ago. About this time I was having a talk with myself saying "Hey, Joe, do you know what the hell you are doing? You have been running around for a long time and now you are getting yourself some responsibilities — are you ready for them?" I was giving myself a pretty thorough going over. Needless to say though of all the "steps" or decisions that I have made in my life, that one was the wisest and the epitome of them all. The one that has given me the most happiness and has been the most satisfying. The beautiful part of it is that I'm sure you feel the same. Today is my big holiday of the year. I wouldn't swap the last five years of my life (even considering the past 18 months) for anything or any other part of my previous life. Let's hope that the rest of our lives together will be just as happy and fruitful.

Yes, I'm still sweating it out from this end of the line too. As far as I know, our boat is not even in the harbor yet, and a storm is in the making. The wind is high, and it is very cloudy, and the white caps are large and plentiful in the bay. We are all just as restless as ever if not a little more so.

I received three letters from you yesterday — one was written May 24 when you were just reading between the lines that I would probably be home by July 4. I still hope and believe I will make it by then at least, for the way the days are dragging now that in itself will be an eternity. In the meantime, you be sure not to overdo yourself fixing things up — out of paper!

I love you,
Joe.

June 1, 1945
11 a.m.

My dearest Honey,

Here I sit still sweatin' out the days to pass — hope it won't be much longer and don't think it will. Just saw Harvey for the last time, and he had just read a letter from Helen. She said she had just seen you and that you were all "built up" — I still hope it's not for a let down and still don't think it is. Boy, the days are getting longer and longer — by the time I reach the States, each day will seem like a month. I'll bet that train ride across the States will seem like a real eternity. Honey this may be my last letter to you — the next communication between us might be of a different nature. I am certainly looking forward to it.

I love you,
Joe.

June 3, 1945
2 p.m.

My dearest Honey,

Every time that I sit down to write, I think that it will be the last time that I write. Again, I think this really will be the last time I get to write to you. I was supposed to board the ship yesterday, but it was put off until early in the morning. So if nothing unforeseen happens, I will climb aboard in the morning and the next scheduled time I'm supposed to step on ground it is supposed to be good clean U.S.A. land. Where we will land is still a question. It will either be Newport News, Virginia or New York, I imagine.

It is the consensus of opinion around here that it is a pretty fast boat — consequently we should be hitting the States pretty soon. It still sounds too good to be true, and I almost find myself refusing to believe it.

We have recently been told that they are now serving milk and fresh meat on the ships returning veterans to the States. We are anxiously awaiting to see if that is true. I don't think there will be very much griping though even if the boys have to live on C-rations. I'm sure I can put up with whatever the fare will be. The days here have certainly been extremely long though, and I suppose the days on the boat will be longer. I believe it will help some though just to know that you are moving in the right direction.

I am surely eager to see you, Honey, I know that you know that without me saying so, but I can hardly wait. Just to know that I might be seeing you even by the time that this letter reaches you is almost unbelievable. I've been playing cards and reading so much since I've been here that I'm fed up — now I guess I'd better go on back to my room and finish packing. Bye, Honey, for now.

I love you,
Joe.

* * *

On June 4, Joe sailed aboard the U.S.S. General Meigs to Virginia, arriving 10 days later. After landing, the 451st boarded a train for Camp Patrick Henry, where they were given a 30-day leave and new orders. Finally reunited after nearly two years apart, Joe and Babe were determined to stay together. Babe packed up Kay and Carol and followed Joe for the rest of his assignments.

After Joe's leave, he reported to Dow Field in Bangor, Maine, where the 451st had been reassigned as part of the Air Transport Command. The men were only in Maine for a few weeks before they received news that the 451st had been officially deactivated following the end of the war in the Pacific. At the end of September, Joe reported to Fort McPheerson, Georgia to be released from active duty.

Officers and their wives attending a farewell party at Dow Field after the 451st was disbanded in August 1945. Joe is sitting on the chair in the right side of the photograph.

Babe's brother Hugh also returned safely from the war and went back to school to become a librarian in San Marcos, Texas. Hugh lived to celebrate his 100th birthday in 2020.

Afterword

After the war, Joe returned to his family in Memphis. Following his interest in returning to school, he applied and was accepted to an orthopedic residency program at the Ochsner's Clinic. The family moved to New Orleans to start the next phase of their lives. In 1947, Babe and Joe welcomed a son, whom they named Joe Wesley King, Jr.

In 1948, they moved to Houston so Joe could complete his orthopedic training at Shrine Crippled Children's Hospital. He started his private practice a year later. In 1951, they bought a house on Buffalo Speedway in Houston, where they raised their family and lived out the rest of their lives. The house is still in the family and is "home" to us all.

Joe went on to have a very successful career in orthopedic medicine. He saw patients in his office, practiced surgery at the Methodist Hospital, was the head of Orthopedic Surgery and a clinical professor at Baylor College of Medicine. He established and directed the Fondren Orthopedic Center and was a medical consultant at the Veterans Administration Hospital, Shrine Crippled Children's Hospital, Hermann Hospital, and Texas Children's Hospital.

He was one of the first doctors for the Houston Astros baseball team in 1962. Over the course of his career, he did extensive research in sports medicine including a thesis on the "pitching arm" of professional baseball pitchers. Throughout his career, he treated many notable patients including Ella Fondren, Nolan Ryan, and Jerry Lewis.

Team doctors Harold Brelsford, Hatch Cummings, and Joe King at the Houston Astros spring training.

As a member of many Orthopedic and Medical Societies, Joe frequently took leadership positions. He served as the Chief of Orthopedic Services at Methodist Hospital in 1955 and the President of the Methodist Hospital Staff. Babe was at his side throughout it all. She was very active in the Houston Medical Auxiliary, often traveling with Joe during these years for different medical meetings. She hosted events and helped raise money for hospital initiatives.

Joe and Babe spent their years together laughing. They took the trips they always talked about, and Joe finally bought the wedding ring he couldn't afford when they married. They watched their family grow and continued to do "projects," mostly at a vacation property in Crystal Beach, Texas on Bolivar Peninsula. They spent weekends working on the beach house together. Joe installed the windows and doors—and the plumbing, a continuation of the King Water Works he started in Italy. Babe painted cabinets and hung curtains. The first two beach houses were blown away by hurricanes, but the third was the family vacation spot where the kids and grandkids learned to swim, fish, and crash golf carts.

In 1977, Joe received a letter from Bob Karstensen, who was also a member of the 451st. Bob and Pete Massare located the surviving group members and formed The Former Members of the 451st. In his letter, Bob told a story about a time that Joe helped him out of a plane after a mission. Bob had been injured when the glass from the nose gunner position hit him in the eyes. Joe was the first one to the plane when it landed. Bob never saw Joe but remembered his help. Pete Massare was the Assistant Operations Officer for the 727th. He was one of the original members and knew Joe from the beginning, remembering him as one of the kindest and finest flight surgeons.

Joe and his former roommate Eli Zinn remained friends throughout their lives. As an original member of the 451st, Zinn said that the whole squadron liked and respected Joe. Years after the war, they fished together at Bolivar, and Zinn attended Joe's funeral in 1979.

On June 6, 1979, at the age of 64, Joe died from complications from surgery. Out of a lifelong habit, he wrote Babe one final letter.

To the loveliest lady There is :—

Thank you for taking care
of me for the last 64 years
and particularly for The past
10 days — Being so nice —
waiting on me hand + foot
(and knee) — Helping to dress
me and all The other things
to numerous to mention — And
being so gracious + nice about
it. I have always loved you
+ always will —

Jim

Joe and Babe, Christmas 1978.

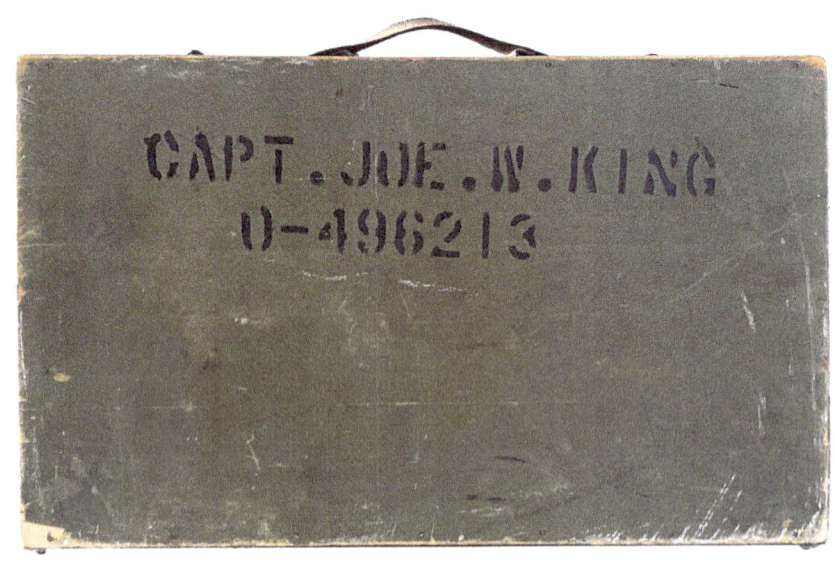

Bibliography

The majority of sources for this book come from Babe King's personal archives of letters, photographs, and Joe's military records.

Letters, Telegraphs, and Postcards: Joe King's War Letters. Privately held by Kelly Mazade, Fort Worth, Texas. The personal WWII Letters of Joe W. King were originally written from 1940 to 1945 and never published. Reprinted by permission of the Estate of Joe W. King. Collection of personal letters with envelopes and postmarks.

Photographs: Family Photos. Privately held by Kelly Mazade, Fort Worth, Texas. The personal photographs of Joe W. King and Olive King were originally taken from 1940 to 1945 and never published. Reprinted by permission of the Estate of Joe W. King. Family photos with names and dates written on the backs of most.

Military Records: Joe W. King's Military Records. Complied by Olive (Black) King. Privately held by Kelly Mazade, Fort Worth, Texas. Babe's collection of Joe's orders.

Cover Quotations: DeBakey, Dr. Michael E. "Joe W. King: In Memoriam," *Inside Baylor Medicine* Vol. X (June/July 1979): 2. Quote from Dr. Michael E. DeBakey, BCM chancellor and president, in Joe W. King's obituary about his longterm colleague that was printed in the Baylor College of Medicine newsletter.

King, Olive (Black). Personal Correspondence from Peter Massare to Joe W. King, 5 December 1977. Privately held by Kelly Mazade, Fort Worth, Texas. 2002–2020. Letter from Peter Massare, co-founder of the Former Members of the 451st, to Joe W. King, inviting him to join the reunion group in 1977.

Stars and Stripes Newspaper Clippings: ©1944/1945, 2021 Stars and Stripes, All Rights Reserved. Reprinted with the permission of the Stars and Stripes archive. Each clipping is individually cited in its corresponding chapter.

Dearest Babe,

Introduction

1973 National Records Fire: National Archives. "The 1973 Fire: National Personnel Records Center." *Archives.gov.* www.archives.gov/personnel-records-center/fire-1973. 8 October 2021. Accessed 2021.

Joe's birth: Mississippi. State Board of Health. Birth Certificate. Bureau of Vital Statistics. Jackson. Certified copy of record of birth.

Joe's family history: Farris, Carol King. McAllen, Texas. Interviews by Kelly Mazade. 2020. Transcripts. Privately held by Mazade, Fort Worth, Texas. 2020. McKechnie, Kay King. Houston, Texas. Interviews by Kelly Mazade. 2020. Transcripts. Privately held by Mazade, Fort Worth, Texas. 2020. Family story.

Joe's skating accident: Ibid. Family story.

Joe's higher education: King, Olive (Black). Officer's Data Sheet. 24 October 1943. Privately held by Kelly Mazade, Fort Worth, Texas. 2002–2020. Form required by the Army Air Corps before deployment that listed Joe's education history along with other personal and professional information.

Babe's family history: Carol King Farris, interview, 2020. Family story.

Babe's higher education: Southwestern, Memphis, Shelby County, Tenn. "Diploma, 4 June 1935." Privately held by Kelly Mazade, Fort Worth, Texas. 2002–2020. Babe's diploma from Southwestern with Bachelor of Arts in Education.

Babe's employment: Kay McKechnie, interview, 2020. Family story.

Joe and Babe's high school: "Tech Class Rosters, Yearbooks and Photos, 1911-1987." Digitized yearbook. Memphis Technical High School. *Memphis Tech High Alumni.* http://memphistechhigh.com: 2020. Babe graduated from Memphis Technical High School in 1931. Joe graduated the following year.

First date: Kay McKechnie, interview, 2020. Family story.

Marriage: Marriage Certificate. 31 May 1940. Compared to other legal documents (birth and death certificates), Joe and Babe's ages are incorrect on their marriage certificate.

Babe's wedding attire: "King-Black Vows Are Said," *(Memphis) The Commercial Appeal,* 1 June 1940. Newspaper clipping is within personal holdings.

Honeymoon in Mexico: Hotel Receipt. 7 June 1940. Receipt from the Hotel "Imperial" in Mexico.

Kay's birth: Arkansas. The Helena Hospital. Birth Certificate. Helena. Birth story from interview with Kay King McKechnie.

Joe joins the military: Farris. "KING, Joe Wesley," 1. Carol Farris's unpublished biography of Joe King.

Joe's enlistment: Officer's Data Sheet, 24 October 1943. Enlistment date written on form.

Joe's draft registration: "U.S. WWII Draft Cards Young Men, 1940-1947." [database online]. Lehi, UT, USA: Ancestry.com Operations, Inc., 2011. Draft registration card sourced from Ancestry.com database.

Assigned to March Field: Special Orders No. 250, 15 September 1942. Restricted Orders from the War Department requiring 1st Lt. Joe Wesley King to report to March Field, Riverside, Calif. on 29 September 1942.

King family car: Automobile Driver's Permit, September 1942. Driver's Permit for March Field Air Base notes that Lt. Joe W. King drove a 1941 Chrysler Conv. Cpe. Grey with red plaid seats 189-111.

Transfer to Randolph Field: Special Orders No. 8, 8 January 1943. Restricted Orders from HQ Fourth AF requiring 1st Lt. Joe Wesley King to report to School of Aviation Medicine, Randolph Field, Texas on 24 January 1943.

Transfer to Santa Ana, Calif.: Special Orders No. 38, 8 March 1943. Restricted Orders from HQ Fourth AF requiring 1st Lt, Joe Wesley King to report to the Branch School of Aviation Medicine, SAAAB, Santa Ana, Calif. on 8 March 1943.

Aviation Medicine Certificate: Aviation Medicine Certificate, 22 April 1943. Certificate from the Air Corps School of Aviation Medicine naming 1st Lt. Joe W. King, M.D. as a qualified Aviation Medical Examiner on 22 April 1943.

Assigned to 15 Bomb Wing: Special Orders No. 71, 17 April 1943. Special Orders from The School of Aviation Medicine requiring 1st Lt. Joe Wesley King to report to the 15 Bomb Wing, AAB, Sioux City, Iowa.

Transfer to Gowen, Field: Special Orders No. 38, 1 May 1943. Special Orders from HQ 15 Bombardment Operation Training requiring 1st Lt. Joe Wesley King to report to the Station Hospital at Gowen Field, Boise, Idaho.

Transfer to Casper, Wyo.: Special Orders No. 146, 26 May 1943. Special Orders from Air Base HQ requiring 1st Lt. Joe Wesley King to report to the Station Hospital at Casper, Wyoming by rail.

Assigned to 451st Bomb Grp.: Special Orders No. 151, 31 May 1943. Special Orders from HQ Army Air Base assigning 1st Lt. Joe Wesley King to the 451st Bomb Grp, AAB, Davis Monthan Field, Tucson, Ariz. by 7 June 1943.

History of 451st Bomb Grp.: Air Force Historical Research Agency. "History of the 451st Bombardment Group (H) - UNCLASSIFIED." 1943-1945. FIIL microfilm, rolls B0595 and B0596. Department of the Air Force, Maxwell Air Force Base, Alabama.
Hill, Mike. *The 451st Bomb Group In World War II A Pictoral History.* Atglen: Schiffer, 2001.
Hill, Mike. *"The Fight'n" 451st Bombardment Group (H)*. Paducah: Turner, 2000.
"History of the 451st Bombardment Group (H)." Online publication. 451st. 451st.org/History/pdf's/History-of-the-451st-Bombardment-Group%20_H_.pdf: 2020." Online publication courtesy of Bob Karstensen in 2006.

Transfer to Dyersburg, Tenn.: Special Orders No. 155, 4 June 1943. Special Orders from HQ Davis Monthan Field attaching 1st Lt. Joe Wesley King and the rest of the 451st to the parent 436th Bomb Grp, AAB, Dyersburg, Tenn. on 4 June 1943.

Transfer to 346th parent group: Hill, "The Fight'n" 451st, 14. The 451st was placed under the parent 436th Bomb Grp, AAB, Dyersburg, Tenn. on 4 June 1943.

History of Dyersburg Army Air Base: Reynolds, Robert. "WWII Dyersburg Army Air Base – Halls, TN, 1942." *Tennessee History: Tennessee Good Old Days.* tennessee-historyblog.wordpress.com/2017/08/22/wwii-dyersburg-army-air-base-halls-tn-1941 22 August 2017. Accessed 2020.

451st AAFSAT Training: Air Force Historical Research Agency. "History of the 451st," roll B0595, Narrative History, Chpt. 1.
Hill, The 451st Bomb Group, 8-9. Hill, "The Fight'n" 451st, 14.

Joe's flight status: Physical Examination for Flying, 4 June 1943. Physical Examination for Flying from Capt. M.J. Hitjito approving 1st Lt. Joe Wesley King for flying on 14 June 1943. Includes letters from Joe requesting permission for flight status and permission letter signed by 1st Lt. M.Y. Kremers, as well as personnel orders requiring Joe to completed regular flight hours.

Beheading story: "Orlandoan Saves Life," *Orlando Evening Star*, 25 June 1943. Page 1. Newspapers.com. www.newspapers.com/image/340892981/. Accessed 3 October 2021. The story Joe tells Babe about seeing a woman beheaded by his wife is true, as evidenced in this article.

Troop train: Strack, Don "Pullman Troop Sleeper and Troop Kitchen Cars." *Utah-Rails.net.* utahrails.net/pass/pullman-troop-sleeper-kitchen.php 14 October 2018. Accessed 2020. Description of troop train.

Transfer to Wendover Field: Special Orders No. 142, 19 June 1943. Special Orders from HQ Army Air Base Dyersburg requiring the 451st to report from Orlando to Wendover Field Utah.

History of Wendover Field: Air Force Historical Research Agency. "History of the 451st," roll B0595, Narrative History, Chpt. 1.
Hill, The 451st Bomb Group, 9-10. Hill, "The Fight'n" 451st, 15.

Medical corps duties: Craven and Cate. "Medical Service of the AAF." Army Air Forces Medical Services in World War II. Digital Book. Office of Air Force History, www.ibiblio.org/hyperwar/AAF/VII/AAF-VII-13.html. Accessed 2020. "Medical History." *451st.org*. https://451st.org/History/pdf's/Medical%201.pdf: 2020. Nanney, James S. Army Air Forces Medical Services in World War II. Digital Book. Air Force History and Museums Program, media.defense.gov/2010/Sep/23/2001330103/-1/-1/0/AFD-100923-014.pdf. 1998.

Spring Street House: "Army Couple Buys Home And All the Furnishings," *(Memphis) Press-Scrimitar*, June 1943. Newspaper clipping within personal holdings.

Wendover plane crash: "Army Bomber Crashes on Utah Desert, Wrecking Freight Train." *The San Bernardino Sun*, Tuesday, August 10, 1943, Page 1. UCR Center for Bibliographical Studies and Research, California Digital Newspaper Collection. cdnc. ucr.edu. Accessed 7 August 2020. Newspaper clipping from front page of The San Bernardino Sun pulled from online newspaper archives.

History of the B-24 airplane: Dorr, Robert F. *B-24 Liberator Units of the Fifteenth Air Force*. Oxford: Osprey, 2000, 6-16.
Dwyer, Larry. "Consolidated B-24 Liberator." *The Aviation History Online Museum*. www.aviation-history.com/consolidated/b24.html. 6 October 1998. Accessed 2020.

Flight suits: Hill, "The Fight'n" 451st, 66. Personal account of Sgt. Robert Karstensen, nose turret gunner, 451st Bombardment Group, 724th Squadron.

Wendover wind storm: Air Force Historical Research Agency. "History of the 451st," roll B0595, Narrative History, Chpt. 2.

Joe's furlough: Special Orders No. 65, 14 September 1943. Special Orders from Fairmont AAF granting 1st Lt. Joe King a 10-day leave of absence on 17 September 1943.

Carol's birth: Tennessee. Department of Public Health. Birth Certificate. Division of Vital Statistics. Shelby. Birth story from interview with Carol King Farris. Newspaper clipping from Memphis Press-Scimitar.

Joe's return to Fairmont: TWA Correspondence, 25 September 1943. Letter from Passenger Agent E.M. Clark of Transcontinental and Western Air, Inc., noting the delayed transportation of Lt. Joe King on 25 September 1943, to his C.O.

History of Fairmont: Air Force Historical Research Agency. "History of the 451st," roll B0595, Narrative History, Chpt. 2.
Hill, The 451st Bomb Group, 12. Hill, "The Fight'n" 451st, 16.

Jimmy Stewart: Signor, Johnny. "703rd Bomb Squadron." *American Air Museum*. www.americanairmuseum.com/unit/1367. 10 April 2015. Accessed 2020.
"Bomber Group Training." *Historic Wendover Air Field Foundation.* http://www. wendoverairbase.com/bomber_groups. Accessed 2020.

Joe's last will and testament: Last Will and Testament, 4 October 1943. Before he left for the war, Joe W. King left the entirety of his estate to Olive B. King.

"3700 men and 62 B-24 Liberators": Hill, The 451st Bomb Group, 12. Hill, "The Fight'n" 451st, 17.

Transfer to Lincoln: "Special Order No. 251." *451st.org*. www.451st.org. 2020. Special Order from HQ FAAF requiring the 451st air echelon to report to AAB Lincoln, Nebraska on 18 November 1943.

History of Lincoln Air Base: Hill, "The Fight'n" 451st, 17.
"World War II." *History of the Former Lincoln Air Force Base*. www.lincolnafb.org/ history.php. Accessed 2020.

Joe's captaincy: Special Orders No. 299, 26 October 1943. Special Orders from the War Department promoting 1st Lt. Joe W. King to Captain on 26 October 1943. His promotion was also featured in the *Memphis Press-Scimitar* on 18 November 1943. Newspaper clipping is within private holdings.

Joe and Young's rescue: Air Force Historical Research Agency. "History of the 451st," roll B0595, Narrative History, Chpt. 4.

Flight to West Palm Beach: Ibid.

Dearest Babe,

En Route to Italy

Journal entries: King, Joe W. "Journal." MS. Multiple Unit Locations during WWII, 9 December 1943-12 March 1944. Privately held by Kelly Mazade, Fort Worth, Texas, 2020. Joe's private journal while traveling to Italy with the 451st Bomb Group from December 9, 1943 to March 12, 1944.

Travel records: Air Force Historical Research Agency. "History of the 451st," roll B0595, Narrative History, Chpt. 5.
Hill, The 451st Bomb Group, 14-20.

Gioia del Colle

History of Gioia del Colle: Hill, The 451st Bomb Group, 20.
Hill, "The Fight'n" 451st, 19-20.

Mission Histories: Zinn, Engineering Officer Eli. San Antonio. Interview by Kelly Mazade. 3 April 2006. Transcript. Privately held by Mazade, Fort Worth, Texas. 2020
Karstensen, Bob. 451st Mission Log. ca. 1980-99. Privately held by Kelly Mazade, Fort Worth, Texas. 2006-2020.
Hill, "The Fight'n" 451st, 195-196. History and mission logs from Mission 1 to Mission 245 were contained in a personal letter from Eli Zinn to author. According to Zinn, he received a copy of the mission logs from Bob Karstensen.

Gioia del Colle Airstrip: Hill, The 451st Bomb Group, 23.

Description of V-Mail: "Mail Call: V-Mail." *National WWII Museum.* www.national-ww2museum.org/war/articles/mail-call-v-mail#. Accessed 2020.

Journal entries 13 January 1944 - 6 March 1944: King, "Journal," 13 January 1944 - 6 March 1944.

Missions #1-10: 451st Mission Log, Mission #1-10, 1980-99. The mission notations were quoted directly from the complied mission log.

Regensburg DUC: Karstensen, Bob. "Narrative History Regensburg, Germany, Mission #10." 451st Ad-Lib (January 1982): 4-6. Image copy. 451st.org. www.451st.org/Ad%20Lib/Pdfs/JAN%201982.pdf: 2020.
General Orders No. 12, 23 April 1944. General Orders from the Headquarter of the 47th Bomb Wing commending the 451st unit for a Distinguish Unit Citation for Mission #10 to bomb Regensburg A/C factory. Narrative History from Karstenen's *Ad-Lib* reunion periodical accessed online.

Regensburg Newspaper Article: "Battle to Finish Luftwaffe Passes Sixth Day: Attacks on Stuttgart, Regensburg Climax Furious 24-Hour Drive." The Stars and Stripes, 26 February 1944. Northern Ireland Edition, archived. starsandstripes.newspaperarchive.com/northern-ireland-stars-and-stripes/1944-02-26 : 2020.

Bibliography

San Pancrazio and Manduria

Move to San Pancrazio and Manduria: Hill, "The Fight'n" 451st, 21.

Ploesti Raids: Hill, "The Fight'n" 451st, 22.
Hill, The 451st Bomb Group, 27-28.

Missions #11-24: 451st Mission Log, Mission #11-24, 1980-99. The mission notations were quoted directly from the complied mission log.

Journal entries 9 March 1944 - 12 March 1944: King, "Journal," 6 March 1944 - 9 March 1944.

Lt. McAlister's Last Words: NARA. "Individual Causality Questionnaire." 451st.org. www.451st.org/MACRs/451st%20MACRs/42-52103/42-52103%20Combined.pdf. 1973. Accessed 2020.

Ploesti Newspaper Article: "Ploesti Oil Fields Hit As Allies Step Up Air Attack to Aid Soviets." The Stars and Stripes, 8 April 1944. London Edition, archived. starsand-stripes.newspaperarchive.com/london-stars-and-stripes/1944-04-08 : 2020.

Castelluccia di Sauri

Reassignment to 49th Bomb Wing: Hill, "The Fight'n" 451st, 22.

Description of base: Hill, The 451st Bomb Group, 30.

Men's Living Quarters: Ibid, 31.

Dispensary Whiskey: Hill, "The Fight'n" 451st, 144. The note about the purpose of the 2 oz shot of whiskey is courtesy of James Greco, a member of the 451st Reunion Facebook Group.

Medical Disposition Board: United States. Army Medical Service, Arnold Lorentz Ahnfeldt, Robert S. Anderson, John Boyd Coates, Calvin H. Goddard, William S. Mullins, *The Medical Department of the United States Army in World War II*, Google Book. University of California: Office of the Surgeon General, Department of the Army, 1973. 871-873.

Mission 108 and 109: Ibid, 50-56.

Mission requirements for return to U.S.: Air Force Historical Research Agency. "History of the 451st," roll B0595, Narrative History, Chpt. 7.

Allies Take Rome Newspaper Article: "Allied Forces Occupy Rome." *The Stars and Stripes*, 6 June 1944. Casablanca Edition, archived. starsandstripes.newspaperarchive. com/casablanca-stars-and-stripes/1944-06-06 : 2020.

France Invaded Newspaper Article: "France Invaded Beach Defenses Pierced 10 Miles." *The Stars and Stripes*, 7 June 1944. Mediterranean Edition, archived. starsand-stripes.newspaperarchive.com/mediterranean-naples-stars-and-stripes/1944-06-07:2020.

100th Mission celebrations: Hill, The 451st Bomb Group, 47.

727th Insignia: "Flying Box Car" *451st.* www.451st.org/Photos/images/451st_Patch.jpg. Accessed 2020.

French Riviera Newspaper Article: "Riviera Coast Secured." *The Stars and Stripes*, 17 August 1944. Mediterranean Edition, archived. starsandstripes.newspaperarchive.com/mediterranean-rome-stars-and-stripes/1945-08-17 : 2020.

Markersdorf DUC: "Distinguished Unit Citation" *451st.* www.451st.org/Awards/PDFs/Citation_2-10-44.pd. Accessed 2020.

Paris Freed Newspaper Article: "Patriots Report Paris Freed." *The Stars and Stripes*, 24 August 1944. Mediterranean Edition, archived. starsandstripes.newspaperarchive.com/mediterranean-naples-stars-and-stripes/1944-08-24 : 2020.

Romanian POWs: Haulman, Daniel. "Operation Reunion and the Tuskegee Airmen" *Tuskegee Airmen.* tuskegeeairmen.org/wp-content/uploads/Operation-Reunion.pdf . Accessed 2020.

Group Dispensary: Air Force Historical Research Agency. "History of the 451st," roll B0595, Narrative History, Chpt. 11.

1944 World Series: "1944 World Series." *Baseball Reference.* www.baseball-reference.com/postseason/1944_WS.shtml. Accessed 2020.

Roosevelt Election Newspaper Article: "Roosevelt Is Winner." *The Stars and Stripes*, 8 November 1944. Mediterranean Edition, archived. starsandstripes.newspaperarchive.com/mediterranean-naples-stars-and-stripes/1944-11-08 : 2020.

Thanksgiving Day football game: Hill, The 451st Bomb Group, 63.

Joe's lecture in the *Ad-Lib*: Air Force Historical Research Agency. "Doc King Talks on Bugs." Ad-Lib, Volume 1, No. 3. 13 February 1945, roll B0595.

VD 8-ball: Air Force Historical Research Agency. "VD 8 Ball to 727th: Big Sphere in Mess Hall." *Ad-Lib*, Volume 1, No. 1. 1 February 1945, roll B0595.

T/S Card: Hill, "The Fight'n" 451st, 144.
King, Olive (Black). T/S Card. December 1978. Privately held by Kelly Mazade, Fort Worth, Texas. 2002–2020. 451st reunion materials sent by Karstensen to Joe in 1977.

Warsaw Newspaper Article: "Red Army Frees Warsaw; Vast Offensive Under Way." *The Stars and Stripes*, 18 January 1945. Mediterranean Edition, archived. starsandstripes.newspaperarchive.com/mediterranean-rome-stars-and-stripes/1945-01-18:2020.

Roosevelt Dies Newspaper Article: "President Roosevelt Does: Hemorrhage Fatal in Georgia." *The Stars and Stripes*, 13 April 1945. London Edition, archived. starsandstripes.newspaperarchive.com/london-stars-and-stripes/1945-04-13 : 2020.

Mussolini Newspaper Article: "Mussolini Executed; 5th Enters Milan; 7th In Munich." *The Stars and Stripes*, 30 April 1945. Mediterranean Edition, archived. starsandstripes.newspaperarchive.com/mediterranean-naples-stars-and-stripes/1945-04-30:2020.

Hitler Dead Newspaper Article: "Hitler Dead, Nazis Report; Doenitz Becomes 'Fuehrer,' Declares War Will Go On." *The Stars and Stripes*, 2 May 1945. Mediterranean Edition, archived. starsandstripes.newspaperarchive.com/mediterranean-rome-stars-and-stripes/1945-05-02 : 2020.

Bibliography

VE-Day Special Newspaper Article: "It's Over Over Here." *The Stars and Stripes*, 5 May 1945. Mediterranean Edition, archived.starsandstripes.newspaperarchive.com/mediterranean-naples-stars-and-stripes/1945-05-08 : 2020.

451st accomplishments and results: Hill, "The Fight'n" 451st, 144.
Hill, The 451st Bomb Group, 73-74.
Ad-Lib Issue 38. *451st.org*. www.451st.org/Ad%20Lib/Pdfs/Issue%2038%20Spring%202004.pdf. Spring 2004. Accessed 2020.

Returning Home

451st' return home: Hill, "The Fight'n" 451st, 55.
Hill, The 451st Bomb Group, 73.
Air Force Historical Research Agency. "History of the 451st," roll B0595, Narrative History, Chpt. 16.

Joe's original travel arrangements: Farris, Jim. McAllen, Texas. Interview by Kelly Mazade. 2020. Transcript. Privately held by Mazade, Fort Worth, Texas. 2020.

Assignment to Dow Field: Hill, The 451st Bomb Group, 73.

Joe's discharge: Special Orders No. 243, 20 September 1945.

Afterword

Orthopedic training in New Orleans: King, Olive Black. "Babe's Notes." MS. Houston, TX, 1981. Privately held by Kelly Mazade, Fort Worth, TX. 2020.

Joe's medical accomplishments: King, Olive Black. "CV of Dr. Joe W. King" MS. Houston, TX, 1981. Privately held by Kelly Mazade, Fort Worth, TX. 2020.
Smith, Dr. Edward T. *A History of Orthopaedic Surgery in Houston, Texas*. Austin: Eakin Press, 1988, 63.

Joe's patients: Carol Farris, interview, 2020.

Leadership roles: The Methodist Hospital. "Dr. King Retires as Chief of Orthopedics," *The Journal* (October 1974): 2.

Bob Karstensen's story: Robert Karstensen, Former Members of the 451st, Marengo, Illinois to Joe King, letter, 24 September 1977; Personal Correspondence. Privately held by Kelly Mazade, Fort Worth, Texas.

Pete Massare's story: Pete Massare, Former Members of the 451st, Rochester, New York to Joe King, letter, 18 November 1977; Personal Correspondence. Privately held by Kelly Mazade, Fort Worth, Texas.

Eli Zinn's story: Eli Zinn, interview, 2006.

People

A

Arnold, 2nd Lt. Elliot *Member of the 726th whom Joe met in Dyersburg. Original member of 451st.* 37, 39, 42-44, 55, 408

B

Baldwin, Dan *Joe's best friend from Memphis, who was the best man in his wedding.* 10, 130-31, 144, 146-47, 153-54, 168, 183, 265, 342-343, 379

Bassett, Maj. *Joe's friend from March Field.* 243, 323, 393

Bell, Capt. *Doctor from the service squadron attached to the 451st, who asked Joe to take care of his outfit while he was on leave.* 264, 355

Bently, Maj. Charlie *Executive Officer of the 724th who took Maj. Reynolds' place and was sent to the hospital for TB.* 157, 276

Bernstein, 1st Lt. Jack *Original member of the 451st and Joe's friend.* 128, 166, 175, 204, 211, 232, 238, 252, 284, 292, 303, 307, 418, 420

Blackmon, Capt. Linnon R. "Blackie" *Pilot in the 451st whom Joe knew in Orlando and Wendover.* 84, 89, 90, 93, 98, 100, 102-5, 109-10, 112, 115, 138, 147, 166, 171, 173, 175, 177, 205, 213, 215, 220-21, 231, 350

Blaschke, Lt. *Member of the 727th; a P.O.W. who was liberated from Romania and stayed with Joe.* 357-58

Boyd, Dr. *Doctor who Joe knew from Campbells, helped Joe get into school in New Orleans.* 331, 314, 325

Bryan, Lt. "Judge" *Friend of Joe's from the 727th who visited Babe when he returned to the States.* 241, 291, 294-95, 316

C

Carter, Dr. Harvey *Doctor Joe knew who was stationed in Naples.* 307-08, 367, 438

Craycroft, Burr and Jean *Joe and Babe's friends from Riverside. Also trained in San Antonio. Member of the 364th Fighter Group.* 19-20, 37, 192, 194, 218, 224, 313, 336, 368

Curtis, 1st Lt. George R. *Member of 724th whom Joe met in Dyersburg; one of Joe's poker buddies.* 126-28, 140, 155, 164, 174-75, 204, 246

E

Eaton, Col. *C.O. of the 451st.* 224, 286

Ellis, Rev. D.A. *Preacher of Bellvue Baptist Church in Memphis who married Joe and Babe.* 49

Epes, Thea *Joe and Babe's friend from Helena.* 256, 381

Eubanks, Jr., General William E. *C.O. of the 2nd Air Force.* 61

Evans *Joe's C.O.* 128-129, 146, 152, 155, 157-58, 171, 174-75, 183, 190-191, 206, 208, 210, 213, 215, 219, 223-25, 232-33, 235-37, 239, 240-45, 283

F

Ferguson *Member of the 727th.* 408, 429

Formanek, 2nd Lt. Arthur A. *Pilot in the 727th.* 132, 142, 195, 208, 244

G

Garrity *One of the men that worked for Joe in the medical staff.* 188, 242

Giovanni *Joe's house boy who was injured.* 396, 307, 312, 326, 330-34, 336, 388

H

Haltom, Capt. Charles C. *Member of the 451st whom Joe knew in Orlando and Wendover. C.O.* 87, 142, 158, 175, 232, 237

Hoppock, Lt. Col. *New colonel whom Joe did not like.* 277, 281, 283-84, 287-88, 296-297, 299, 324, 344, 347, 352

Hughes Twins *Doctors whom Joe knew in the States that were stationed in Naples, Italy and attended a medical meeting in Bari with Joe.* 307, 351, 367-68, 384, 438

I

Inez *Babe's friend from Memphis.* 73

Ingham, Mrs. *Joe and Babe's landlord in Riverside.* 17-18

J

Johnson, 1st Lt. Eldridge W. *Doctor whom Joe met in Orlando.* 147-48, 230, 316, 332-33, 346, 368, 383

K

Kinard, Col. *Colonel who helped Joe get a dispensary tent in Wendover.* 38, 60-61

Koenig, Emil *Doctor friend of Joe's.* 51, 144, 194-95, 206, 287, 315

Kremers, Capt. Marshall Y. *Flight Surgeon for the 726th whom Joe met in Santa Ana.* 25-26, 32, 43, 59, 62-64, 87, 90, 147, 154, 201, 206, 208, 220, 222, 225, 231, 236, 262, 270, 274, 286, 288, 289, 295, 300, 310-11, 316, 318-19, 321, 323, 341, 380, 389, 398

L

Larson *Pilot in the 727th.* 216, 224, 232

Lather, Maj. *C.O.* 374, 378, 409, 423

LaVigne, 1st Lt. Richard J. *Doctor who attended the School of Aviation Medicine with Joe.* 103

Leonardo *Joe's 12 year old Italian "valet" house boy.* 268

Lewis, 1st Lt. Robert *Doctor who attended the School of Aviation Medicine with Joe.* 27, 58

Love, Lt. *Member of 727th.* 171, 207

M

Mahon *Member of 727th who took Evans' place as C.O., whom Joe liked very much.* 239, 241, 245-46, 254, 256, 276-78, 282, 286-92, 294-95, 321, 341, 347, 354

Manoogian, 1st Lt. Morris *Original member of the 451st.* 93, 408

Marco, 1st Lt. Herbert *Original member of the 451st.* 86, 89-90, 273

Martino *Joe's Italian house boy.* 407

Mason, Sergeant John R. *Medical non-commissioned officer who served under Joe in Italy.* 156, 162, 167-69, 175-76, 188, 242, 303, 399, 431

Massare, Lt. Peter A. *Assistant Operations officer of the 727th who Joe knew from the beginning of the war and co-founder of the Former Members of the 451st.* 207, 312, 477

McAllister, Wilfred B. *Pilot of the Craven Raven whom Joe met in Wendover. Best friends with Pep Prewitt. Killed in action April 5, 1944.* 84, 136, 142, 164-65, 179, 206

McFarland, Capt. Ward J. M.C. *Flight Surgeon for the 725th. Original member of 451st.* 32, 45, 59, 62, 103, 132, 136, 167, 201, 207, 222, 316

O

Oleen, Robert A. *Bombardier of the 727th.* 107

P

Peterson, Dr. *Dentist of the 451st.* 236, 250-52, 255, 341, 344-45, 411

Posey, Dr. Frank *Joe's friend.* 21, 230, 337

Prewitt, Terrell G. "Pep" *Member of the 451st whom Joe met in Wendover.* 92, 114, 173, 188, 205, 208, 232, 238, 244

Pyle, Ernie *War correspondent.* 422

Q

Quinn, Capt. Henry F. *Flight Surgeon for the 724th.* 32, 59, 62, 64, 222, 236, 314, 316, 424

R

Reese, Harold *One of Joe's roommates.* 219, 223, 247, 283-84, 288, 298, 303, 306, 318, 343, 345, 375, 383, 407, 410-411

Roach, Lt. *Member of the 451st whom Joe met in Wendover; one of Joe's poker buddies.* 84, 132, 204, 211, 213, 225, 232, 234, 241

Roosevelt, Pres. Teddy 48, 318, 411-12

Rosenbaum, Cpl. Raymond J. "Rosie" *Joe's medical technician in Italy.* 170, 175, 303, 350

Family Members

Some family are mentioned throughout the book. No page numbers are listed.

Joe Wesley King

Olive "Babe" Black King

> **Kay King McKechnie** *Joe and Babe's daughter.*

> **Carol King Farris** *Joe and Babe's daughter.*

> **Joe Wesley King, Jr.** *Joe and Babe's son.*

Alvin Coy King *Joe's father.*

Mabel Westmoreland King *Joe's mother.*

> **Adrian King** *Joe's brother.*

> **Otha King** *Joe's brother.*

> **Sally "Sis" King Fowler** *Joe's sister.*

> **Royce Fowler** *Joe's brother-in-law.*

John King *Joe's uncle.*

Hugh Black, Sr. *Babe's father.*

Ola Blancet Black *Babe's mother.*

> **Hugh Black, Jr.** *Babe's younger brother who served in France during WWII.*

Entertainment

A Guy Named Joe, starring Irene Dunn, 1943

A Tree Grows in Brooklyn, by Betty Smith, 1943

And the Angels Sing, starring Betty Hutton, 1944

Air Force, 1943

As Thousands Cheer, 1943

Barretts of Wimpole Street, starring Katharine Cornell, Brian Aherne, 1931

Battle of Britain, 1943

Bombers Moon, 1943

Casanova Brown, starring Gary Cooper, 1944

Christmas Time, starring Deanna Durben, 1944

Coney Island, starring Betty Grable, 1943

Dangerous Blondes, 1943

Dixie, starring Bing Crosby, 1943

Fired Wife, 1943

Flame of the Barbary Coast, starring John Wayne and Ann Dvorak, 1945

Frenchman's Creek, starring Joan Fontaine, 1944

Frontier Badsman, 1943

Going My Way, starring Bing Crosby, 1944

Guadalcanal Diary, 1943

"I Get the Blues When It Rains," by Carl Haworth, 1928

"I Love You Truly," by Carrie Jacobs-Bond, 1901

Impatient Years, starring Jean Arthur, 1944

Joe Smith, American, starring Robert Young and Marsha Hunt, 1942

Johnny Come Lately, starring Jimmie Cagney, 1943

Kismet, 1944

"La Traviata," 1853

Lady Takes a Chance, starring John Wayne and Jean Arthur, 1943

Lily Mars, starring Judy Garland, Van Heflin, 1943

Terms and Locations

T

U

V

W